MATERIALS SCIENCE AND TECHNOLOGIES

CERIUM

MOLECULAR STRUCTURE, TECHNOLOGICAL APPLICATIONS AND HEALTH EFFECTS

Materials Science and Technologies

Additional books in this series can be found on Nova's website
under the Series tab.

Additional e-books in this series can be found on Nova's website
under the e-book tab.

MATERIALS SCIENCE AND TECHNOLOGIES

CERIUM

MOLECULAR STRUCTURE, TECHNOLOGICAL APPLICATIONS AND HEALTH EFFECTS

ALEKSEY IZYUMOV
AND
GULNAZ PLAKSIN
EDITORS

nova publishers
New York

Library of Congress Cataloging-in-Publication Data

Cerium : molecular structure, technological applications and health effects / editors, Aleksey Izyumov and Gulnaz Plaksin.
p. cm.
Includes index.
ISBN 978-1-62257-670-8 (hardcover)
1. Cerium group--Research. I. Izyumov, Aleksey. II. Plaksin, Gulnaz.
QD172.R2C456 2011
546'.412--dc23

2012030294

Published by Nova Science Publishers, Inc. † New York

CONTENTS

PREFACE

Cerium is the most abundant of the rare earth elements, making up about 0.0046% of the Earth's crust by weight. It is found in a number of minerals, the most important being monazite and bastnasite. Commercial applications of cerium include catalysts, additives to fuel to reduce emissions and to glass and enamels to change their color. In this book, the authors study the molecular structure, technological applications and health effects of cerium. Topics include the synthetic and structural features of ceriopolyoxotungstates; cerium-based corrosion inhibitors; tetravalent cerium chemistry; comparative toxicity of cerium and other rare earth elements (REEs) in plant and invertebrate test systems; cerium-based catalysts for the selective catalytic reduction of NOx with NH3; cerium oxide and its reduction in composite materials under hydrogen containing atmosphere; cerium as fertilizer for soils in Russia; and the stability and reactivity of cerium IV complexes with aliphatic organic compounds.

Chapter 1 – Polyoxotungstates constitute a large class of anionic tungsten-oxide clusters with rich solution chemistry, highly diverse topologies and numerous applications in key areas within the materials science. These clusters have been thoroughly reacted with lanthanoids, and in the specific case of cerium, it is only in the last decade that a significant increase in the number of structurally characterized ceriopolyoxotungstates is observed despite of the fact that the first structural report on such compounds dates from the early seventies. Cerium has primarily been combined with lacunary polyoxotungstates acting as polydentate O-donor rigid ligands to generate a variety of molecular assemblies. These range from several small 1:2 sandwich-type clusters or 1:1 monomeric building blocks derived as dimers or chains to some giant, nanosized ceriopolyoxotungstates that contain over sixteen cerium and one hundred tungsten atoms per cluster and are amongst the largest polyoxotungstates known. Because of its high oxophilicity, cerium has also been used as stabilizing agent to isolate polyoxotungstates with unprecedented topologies or as linker between archetypal, plenary clusters to construct purely inorganic or hybrid organic-inorganic extended frameworks with different dimensionalities. This chapter reviews these aspects with focus on synthetic and structural features.

Chapter 2 – Cerium is recognised as a good cathodic inhibitor for a range of metals. The corrosion inhibition of cerium as a simple salt as well as in cerium-organic compounds has been the subject of extensive research since the mid 1980’s. This chapter examines the authors’ current understanding of the fundamental mechanisms of corrosion inhibition using cerium salts. It then explores how the chemistry has been adapted for industrial inorganic coating applications called conversion coatings. In finishes with a review of how multifunctionality is built into cerium compounds using various ligands and how these compounds are incorporated into paint systems used for corrosion protection.

Chapter 3 – This chapter gives an overview on the chemistry of tetravalent lanthanide compounds, especially those of tetravalent cerium. Following a brief Introduction it covers the tetrahalides, dioxides and other lanthanides(IV) salts. Coordination compounds of cerium in the oxidation state +4 include halogeno complexes and complexes of oxo acids, *β*-diketonates and related Schiff-base complexes as well as porphyrinates and related complexes. The chapter also covers the chemistry of organocerium(IV) complexes as well as "pseudo-organometallics" of tetravalent cerium. The latter comprise mainly the alkoxides and siloxides as well as various amido and amidinato complexes of cerium in the oxidation state +4. True organocerium(IV) complexes are thus far restricted to a handful of cyclopentadienyl derivatives and some cerocene sandwich complexes. Cerium(IV) complexes comprising specially designed *N*-heterocyclic ligands comprise the latest addition to the field.

Chapter 4 – The widespread and growing use of REEs in a number of technological applications requires a comparative evaluation of REE-associated effects on environmental health, in view of optimizing the choice between the technological requirements for individual REEs and their expected effects on environmental and human health. A review of the published literature in the last decade points to an imbalance between the dataset accumulated for Ce and La, and a definite scarcity of information on the toxicity of many REEs. Moreover, only few comparative studies have focused on the relative toxicities of several REEs.

Along with a well-established literature in toxicity testing by means of plant and invertebrate bioassay systems, a limited number of independent studies have investigated the Ce- and REE-associated effects in plant and invertebrate bioassays. The reported effects were evaluated in a wide range of endpoints, including growth and viability, cytogenetic effects, developmental defects, and in either triggering or counteracting oxidative stress. Indications for a hormetic, non-monotonic concentration-related trend were provided by some recent studies, providing mechanistic hints for the coexistence of toxic and beneficial effects (e.g. on plant growth), and of antioxidant/prooxidant effects. The current information should prompt further investigations providing a reliable classification and ranking of REE-associated relative toxicities.

Along with a well-established literature in toxicity testing by means of plant and invertebrate bioassay systems, a limited number of independent studies have investigated the Ce- and REE-associated effects in plant and invertebrate bioassays. The reported effects were evaluated in a wide range of endpoints, including growth and viability, cytogenetic effects, developmental defects, and in either triggering or counteracting oxidative stress. Indications for a hormetic, non-monotonic concentration-related trend were provided by some recent studies, providing mechanistic hints for the coexistence of toxic and beneficial effects (e.g. on plant growth), and of antioxidant/prooxidant effects. The current information should prompt further investigations providing a reliable classification and ranking of REE-associated relative toxicities.

Chapter 5 – Selective catalytic reduction of NO_x with NH_3 (NH_3-SCR) is one of the most promising NO_x control technologies for lean burn engines, particularly for diesel engines, to meet the increasingly stringent standards for NO_x emission worldwide. Due to some inevitable disadvantages of the well-established vanadium-based catalyst (V_2O_5-WO_3

(MoO_3)/TiO_2) for NH_3-SCR of NO_x from stationary sources, including the narrow operation temperature window, the large N_2O formation at high temperatures, the easy sublimation and biological toxicity of vanadium pentoxide, environmentally-benign and vanadium-free catalysts for NH_3-SCR reaction are requisite to be developed urgently. Recently, cerium-based oxides with high oxygen storage capacity and excellent redox property attracted much attention for their use as environmentally-benign NH_3-SCR catalysts. In this chapter, the authors will comprehensively introduce the authors' recent research results about the cerium-based catalysts for NH_3-SCR process, including the supported type CeO_x/TiO_2 catalyst prepared by conventional impregnation method and mixed oxide $CeTiO_x$ catalyst prepared by a novel homogeneous precipitation method. The effect of WO_x addition into the $CeTiO_x$ catalyst will also be studied in detail, which can improve the NH_3-SCR activity in the whole temperature range and enhance the N_2 selectivity in the high temperature range simultaneously. The total substitution of titanium species by tungsten species in $CeWO_x$ catalyst can further improve the SCR activity and N_2 selectivity to a certain extent, with nearly 100% NO_x conversion in a wide temperature range from 250 to 425 °C even under an extremely high gas hour space velocity (GHSV) of 500,000 h^{-1}. The excellent catalytic performance of this catalyst is attributed to a synergistic effect between Ce and W species. The cerium-based catalysts introduced in this chapter can be used as potential candidates for practical application in the deNO_x process for diesel engines, which is an important developing direction for the technological applications of cerium materials.

Chapter 6 – Ce-based materials have attracted much attention because of their applications as electronic ceramics, membranes, electrodes in fuel cells, and catalysts. The redox activity of cerium cations, which is the basis of its application as the catalysts, was usually observed in undoped cerium oxide at elevated temperatures and low oxygen partial pressure. One of the questions remained to be answered is whether cerium oxide in a composite material would reduce during long-term exposure to hydrogen containing atmosphere. If it is, how wide extend would be. To explore this aspect, the redox activity of the composite systems $La_{0.6}Sr_{0.4}CoO_3$ - CeO_2 (LSCCe), $La_{0.95}Ni_{0.6}Fe_{0.4}O_3$ - CeO_2 (LNFCe), and $La_{0.8}Sr_{0.2}MnO_3$ - CeO_2 (LSMCe) fabricated through the reactive sintering in air at 1350 °C was studied under hydrogen containing atmosphere. During the reactive sintering the cross-dissolution of cations at the phase interfaces occurs and results in the modification of the initial phases, forming cerium oxides with different degree of oxygen deficiency. The degree of the cross-dissolution at the phase interfaces increases in the following sequence: LSMCe → LSCCe → LNFCe. The results obtained in the present study demonstrate that reduction of modified cerium oxides in composite materials with highly reducible transition metal cations occurs under hydrogen atmosphere at the initial stage of reduction. During long-term exposure to H_2-Ar atmosphere the involvement of cerium oxide in the reduction process seems to depend on its lattice parameter that determines the length of the pathway for oxygen ion transport through the lattice. Kinetics of the reduction of composite materials at 800 °C has been discussed in more details.

Chapter 7 – The study of lanthanides in soils is not only of theoretic and soilgenetic importance but also of applied value, because lately lanthanides-containing substances and wastes are used as microfertilizers. The interest in rare earth elements in soils increased in the 1990s, when they found wide use (and cerium in particular) as microfertilizers for some crops in China, which increased the yield and quality of grain. Ce-containing fertilize usage will be baseless without information about Ce-content in soils. Revelation of regions with negative

Ce-anomalies is the most important, where Ce-containing fertilize usage is the most effective. Neither positive nor negative geochemical anomalies of natural or technogenic rare earth metals have been revealed in Russia until now.

The recent progress in the study of rare earth elements was related to the use of the expensive method of inductively coupled plasma mass spectroscopy (ICP MS). The expensive method of neutronactivation analysis using a nuclear reactor with the use of a gamma spectrometer allows determining the different dispersed elements in soils even at low concentrations.

The authors used for determining Ce contents the X-ray radiometric modification of the energydispersive X-ray fluorescence method, in which the sample is excited by a high-energy radioisotope source ^{241}Am rather than by the X-ray tube radiation. This method has made it possible to obtain data on Ce geochemistry in soils of humid landscapes quite expressively and cheaply.

Podzolic soils are specified by substantial negative cerium anomalies. The authors' studies of podzol and podzolic soil in Pinega (Arkhangelsk oblast), soddy-podzolic soil in town Chusovoi (Perm region) revealed a negative cerium anomalies. This situation is typical for taiga zone, where lanthanides are removed from soil horizons.

In Buryatia, Russia study the effectiveness of introducing lanthanides into the soil is produced. There are many negative Ce-anomalies in Russia. Ce-deficit in these soils may be removed by addition the Ce-containing wastes, including phosphogypsum.

Chapter 8 - Cerium is the most stable lanthanide in the (+4) oxidation state. Consequently, cerium(IV) is commonly used as a one-electron oxidant in aqueous organic synthesis, aqueous polymer synthesis, and water oxidation. Since cerium (IV is a fairly strong Lewis acid, the hydrated ion $[Ce(H_2O)_n]^{4+}$ is easily hydrolyzed and polymerized except at very low pH condition, leading to the loss of oxidizing power. Thus $[Ce(H_2O)_n]^{4+}$ ion exists only in concentrated perchloric acid solution. On the other hand, it has been pointed out in many previous reports that cerium(IV) has specific reactivity in acidic media. In cerium(IV)-driven water oxidation, no oxygen evolves when cerium(IV) is replaced with other strong oxidants, although most of the ruthenium-based O_2-evolving catalysts used cerium(IV) as an oxidant. In acidic aqueous media, cerium(IV) oxidizes aldehydes, ketones and alcohols at the α-carbon atom via the radical-like H-atom abstraction mechanism. In fact, electronic structures of some cerium(IV) complexes are recognized as an 'open-shell singlet state'. These indicate the 'non-innocent' character of cerium(IV) in the oxidation. This chapter focuses on the detailed electronic structure and the specific reactivity of cerium(IV) in the acidic solution.

Chapter 9 - An approach to investigation of thermodynamic and kinetic stability of variable-valence metal complexes is developed. Concepts of the generalized inverse problems of quantitative instrumental analysis and the complex formation equilibria are intro- duced. The procedure for this investigation is regarded as a sequential procedure for solving these two inverse problems. The classical methods for investigating the com- plex formations in solution are analyzed in terms of these inverse problems. Their generalizations are given to determine together with the formation constants also redox- decomposition rate constants of variable-valence metal complexes, and the rate law of numerous redox processes for which the rate-determining step is the decay of an inter- mediate complex. These generalized methods are applied to a comparative study of the complexation and intramolecular redox decomposition of cerium(IV) hydroxo com- plexes with some dicarboxylic, oxycarboxylic acids, and aliphatic polyatomic alcohols in acidity dependence (pH 0.7–3.4) of a sulfate medium at the ionic strength $I = 2$, temperatures of 10.0–43.0◦C by UV-Vis spectroscopy, and pH measurements. The region of existence, composition, ligand speciation, stability constants of the complexes $[C\ eOH\ (H{-}xR)n\ ]^{3-nx}$, rate constants, and activation energies of the intramolecular redox decomposition were determined for these complexes on a sulfate background. The rate law for the redox process that occurred in these studied systems is derived on the basis of calculated constants and the set of all the chemical and algebraic equations describes the equilibria within these systems. The quantitative model for this process is created. The existence of the inverse linear correlation between the logarithms of stability constants and logarithms of the redox-decomposition rate constants for $[C\ eOH\ H{-}2R]+$ complexes with dibasic ligands $R = H2L$, $R = H\ (OH\ )L$, and $R = L(OH)_2$ was established. It is shown that the kinetic stability for the corresponding bis-complexes $[C\ eOH\ (H{-}2R)2\ ]-$ does not differ from the kinetic sta- bility of the mono-complexes. The constancy of the reactivity for the complexes $[C\ eOH\ H{-}xR]^{3-x}$ with $x = 2, 3, 4$ in the sulfate medium and its independence from their thermodynamic stability are also demonstrated. We discuss the computational aspects of solving the regarded generalized problems.

In: Cerium
Editors: Aleksey Izyumov and Gulnaz Plaksin
ISBN: 978-1-62257-670-8

Chapter 1

CERIOPOLYOXOTUNGSTATES: FROM MONOMERIC SPECIES TO NANOSIZED MOLECULAR ASSEMBLIES

Santiago Reinoso**, Leire San Felices and Juan M. Gutiérrez-Zorrilla
Departamento de Química Inorgánica, Facultad de Ciencia y Tecnología, Universidad del País Vasco UPV/EHU, Bilbao, Spain

ABSTRACT

Polyoxotungstates constitute a large class of anionic tungsten-oxide clusters with rich solution chemistry, highly diverse topologies and numerous applications in key areas within the materials science. These clusters have been thoroughly reacted with lanthanoids, and in the specific case of cerium, it is only in the last decade that a significant increase in the number of structurally characterized ceriopolyoxotungstates is observed despite of the fact that the first structural report on such compounds dates from the early seventies. Cerium has primarily been combined with lacunary polyoxotungstates acting as polydentate O-donor rigid ligands to generate a variety of molecular assemblies. These range from several small 1:2 sandwich-type clusters or 1:1 monomeric building blocks derived as dimers or chains to some giant, nanosized ceriopolyoxotungstates that contain over sixteen cerium and one hundred tungsten atoms per cluster and are amongst the largest polyoxotungstates known. Because of its high oxophilicity, cerium has also been used as stabilizing agent to isolate polyoxotungstates with unprecedented topologies or as linker between archetypal, plenary clusters to construct purely inorganic or hybrid organic-inorganic extended frameworks with different dimensionalities. This chapter reviews these aspects with focus on synthetic and structural features.

* E-mail: santiago.reinoso@ehu.es.

1. Brief Notes on Polyoxometalates

Polyoxometalates (POMs) are anionic, oxo-bridged transition-metal clusters with outstanding structural, electronic and reactive versatility. These features make POMs a unique class of compounds that have found remarkable applications in very diverse fields, such as catalysis, materials science or medicine [1].

The large POM family includes a wide variety of discrete, molecular species ranging from small dimetalates to complex clusters comparable to a protein in size [2]. According to their composition, they can be classified in isopolyoxometalates or heteropolyoxometalates. The former exclusively contain the so-called *addenda* metals (M) and oxygen atoms acting as bridging or terminal oxo ligands (often also as hydroxo, especially when bridging), whereas the latter show additional elements playing the role of heteroatoms. Addenda metals require a suitable combination of relatively small size and high charge to facilitate condensation processes and also empty *d* orbitals accessible for back-donation in the formation of M–O bonds (terminal M–O bonds show indeed a strong double-bond character). Addenda metals are thus virtually restricted to elements of groups 5 and 6 in high oxidation states (especially V, Mo and W, polyoxoniobates being less extended but rapidly growing [3]), although a few examples of noble-metal POMs have been recently reported [4]. Heteroatoms (X) are usually p-block elements (either metallic or non-metallic), but first-row transition metals also occupy heteroatomic positions in some cases.

From a structural point of view, POMs can be described as discrete assemblies of condensed MO_n polyhedra, usually distorted octahedra ($n = 6$) and less frequently distorted tetrahedra ($n = 4$) or square-pyramids ($n = 5$). General rules governing the structures are: i) polyhedra share edges or vertexes (face-sharing is very scarce) in such a way that each polyhedron must show either one or two *cis*-related unshared oxygens to minimize the strong *trans*-effect of the terminal M–O bonds; and ii) addenda metals in each polyhedron are shifted toward the terminal M–O bonds as a result of their π contribution.

Polyoxometalates show extremely rich solution chemistry. They are formed by self-assembly processes via complex acidic condensation equilibria that are still not-well understood. These equilibria are affected by several other factors besides the pH, like temperature, reactant concentrations, ionic strength, or even counter-cations. Different POMs can coexist in solution for a given pH, in such a way that slight variations on the above-mentioned parameters can promote a particular species becoming predominant, being selectively isolated or even transforming into a radically different cluster.

Isopolyoxotungstates clearly illustrates this, as shown in Figure 1. Acidification of a tungstate solution results in the formation of heptatungstate $[W_7O_{24}]^{6-}$ at slightly acidic pH values. This cluster of seven edge-shared WO_6 octahedra with ideal C_{2v} symmetry is in equilibrium with paratungstate $[H_2W_{12}O_{42}]^{10-}$. The latter is the most commonly isolated species and it is formed by corner-sharing of two angular $\{W_3O_{18}\}$ and two triangular $\{W_3O_{13}\}$ trimers of edge-shared WO_6 octahedra. When the pH is lowered to moderately acidic conditions, a not-yet fully determined polyoxotungstate is formed, which transforms into the metatungstate $[H_2W_{12}O_{40}]^{6-}$ cluster composed of four of the above-mentioned $\{W_3O_{13}\}$ trimers sharing corners in an ideal T_d symmetry. The predominant species at highly acidic conditions is decatungstate $[W_{10}O_{32}]^{4-}$, formed by two fused $\{W_5O_{18}\}$ subunits that can be seen as monovacant derivatives of the Lindqvist-type $[W_6O_{19}]^{2-}$ hexatungstate. The latter is an octahedral assembly of six edge-shared WO_6 octahedra.

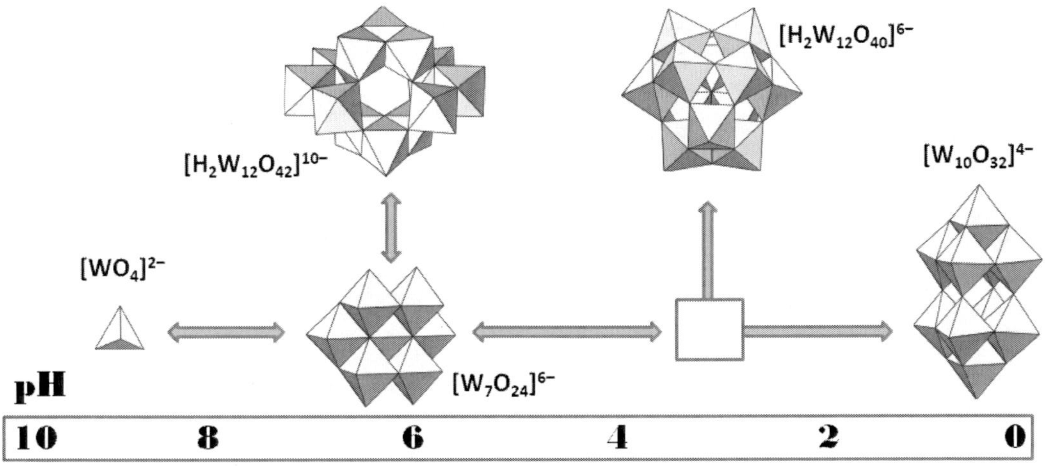

Figure 1. Isopolyoxotungstates formed by acidification of aqueous tungstate solutions.

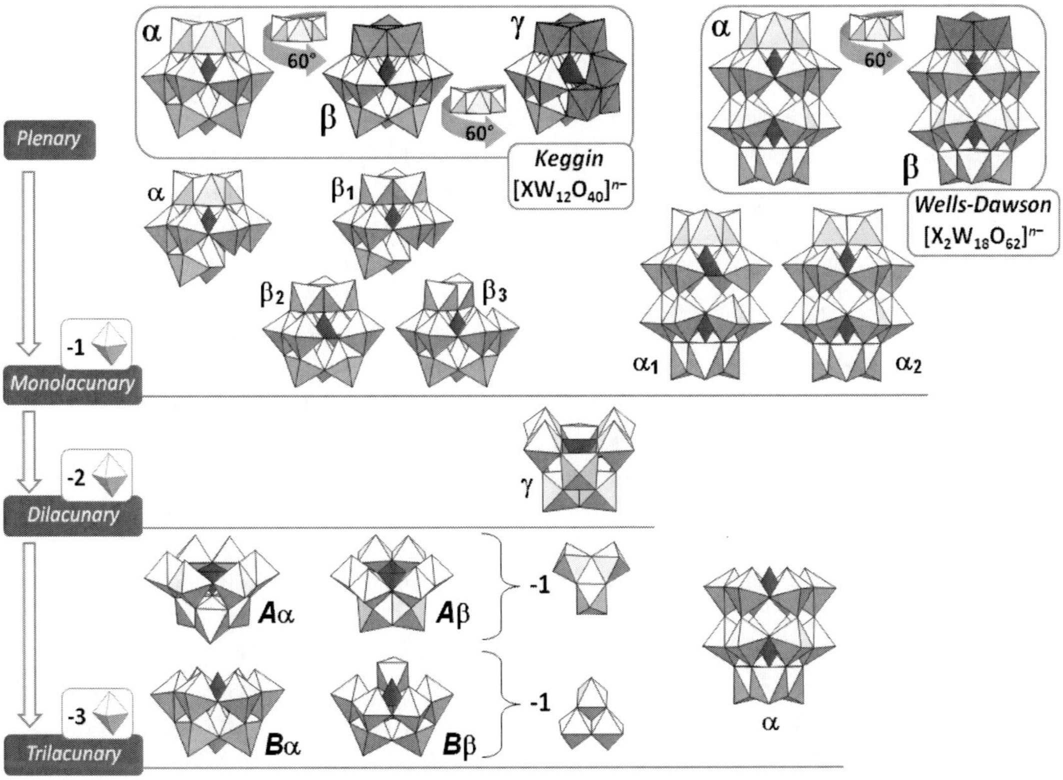

Figure 2. Isomers and mono- to trilacunary derivatives of Keggin $[XW_{12}O_{40}]^{n-}$ and Wells-Dawson $[X_2W_{18}O_{62}]^{n-}$ heteropolyoxotungstates isolable as independent polyoxometalates.

On the other hand, Keggin-type $[XW_{12}O_{40}]^{n-}$ anions and their derivatives can be cited as the most archetypal examples of heteropolyoxotungstates. The Keggin-type framework is virtually identical to that of the metatungstate but encapsulating a central XO_4 tetrahedral heterogroup that shares one vertex with each of the $\{W_3O_{13}\}$ trimers. This assembly is known as α-isomer, but four additional isomers can be postulated by 60^0 rotation of one to four $\{W_3O_{13}\}$ trimers around their ternary axis, namely β, γ, δ, and ε with C_{3v}, C_{2v}, C_{3v} and T_d symmetries, respectively. Their solution stability is lowered as the number of rotated trimers increases, in such a way that only α to γ isomers have been observed for polyoxotungstates.

Lacunary species of these Keggin-type isomers can be derived by removal of a variable number of WO_6 octahedra from the plenary skeletons, resulting in clusters with vacant addenda metal sites. Nine different lacunary clusters are readily isolable in the solid state as independent polyoxotungstates, as shown in Figure 2. Three of these species are derived from the α-isomer: the monovacant $[XW_{11}O_{39}]^{n-}$ and the A or B trilacunary forms by removal of one corner-shared $\{W_3O_{15}\}$ or one edge-shared $\{W_3O_{13}\}$ fragments, respectively. In contrast, three different monovacant forms are obtained for the β-isomer because addenda metal positions are no longer equivalent: β_1, β_2 and β_3 with vacant sites located in the $\{W_3O_{15}\}$ triad opposite to the rotated trimer, the so-called central belt and the rotated trimer itself, respectively. As for the α-isomer, trilacunary A and B forms are also known. Finally, the $[XW_{10}O_{36}]^{n-}$ cluster represents the only lacunary derivative of the γ-isomer and the only divacant cluster known. It is obtained by removal of the edge-shared $\{W_2O_{10}\}$ fragment involving one WO_6 octahedron on each of the rotated trimers.

Plenary and lacunary Keggin-type polyoxotungstates are metastable at different pH ranges: trilacunary species are generally formed at moderately basic pH values (pH ≈ 8 – 10), whereas monolacunary clusters predominate at moderately acidic conditions (pH ≈ 4 – 6) and they lead to plenary anions in highly acidic media (pH < 2). Thanks to their kinetically slow equilibria, they can be isolated separately by either controlled acidification of tungstate solutions in the presence of heteroatomic sources or controlled hydrolysis of plenary anions, but these two routes do not necessarily converge in the same species. Moreover, not all lacunary anions are accessible for all heteroatoms; for example, while the four β-monovacant clusters are known for group 14 heteroatoms (Si^{IV}, Ge^{IV}), they have not been isolated for the group 15 ones (P^{V}, As^{V}) because of rapid isomerization to the α-form. On the other hand, there are several other lacunary frameworks that have not been obtained as independent clusters, but have been observed as constituent subunits of more complex assemblies.

Condensation of two trilacunary $[A\text{-}\alpha\text{-}XW_9O_{34}]^{n-}$ subunits in an eclipsed conformation via corner-sharing gives rise to the Wells-Dawson $[X_2W_{18}O_{62}]^{n-}$ anion, which constitutes the parent cluster of another representative heteropolytungstate family. It is mainly known for group 15 heteroatoms and it also shows one isomeric form (β) and lacunary derivatives. The latter consist in two monolacunary $[X_2W_{18}O_{61}]^{n-}$ species with vacant sites at the $\{W_3O_{13}\}$ cap (α_2) or at the central double-belt (α_1) and a trilacunary one obtained by removal of one of the caps. There is also a hexalacunary cluster $[X_2W_{12}O_{48}]^{n-}$, showing six vacant positions all on the same side of the Wells-Dawson skeleton.

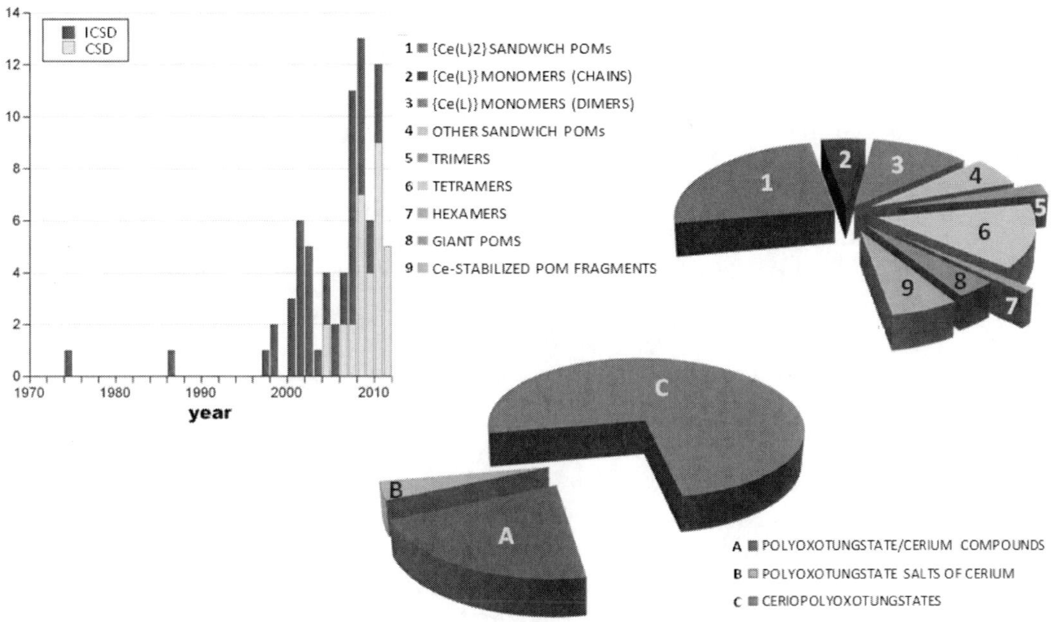

Figure 3. Temporal distribution of structures containing cerium and polyoxotungstates that are deposited in the ICSD and CSD databases (left) and diagrams showing a possible classification based on their nature (right).

2. Introduction to Ceriopolyoxotungstates

Combination of lacunary POMs and 4f metals constitutes a powerful tool for designing complex POM architectures, and at the same time, introducing new properties (additional catalytic centres, photoluminescence or magnetism). This chemistry has become a remarkable focus of interest, so that 4f-metal-containing POMs currently represent one of the largest and most active groups among the vast POM family [5]. One of the seminal works for these POMs refers to ceriopolyoxotungstates and dates from the early seventies, when Peacock and Weakley determined that the correct formulations of two previously reported species were decatungstocerate(III) and –(IV) and prepared analogues with several other 4f metals [6]. They also explored the reactivity of 4f metals (including Ce^{III} and Ce^{IV}) toward monolacunary Keggin- and Wells-Dawson type anions (L) and isolated a family of $\{4f(L)_2\}$-type complexes. The nature of decatungstocerate(IV) was established a few years later by single-crystal X-ray analyses [7], which showed it to be also a $\{4f(L)_2\}$-type anion where L was the monovacant pentatungstate. Subsequently, $\{4f(L)_2\}$-type anions are currently named as Peacock-Weakley sandwich anions. This work not only constitutes the first structural report on a ceriopolyoxotungstate, but also on a 4f-metal-containing polyoxotungstate.

The number of structurally characterized compounds combining polyoxotungstates and cerium has obviously increased during these four decades, but with an irregular profile. A search in the usual structural databases by combination of the compositional elements (Ce, W and O) results in 46 different POM entries deposited in the Inorganic Crystal Structure Database [8] and other 32 POM entries retrieved from the Cambridge Structural Database [9] up to December 2011. The authors found 5 more compounds that should have also been included in the ICSD database: Kazanskii's $[Ce^{IV}(\alpha_2\text{-}P_2W_{17}O_{61})_2]^{16-}$ [10], Niu's $[Ce^{III}(\alpha\text{-}SiW_{11}O_{39})_2]^{13-}$ [11], Kholdeeva and Hill's $[Ce^{IV}(H_2O)_3(\alpha\text{-}SiW_{11}O_{39})]_2^{8-}$ [12], and Nogueira's $[Ce^{IV}_2(\alpha(1,4)\text{-}PW_{10}O_{38})(\alpha\text{-}PW_{11}O_{39})_2]^{17-}$ [13] and $[Ce^{III}(H_2O)_3(\alpha\text{-}BW_{11}O_{39})]_\infty^{6-}$ [14]. The authors also noticed that Xu's $[Ce^{IV}(\alpha\text{-}PW_{11}O_{39})_2]^{10-}$ [15] is simultaneously present in both databases.

Although Kazanskii's sandwich anion dates from 1979, the histogram of Figure 3 shows that it took more than ten years for the structure of a second ceriopolyoxotungstate to be deposited, that of Knoth's $[\{(Ce^{IV}_3O_3(H_2O)_2\}(A\text{-}\alpha\text{-}PW_9O_{34})_2]^{12-}$ [16]. The decade of the nineties did not see a major improvement with only two more structurally characterized ceriopolyoxotungstates, Pope's $[As^{III}_{12}Ce^{III}_{16}(H_2O)_{36}W_{148}O_{524}]^{76-}$ [17] and Yamase's $[\{Ce^{III}_3(CO_3)\}(B\text{-}\alpha\text{-}Sb^{III}W_9O_{33})(W_5O_{18})_3]^{20-}$ [18], and the structural redetermination of the original Peacock-Weakley $[Ce^{IV}(W_5O_{18})_2]^{8-}$ [19]. It is only with the change of century that reports on ceriopolyoxotungstates move from being scarce to show certain regularity across the following years. Recently, the introduction of organic molecules in cerium/polyoxotungstate systems either as cations or as ligands has resulted in a significant increase of structural studies on such compounds, so that this particular field appears to be currently driven by this strategy. Despite of their almost forty years of history, ceriopolyoxotungstates still represent a young, active field with plenty of room for research.

We found a total number of 82 structural determinations in our database searches, but as shown in Figure 3, not all correspond to ceriopolyoxotungstates in the strict sense of this term, i.e. polyoxotungstates where cerium plays a key role in the POM skeleton. Almost one third of the structural reports correspond to compounds where well-defined, plenary isopolyanions (e.g. metatungstate, decatungstate) or heteropolyanions (e.g. Keggin and Wells-Dawson type species) are combined with cerium complexes bearing different types of ligands, pyridincarboxylate derivatives being remarkably abundant. These complexes act either as pendant moieties grafted on the POM surface to give the so-called "decorated" polyoxotungstates [20] or as linking subunits to connect POMs in mono- [21], bi- [22] or tridimensional [23] extended lattices. There are also a few compounds where POMs play the role of pillars for layered cerium-organic architectures or become trapped guests in the voids of three-dimensional cerium-organic frameworks [24]. Some of these structures are displayed in Figure 4 as illustrative examples. The number of polyoxotungstate salts with cerium complexes as counter-cations is merely symbolic [25].

Focusing on ceriopolyoxotungstates, there is a clear predominance of POM assemblies constructed from a variable number of lacunary subunits brought together by coordination of cerium atoms on the vacant sites, and also a few other species where cerium exclusively acts as stabilizing agent of novel POM fragments. Among the former, ceriopolyoxotungstates containing monolacunary subunits represent the largest group because they involve $\{Ce(L)_2\}$ anions and $\{Ce(L)\}$ monomeric building-blocks derived as dimeric species or monodimensional arrangements. There is a wider variety of assemblies for those ceriopolyoxotungstates involving lacunary subunits with two or more vacant sites. Defining the nuclearity as the number of lacunary subunits, they range from low-nuclear, relatively small species like trimers to giant assemblies with nuclearities higher than ten that contain over one hundred tungsten atoms. It is worth highlighting that, while tetramers are particularly abundant (several species derived from POM-cryptates that are constructed from four condensed lacunary fragments have been included in this group), there is an evident lack of medium to large sized POM assemblies showing nuclearities from five to nine. Some synthetic and structural aspects of the mentioned ceriopolyoxotungstates will be commented in detail below.

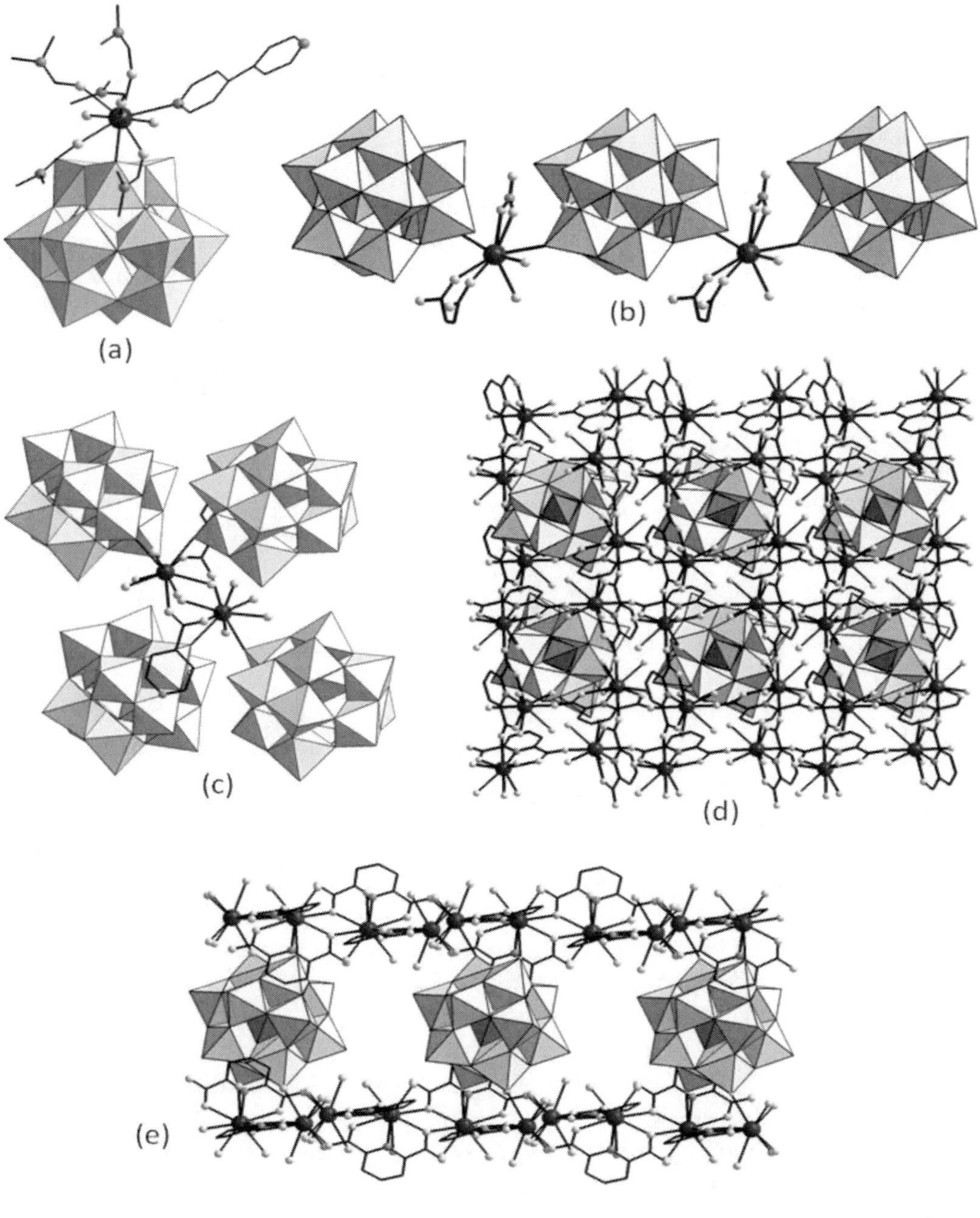

Figure 4. (a) Decorated $[\{Ce^{III}(DMF)_5(H_2O)_2(Hbpy)\}(H_2W_{12}O_{40})]^{2-}$ molecular anion (DMF: dimethylformamide, bpy: 4,4'-bipyridine) [20b]. (b) Monodimensional $[\{Ce^{III}(NO_3)(H_2O)_3(mal)\}(H_2W_{12}O_{40})]^{6-}$ assembly (mal: maleate) [21d]. (c) Fragment of the bidimensional $[\{Ce^{III}$(4-PYC)$(H_2O)_5\}_2\{H_2W_{12}O_{40})]$ assembly (4-PYC: pyridine-4-carboxylate) [22b]. (d) $[\{Ce^{III}$(2,6-PYDC)$(H_2O)_4\}_4\{\alpha\text{-}SiW_{12}O_{40}\}]$ compound consisting in a tridimensional cerium metal-organic framework with Keggin anions as guests (2,6-PYDC: pyridine-2,6-dicarboxylate) [24c]. (e) $[\{Ce^{III}$(2,6-PYDC)$(H_2O)_3\}_4\{\alpha\text{-}SiW_{12}O_{40}\}]$ compound consisting in Keggin-pillared cerium-metalorganic layers [24c]. Hydrogen atoms omitted for clarity.

3. CERIOPOLYOXOTUNGSTATES CONTAINING MONOLACUNARY SUBUNITS

Removal of $(M{=}O)^{n+}$ groups from POM skeletons to generate vacant sites on addenda metal positions endows the POM defect shell with enhanced reactivity towards electrophiles and allows these lacunary species for acting as rigid, highly nucleophilic multidentate O-donor ligands. In the simplest case of monolacunary POMs, creation of the vacant site results in the four surrounding oxygen atoms moving from having bridging nature to a terminal one and also in one central oxygen from the XO_4 group becoming accessible. These four, now "unshared" oxygens are almost coplanar and define a slightly distorted square. Thus, monolacunary POMs can act as pentadentate ligands to incorporate almost any transition metal in the vacant site with a water molecule usually occupying the sixth position, in such a way that mixed-metal POMs with the parent, plenary framework are regenerated.

In contrast, 4f metals cannot be fully incorporated in vacant sites because of their large size, i.e. they cannot access the central oxygen, and therefore, they place over the vacant site with monolacunary species acting as tetradentate ligands in close analogy to an inorganic porphyrin [26]. Having coordination numbers higher than seven, the 4f metal in these {4f(L)} monomeric building blocks still shows a number of coordination sites available for interaction with additional POM subunits.

3.1. Peacock-Weakley $\{Ce(L)_2\}$ Sandwich-Type Species

Peacock-Weakley ceriopolyoxotungstates are sandwich-type $\{Ce(L)_2\}$ species formed when cerium is capped with a second monolacunary subunit acting also as tetradentate O-donor ligand. Both ligands arrange in an alternate fashion, resulting in cerium centers showing axially-compressed $CeO_4(O')_4$ square-antiprismatic coordination geometries. This type of assembly can be viewed as an all-inorganic analogue of a cerium-porphyrinato or phthalocyaninato double-decker sandwich complex [27]. Recent studies have shown that late-4f-containing Peacock-Weakley species with different monolacunary subunits show indeed magnetic properties closely related to those of such double-decker complexes [28]. Structurally characterized examples are given in Table 1, whereas anions showing different types of monolacunary ligands are depicted in Figure 5.

Since the monolacunary Lindqvist-type pentatungstate fragment has never been observed as an independent cluster, decatungstocerates are synthesized in one-pot procedures by acidic condensation of the tungstate oxoanion or partial degradation of other polyoxotungstates in the presence of a cerium salt. Peacock and Weakley prepared several members of the $[4f(W_5O_{18})_2]^{n-}$ family as different alkaline salts (including Ce^{III} and Ce^{IV} POMs) by addition of the lanthanide nitrate (sulfate for Ce^{IV}) to a hot, aqueous solution of sodium tungstate (1:10 ratio) acidified to a pH value in the range 6.5 – 7.5 [6]. Others have completed this family [28, 29]. The authors found that these species are stable in solutions with pH ranging from 5.5 to 8.5 and that decatungstocerates show the highest stability. The Ce^{IV}-derivative could be electrolytically reduced to the Ce^{III}-species, whereas the latter could also be chemically oxidized to the former with H_2O_2 or $S_2O_8^{2-}$ but with partial degradation.

Table 1. Structurally characterized compounds containing Peacock-Weakley {Ce(L)$_2$} sandwich ceriopolyoxotungstates

Compound	Ref.
Lindqvist-type Subunits	
$Na_8[Ce^{IV}(W_5O_{18})_2]\cdot 30H_2O$	[19]
$Na_9[Ce^{III}(W_5O_{18})_2]\cdot NaCl\cdot 30H_2O$	[30]
$Na_9[Ce^{III}(W_5O_{18})_2]\cdot 34H_2O$	[31]
$(NH_4)_3[Na_5(H_2O)_{17}Cl][Na_2(H_2O)_8][Ce^{III}(W_5O_{18})_2]\cdot 6H_2O$	[32]
Keggin-Type Subunits	
$K_{4.5}Na_4(H_3O)_{3.5}[Ce^{IV}(\alpha\text{-}SiW_{11}O_{39})_2]\cdot 23H_2O$	[33]
$K_{8.5}H_{4.5}[Ce^{III}(\alpha\text{-}SiW_{11}O_{39})_2]\cdot 26H_2O$	[11]
$K_{13}[Ce^{III}(\beta_2\text{-}SiW_{11}O_{39})_2]\cdot 25.5H_2O$	[34]
$[(CH_3)_2NH_2]_{10}[Ce^{IV}(\alpha\text{-}PW_{11}O_{39})_2]\cdot 14H_2O$	[35]
$(NH_4)_2[(CH_3)_4N]_6Na_2[Ce^{IV}(\alpha\text{-}PW_{11}O_{39})_2]\cdot 14H_2O$	[15]
$[(CH_3)_2NH_2]_{11}[Ce^{III}(\alpha\text{-}PW_{11}O_{39})_2]\cdot 14H_2O$	[35]
$[H_9\{Cu(en)_2\}\{Ce^{III}(\alpha\text{-}PW_{11}O_{39})_2\}]\cdot 6H_2O^a$	[36]
$H_6[Na_2(en)_2(H_2O)_5][Cu(en)_2(H_2O)]_2[Cu(en)_2][\{Cu(en)_2(H_2O)\}\{Cu(en)_2\}\{Ce^{III}(\alpha\text{-}PW_{11}O_{39})_2\}]_2\cdot 16H_2O^a$	[37]
$H_{14}[\{Cu(en)_2\}_2\{Ce^{III}(\alpha\text{-}PW_{11}O_{39})_2\}]_2\cdot 2en\cdot 21H_2O^a$	[38]
$(H_2en)_2[Cu(en)_2][\{Cu(en)_2\}_{1.5}\{Cu(en)(bpy)\}\{Ce^{III}(\alpha\text{-}PW_{11}O_{39})_2\}]\cdot 2H_2O^a$	[39]
$(HTMA\text{-}CH_3)_2K_2Na_7[Ce^{III}(\alpha\text{-}As^{V}W_{11}O_{39})_2]\cdot HTMA\cdot 19H_2O^a$	[40]
Wells-Dawson-Type Subunits	
$K_{16}[Ce^{IV}(\alpha_2\text{-}P_2W_{17}O_{61})_2]\cdot 50H_2O$	[10]
$(NH_4)_{17}[Ce^{III}(\alpha_1\text{-}P_2W_{17}O_{61})(\alpha_2\text{-}P_2W_{17}O_{61})]\cdot nH_2O^b$	[41]
$K_{19}[Ce^{IV}(\alpha_2\text{-}H_4PW_{17}O_{61})_2]\cdot 22H_2O^c$	[42]
$K_{19}[Ce^{IV}(\alpha_2\text{-}H_4As^{V}W_{17}O_{61})_2]\cdot 21.5H_2O^c$	[42]
Mixed Subunits	
$K_{8.5}H_{3.5}[Ce^{III}(W_5O_{18})(\alpha\text{-}BW_{11}O_{39})]\cdot 25H_2O$	[43]

[a] en: ethylenediamine; bpy: 2,2'-bipyridine; HTMA: Hexamethylenetetraamine. [b] formula as reported; deposited structural data for this anion correspond to a potassium salt with structural formula $CeK_{10}O_{181}P_4W_{34}$. [c] formulae as reported; deposited structural data for these anions correspond to mixed ammonium/potassium salts.

Improved X-ray structural studies on the $[Ce^{IV}(W_5O_{18})_2]^{8-}$ POM were however performed in crystals obtained from an attempted synthesis of a Ce^{III}-containing tungstoantimonate(III) by reacting $CeCl_3$ with the $[B\text{-}\alpha\text{-}HSbW_9O_{33}]^{8-}$precursor in hot water [19]. These studies show ideal D_{4d} symmetry and Ce^{IV}–O bond lengths in the 2.34 – 2.38 Å range for decatungstocerate(IV).

The first structural determination of the $[Ce^{III}(W_5O_{18})_2]^{9-}$ POM was also performed on crystals obtained from an attempted synthesis of a Ce-containing tungstoantimonate(III). Gouzerh and coworkers isolated this POM from addition of $Ce(NO_3)_3$ to a hot, aqueous solution of Na_2WO_4 and Sb_2O_3 acidified to pH 7 [30]. As expected, the anion also shows D_{4d} symmetry and longer Ce^{III}–O bond lengths ranging from 2.50 to 2.53 Å. It is worth mentioning that the sodium ions form clusters as well: di-aquo-bridged dimers linking the decatungstocerates and a cluster analogous to pentatungstate but replacing tungsten by sodium, bridging and terminal oxygens by water molecules and the central oxygen by a chloride. Yuan et al. reported the structure of a mixed ammonium-sodium salt of this POM obtained by heating an aqueous NaCl solution of sodium decatungstate and $Ce(NO_3)_3$ with the pH adjusted to 4.8 [32]. The authors claimed this compound as the first extended lattice containing decatungstocerates based on the fact that they were linked in a one-dimensional assembly by sodium dimers, but this motif is exactly the same previously observed by Gouzerh. The structure shows the pentasodium clusters as well, and furthermore, both Gouzerh's and Yuan's compounds crystallize in the same orthorhombic space group (*I m m* 2) with almost identical cell parameters. Thus, the latter can be simply considered an ammonium isomorph of the former.

Peacock and Weakley also explored the reactivity of 4f metals toward monolacunary α-Keggin- (X = Si, P) and -Wells-Dawson (X = P) anions and isolated a family of $\{4f(L)_2\}$ species by alkaline degradation of plenary clusters in the presence of 4f metals. The Ce^{III} species were prepared as potassium salts by adding an aqueous solution of potassium acetate (pH = 7) to a hot solution of the corresponding heteropolytungstic acid and $Ce(NO_3)_3$, whereas the Ce^{IV} ones were obtained by oxidation of the former with potassium persulfate. For titration studies they used however salts of monolacunary anions as POM precursors. These studies showed that well-defined $\{4f(L)\}$ monomers are formed prior to $\{4f(L)_2\}$ species but they could only isolate the latter from solutions of any composition despite the higher stability of the former. In contrast to anions containing Wells-Dawson fragments, those showing Keggin-type subunits are stable at very low pH, so that acids can be prepared by cation exchange.

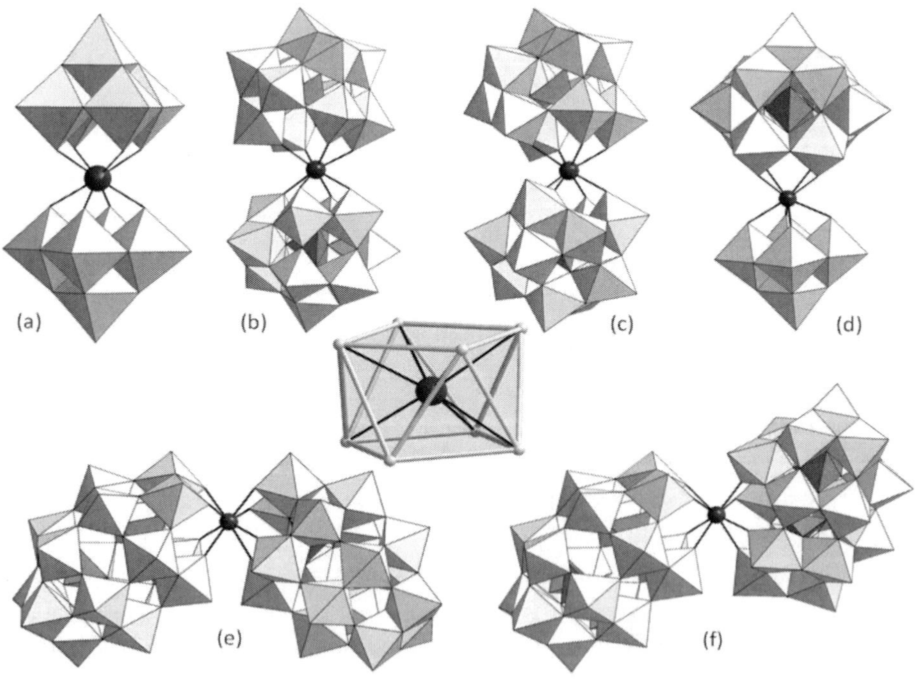

Figure 5. Molecular structures of Peacock-Weakley sandwich ceriopolyoxotungstates containing the monolacunary subunits: (a) Lindqvist; (b) α-Keggin (L-enantiomer); (c) β_2-Keggin (A-configuration); (d) mixed Lindqvist/α-Keggin; (e) Semivacant α_2-Wells-Dawson (*syn* conformer); d) mixed α_1- and α_2-Wells-Dawson (*anti* conformer). Translucent polyhedron illustrates the $CeO_4(O')_4$ coordination geometry.

Since no structural details were provided, other groups continued this work [44] but several 4f-derivatives based on Keggin subunits still lack structural reports. Cerium is probably the 4f metal for which the highest number of structures can be found, more specifically, $[Ce^{IV}(\alpha\text{-}XW_{11}O_{39})_2]^{n-}$ anions with X = Si^{IV} [33] and P^{V} [35, 15], $[Ce^{III}(\alpha\text{-}XW_{11}O_{39})_2]^{n-}$ species with X = Si^{IV} [11], P^{V} [35] and As^{V} [40] and the $[Ce^{III}(\beta_2\text{-}SiW_{11}O_{39})_2]^{13-}$ POM [34]. These anions are all chiral assemblies possessing ideal C_2 symmetry in which the two Keggin fragments are twisted by 40 – 45^0. They show two possible enantiomers derived from a hypothetical C_{2v}-symmetric assembly with eclipsed Keggin fragments by 45^0 clockwise (D) or counter clockwise (L) rotation. The Ce–O bond lengths are less regular compared to decatungstocerates, ranging from 2.30 to 2.50 Å for Ce^{IV} and from 2.38 to 2.56 Å for Ce^{III}. Fedotov et al. performed solution ^{183}W-NMR studies on $[4f(\alpha\text{-}PW_{11}O_{39})_2]^{11-}$ POMs, concluding that larger 4f metals (La to Eu) should have CeO_8 pseudo-cubic coordination according to the six-line pattern corresponding to a species with C_{2v} (twist of 0^0) or C_{2h} (twist of 180^0) symmetry, whereas late 4f metals showed the expected square-antiprismatic geometry [45]. Similar trends were encountered by Francesconi and coworkers [46] in their ^{31}P- and ^{183}W-NMR studies on the $[4f(\alpha\text{-}SiW_{11}O_{39})_2]^{10-}$ family.

Anions with group 15 heteroatoms have been synthesized starting from trilacunary $[XW_9O_{34}]^{9-}$ precursors: Fan et al. [15] reacted $[B\text{-}\alpha\text{-}PW_9O_{34}]^{9-}$ with $(NH_4)_2Ce(NO_3)_6$ and 8-hydroxyquinoline in a hot sodium acetate buffer to isolate $[Ce(\alpha\text{-}PW_{11}O_{39})_2]^{10-}$ (8-hydroxyquinoline showed to be essential for crystallization); Naruke and coworkers [35] prepared this and the Ce^{III} POM as $[(CH_3)_2NH_2]^+$ salts by adding $[A\text{-}\alpha\text{-}PW_9O_{34}]^{9-}$ to an aqueous solution of $(NH_4)_4Ce(SO_4)_4$ or $Ce(NO_3)_3$ (pH = 1.5) and heating the mixture to 60 ^{0}C (other Ce^{IV} salts gave unidentified solids); and Wang and coworkers [40] used hexamethylentetraamine to crystallize the arsenic POM by adding HTMA, $CeCl_3$ and Na_2WO_4 to an aqueous solution of $[B\text{-}\alpha\text{-}AsW_9O_{34}]^{9-}$ and refluxing the mixture with the pH adjusted to 5.0 – 5.5 with KOH. Naruke achieved spontaneous resolution of the L-enantiomeric form in crystals of $[Ce(\alpha\text{-}PW_{11}O_{39})_2]^{10-}$ thanks to using the achiral $[(CH_3)_2NH_2]^+$ cation.

Recently, the $[Ce(\alpha\text{-}PW_{11}O_{39})_2]^{11-}$ POM has been used as the inorganic building-block for constructing hybrid inorganic-metalorganic compounds [36-39]. These are 1-dimensional assemblies in which $[Cu(L)_2]^{2+}$ bridging moieties (L = ethylenediamine, 2,2'-bipyridine) link two adjacent POMs by axial coordination to two terminal oxygens, resulting in axially elongated CuN_4O_2 chromophores. Pendant $[Cu(L)_2]^{2+}$ or $[Cu(L)_2(H_2O)]^{2+}$ groups grafted on the POMs are also observed in some cases. These hybrids are all synthesized by hydrothermal techniques. As an illustrative procedure [36], a suspension of $Na_9[A\text{-}\alpha\text{-}PW_9O_{34}]$, $CuCl_2$, $Ce_2(SO_4)_3$ and ethylenediamine in water was sealed in a Teflon-lined autoclave and kept at 120 ^{0}C for 5 days, resulting in single crystals of $[H_9\{Cu(en)_2\}\{Ce^{III}(\alpha\text{-}PW_{11}O_{39})_2\}]$ (Figure 6).

Kortz and coworkers synthesized several $[4f(\beta_2\text{-}SiW_{11}O_{39})_2]^{13-}$ POMs including the Ce^{III}-derivative by addition of $K_8[\beta_2\text{-}SiW_{11}O_{39}]$ to an aqueous KCl solution of $(4f)Cl_3$, followed by mild heating after adjusting the pH to 5 [34]. AlDamen et al. complemented this study with some additional examples [28]. Since the β_2-fragment possesses intrinsic chirality, two possible configurations are possible: one with the same β_2-enantiomeric forms (A) and another one showing different β_2-enantiomers (B). The former results in a C_2-symmetric assembly and it is observed for early 4f metals including Ce, whereas the latter implying C_1 symmetry corresponds to late metals from Dy to Lu (gradual transition takes place from Sm to Tb). The authors found that the twist angle between the Keggin subunits and the interplanar distance between the two squares defined by the coordinating oxygens correlate with the 4f size, so that an increase of the former and a decrease of the latter take place as the 4f metal contracts. Naruke performed analogous correlation analyses on $[4f(\alpha\text{-}PW_{11}O_{39})_2]^{n-}$ POMs [35] and they observed that, in contrast to $[4f(\beta_2\text{-}SiW_{11}O_{39})_2]^{13-}$, the twist angle decreases with the 4f radius for this family. They rationalized this by means of electronic repulsion between coordinating oxygens, which is lowered by twist-angle decrease in α-Keggin derivatives and by gradual transformation from A- to B-configuration in the β_2-ones.

Pope and coworkers studied the interaction between Wells-Dawson monolacunary anions and cerium by reacting salts of the POM precursors with $Ce(NO_3)_3$ (2:1 ratio) in hot lithium acetate buffer (α_1-isomer) or water at pH 6 (α_2-isomer) [41]. For the α_2-precursor, they obtained the expected C_2-symmetric assembly with *syn*-conformation (twist angle between lacunary fragments of 45^0) similar to that of Kazanskii's $[Ce^{IV}(\alpha_2\text{-}P_2W_{17}O_{61})_2]^{16-}$ [10]. This type of assembly is also known for some other 4f metals [47]. For the α_1-species however, they observed precursor isomerization and isolated the mixed-ligand $[Ce^{III}(\alpha_1\text{-}P_2W_{17}O_{61})(\alpha_2\text{-}P_2W_{17}O_{61})]^{17-}$ sandwich with unprecedented *anti*-conformation, i.e. twisting angle of 135^0. There is only one other example of mixed-ligand $\{Ce(L)_2\}$ anion, Yamase's $[Ce^{III}(W_5O_{18})(\alpha\text{-}BW_{11}O_{39})]^{12-}$ [43]. Given the ligand congestion in a hypothetical $[4f(\alpha_1\text{-}P_2W_{17}O_{61})_2]^{n-}$ sandwich, it is not surprising that examples of such anions for trivalent 4f metals have not been reported so far [48]. Based on their solution ^{31}P-NMR studies and previous results on the $[4f(P_2W_{17}O_{61})_2]^{17-}$ family [46], coexistence of *syn* and *anti* conformers due to slow interconversion arising from ligand congestion was concluded for the mixed-ligand species, whereas fast intramolecular rotation facilitated by large 4f sizes and low steric hindrance was proposed to rationalize Fedotov's and Francesconi's observations.

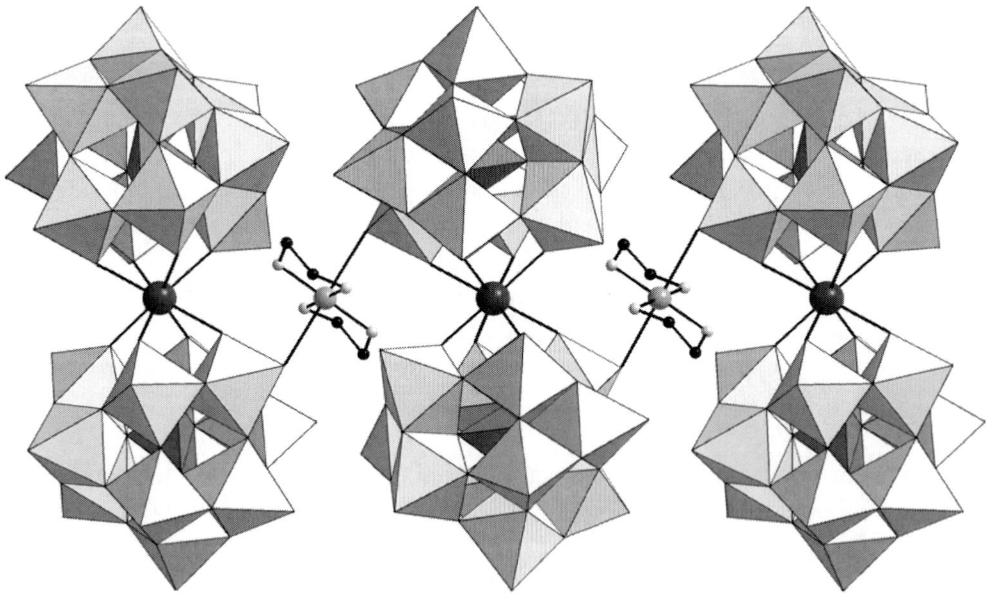

Figure 6. The mono-dimensional assembly of the $[H_9\{Cu(en)_2\}\{Ce^{III}(\alpha\text{-}PW_{11}O_{39})_2\}]$ hybrid compound.

Cerium coordination was the strategy applied by Pope to isolate a derivative of the $[H_4X^VW_{17}O_{61}]^{11-}$ anion (X = P, As) as crystals suitable for X-ray diffraction in order to confirm its semivacant Wells-Dawson nature, i.e. Wells-Dawson framework with one empty X position. Reaction of $Ce(NO_3)_3$ and $[H_4X^VW_{17}O_{61}]^{11-}$ (1:2 ratio) in warm water resulted in the C_2-symmetric $[Ce^{IV}(\alpha_2\text{-}H_4XW_{17}O_{61})_2]^{19-}$ sandwich anion with the expected *syn*-conformation [42]. Structural analyses confirmed the postulated Wells-Dawson framework for the $\{H_4X^VW_{17}O_{61}\}$ fragment with the vacant site on the $\{W_3O_{13}\}$ cap (α_2-isomer). The empty X position is the one not involved in the vacant site and all the four internal oxygens surrounding this tetrahedral position are protonated.

3.2. {Ce(L)} Monomers Derived as Chains

Although Peacock and Weakley could not isolate the {4f(L)} complexes they identified in solution, self-assembled derivatives of these monomers have been lately characterized in the solid state. Pope and coworkers prepared the $[Ce^{III}(H_2O)_3(\alpha\text{-}SiW_{11}O_{39})]_\infty^{5-}$ polymer by reaction of c.a. fourfold excess of $Ce(NO_3)_3$ with $K_8[\alpha\text{-}SiW_{11}O_{39}]$ in hot water [49]. Cerium excess is essential to avoid formation of the Peacok-Weakley sandwich. The resulting anion (Figure 7a) is a 1-dimensional assembly of $\{Ce(\alpha\text{-}SiW_{11}O_{39})\}$ building-blocks with zig-zag arrangement in which the coordination sphere of each cerium is completed by two terminal oxygens from two adjacent monomers ($Ce\text{–}O_{POM}$ = 2.42 – 2.52 Å) and three water molecules ($Ce\text{–}O_w$ = 2.47, 2.60, 2.98 Å). Hence, nona-coordinated cerium atoms show distorted monocapped square-antiprismatic $CeO_4(O')(O'')(O_w)_3$ geometry with a water molecule occupying the capping position. The polymeric anion dissociates in monomeric entities in solution according to the six-line ^{183}W-NMR pattern corresponding to a C_s-symmetric species.

Hu et al. reported a derivative of this anion containing an additional Ce^{III} in the cleft between two adjacent Keggin fragments further stabilizing the assembly [50]. In this case, the skeletal cerium lacks the capping water molecule, whereas the additional cerium is also in square-antiprismatic environment coordinated to two terminal oxygens and six water molecules (Figure 7b). Nogueira and coworkes isolated the $[Ce^{III}(H_2O)_3(\alpha\text{-}HBW_{11}O_{39})]_\infty^{5-}$ analogue by addition of $Ce(NO_3)_3$ dissolved in acetic acid to a hot, aqueous solution of Na_2WO_4 and $B(OH)_3$ with pH adjusted to 7 [14]. The polymeric anion with phosphotungstate Keggin fragments has also been reported, and like in Hu's derivative, the capping water in skeletal cerium atoms is absent leading to a square-antiprismatic $CeO_4(O')_1(O'')_1(O_w)_2$ geometry [51]. Polymeric assemblies of {4f(L)} subunits are not exclusive for Ce-containing species; they are known for several other 4f metals but the arrangement of monomeric building blocks is highly-dependent on the lanthanide, so that different chain-types have been described [52]. It is worth mentioning that no polymeric derivatives of {4f(L)} species containing pentatungstate or Wells-Dawson fragments have been identified so far to our knowledge.

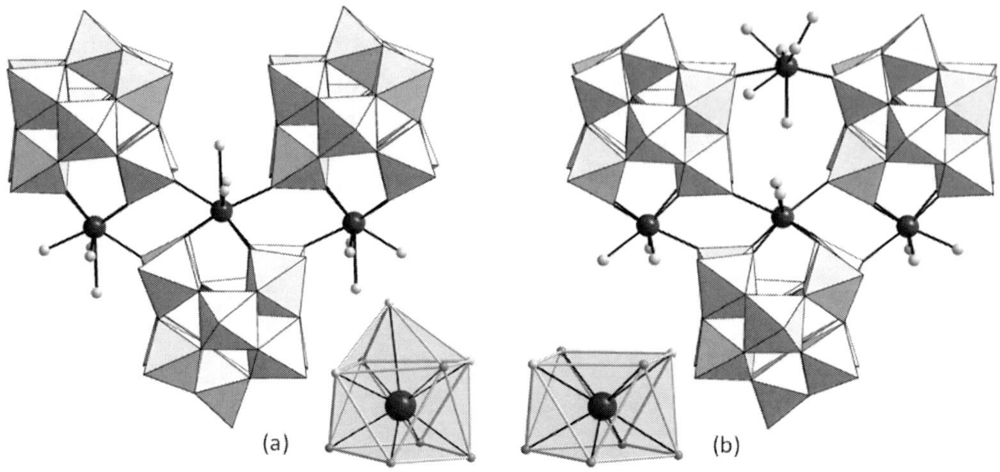

Figure 7. The mono-dimensional $[Ce^{III}(H_2O)_3(\alpha\text{-}SiW_{11}O_{39})]^{5-}$ (a) and $[\{Ce^{III}(H_2O)_6\}Ce^{III}(H_2O)_2(\alpha\text{-}SiW_{11}O_{39})]^{2-}$ (b) assemblies. Translucent polyhedra illustrate the different coordination spheres of the backbone cerium atoms (water molecules depicted in light grey).

Table 2. Structurally characterized compounds containing $\{Ce(L)\}_2$-type dimers

Compound	Ref.
Wells-Dawson-Type Subunits	
$(NH_4)_{10}K_2Li_2[Ce^{III}(H_2O)_4(\alpha_1\text{-}P_2W_{17}O_{61})]_2\cdot10H_2O$[a]	[53]
$(NH_4)_{14}[Ce^{III}(H_2O)_4(\alpha_2\text{-}P_2W_{17}O_{61})]_2\cdot nH_2O$[b]	[54]
$K_4Na_2H_2[\{Ce^{III}(H_2O)_8\}\{Ce^{III}(H_2O)_4(\alpha_2\text{-}P_2W_{17}O_{61})\}]_2\cdot10H_2O$	[55]
$(H_3O)_2[Ce^{III}{}_2(H_2O)_{14}\{Ce^{III}(H_2O)_3(\alpha_2\text{-}As_2W_{17}O_{61})\}]_2\cdot26H_2O$	[56]
Keggin-Type Subunits	
$Cs_8[Ce^{IV}(H_2O)_3(\alpha\text{-}SiW_{11}O_{39})]_2\cdot24H_2O$	[12]
$K_2H_2[Ce^{III}(H_2O)_7\{Ce^{III}(H_2O)_4(\alpha\text{-}SiW_{11}O_{39})\}]_2\cdot28H_2O$	[57]
$[Ce^{III}K(H_2O)_{12}][Ce^{III}(H_2O)_6]_2[Ce^{III}(H_2O)_4(\alpha\text{-}HBW_{11}O_{39})]_2\cdot20H_2O$	[23]
$[H_5\{Ag^I{}_3(H_2O)_2\}\{Ce_2{}^{III}(H_2O)_{12}\}\{Ce^{III}(H_2O)_4(\alpha\text{-}H_2W_{11}O_{39})\}_2]\cdot8H_2O$	[58]
$H[Cu(en)_2(H_2O)]_4[Cu(en)_2]_{1.5}[\{Ce(H_2O)(\alpha\text{-}SiW_{11}O_{39})\}_2(\mu\text{-}CH_3CO_2)_2]\cdot11H_2O$[c]	[59]

[a] formula as reported; deposited structural data for this anion correspond to a salt with structural formula $H_{124}Ce_2K_{0.5}N_{13.50}O_{157}P_4W_{34}$. [b] formula as reported; deposited structural data for this anion correspond to a salt with formula $(NH_4)_{13}K[Ce(H_2O)_4(\alpha_1\text{-}P_2W_{17}O_{61})]_2\cdot18H_2O$. [c] en: ethylenediamine.

3.3. {Ce(L)} Monomers Derived as Dimers

Besides as polymeric assemblies, monomeric {4f(L)} species have also been isolated as dimeric entities in the solid state. Structurally characterized compounds containing such $\{Ce(L)\}_2$ species are summarized in Table 2 and structures of three representative types of dimeric assemblies are depicted in Figure 8.

Following with their studies on the interaction between Wells-Dawson monolacunary anions and cerium, Pope and coworkers used a slight modification of the synthetic procedure of $[Ce^{III}(H_2O)_3(\alpha\text{-}SiW_{11}O_{39})]_\infty^{5-}$ [49] to prepare the corresponding {4f(L)} species. For the chiral α_1-monolacunary Wells-Dawson, they isolated the D,L-meso form of the $[Ce^{III}(H_2O)_4(\alpha_1\text{-}P_2W_{17}O_{61})]_2^{14-}$ dimer by addition of an almost fourfold excess of $Ce(NO_3)_3$ to a suspension of $[\alpha_1\text{-}P_2W_{17}O_{61}]^{10-}$ precursor in a lithium acetate buffer at room temperature [53]. The $[Ce^{III}(H_2O)_4(\alpha_2\text{-}P_2W_{17}O_{61})]_2^{14-}$ anion was obtained by reaction of a fourfold excess of $Ce(NO_3)_3$ with the $[\alpha_2\text{-}P_2W_{17}O_{61}]^{10-}$ precursor in hot water, followed by addition of NH_4Cl, which showed to be essential in the synthesis [54]. Both solid-state dimeric assemblies are constructed by linkage of two {Ce(L)} subunits via cerium coordination to a terminal oxygen from the adjacent Wells-Dawson fragment. Given the different relative positions of the vacant sites in both Wells-Dawson skeletons, *belt-to-belt* and *cap-to-cap* coordination modes for the α_1- and α_2-fragment derivatives were observed, respectively. In both cases, the cerium coordination sphere is completed with four water molecules leading to distorted monocapped square-antiprismatic $CeO_4(O')(O_w)_4$ geometries with a water molecule occupying the capping position. The Ce–O bond lengths range from 2.44 – 2.52 Å, whereas the Ce–O' ones involving adjacent subunits are significantly longer (2.54 – 2.67 Å) and comparable to the $Ce\text{–}O_w$ bond lengths.

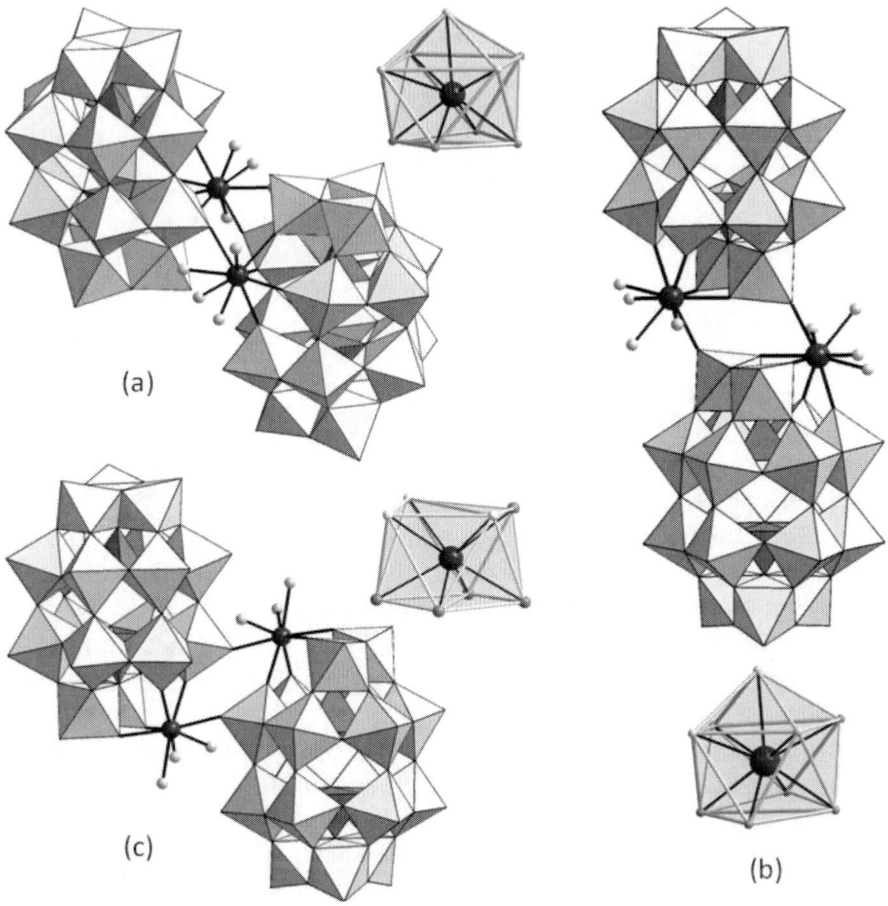

Figure 8. Structures of the *belt-to-belt* $[Ce^{III}(H_2O)_4(\alpha_1\text{-}P_2W_{17}O_{61})]_2^{14-}$ (a), *cap-to-cap* $[Ce^{III}(H_2O)_4(\alpha_2\text{-}P_2W_{17}O_{61})]_2^{14-}$ (b) and *cap-to-belt* $[\{Ce^{III}(H_2O)_3(\alpha_2\text{-}As_2W_{17}O_{61})\}]_2^{14-}$ (c) dimeric entities. Translucent polyhedra illustrate the different cerium coordination spheres (water molecules depicted in light grey).

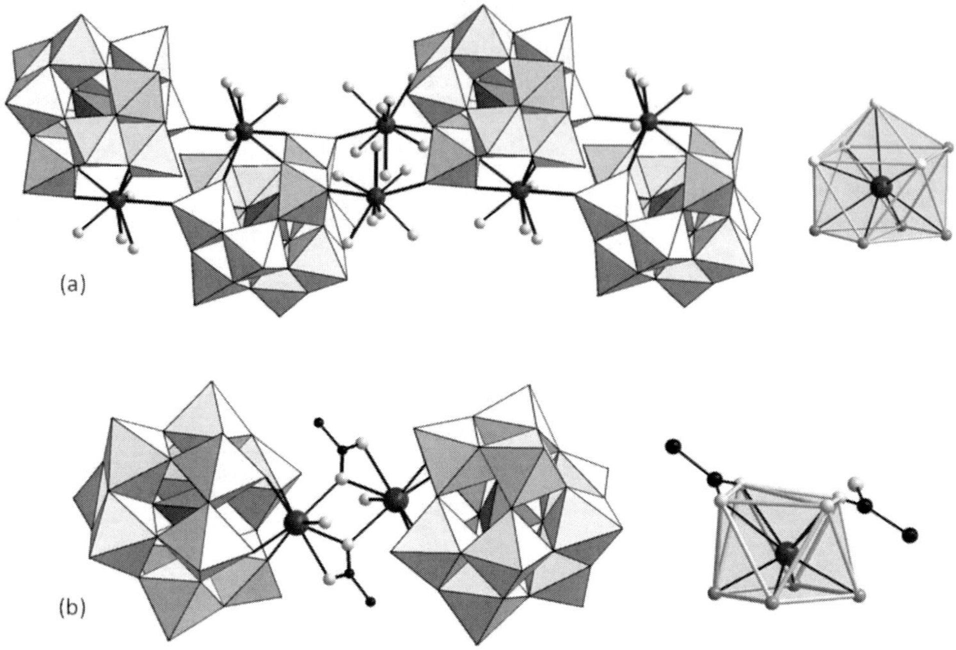

Figure 9. Structures of the monodimensional $[Ce^{III}(H_2O)_7\{Ce^{III}(H_2O)_4(\alpha\text{-}SiW_{11}O_{39})\}]_2^{4-}$ anion (a) and the $[\{Ce^{III}(H_2O)(\alpha\text{-}SiW_{11}O_{39})\}_2(\mu\text{-}CH_3CO_2)_2]^{12-}$ (b) dimer. Translucent polyhedra illustrate the different skeletal cerium coordination spheres (water molecules depicted in light grey).

They also performed exhaustive ^{183}W- and ^{31}P-NMR studies on these systems and concluded that the D,L-$[Ce^{III}(H_2O)_4(\alpha_1\text{-}P_2W_{17}O_{61})]_2^{14-}$ anion dissociates in solution based on the concentration dependency observed for the ^{31}P chemical shifts and estimated the dimerization constant. Further evidence of dissociation was given by splitting of the ^{31}P signals upon addition of a chiral aminoacid (L-proline) to the system, which was attributed to formation of diastereomeric $\{\alpha_1\text{-Ce(L)}\}$/proline pairs. The $\{\alpha_1\text{-Ce(L)}\}$/proline binding constants were also estimated for both diastereomeric pairs from the splitting dependence on the $\{\alpha_1\text{-Ce(L)}\}$/proline ratio owing to the fact that cerium paramagnetism sufficiently magnifies their chemical shift differences. Equilibria for the $[Ce^{III}(H_2O)_4(\alpha_2\text{-}P_2W_{17}O_{61})]_2^{14-}$ dimer showed to be more complicated because, in addition to fast dimer-monomer exchange, the latter was found to be also in slow exchange with the corresponding $\{Ce(L)\}_2$ species through a proposed Ce/$\{Ce(L)\}_2$ ion pair and the corresponding constant was estimated. Splitting of several ^{183}W signals upon addition of L-proline caused by coordination of the latter and subsequent loss of the C_s symmetry in the monomer provided further evidence for dimer dissociation.

A *cap-to-belt* variation of the $[Ce^{III}(H_2O)_4(\alpha_2\text{-}P_2W_{17}O_{61})]_2^{14-}$ dimer was obtained by Shi et al. from addition of $Ce(NO_3)_3$ to a hot, aqueous solution of $[P_2W_{18}O_{62}]^{6-}$ after adjusting the pH to 5.8 with Na_2CO_3 [55]. In this example, Ce–O linkages between monomeric subunits do not involve terminal oxygens from the POM defect caps, but from the double belts. This linking fashion is the one encountered for the $[4f(H_2O)_n(\alpha_2\text{-}P_2W_{17}O_{61})]_2^{14-}$ dimers containing mid-late 4f metals (Nd to Lu), whereas the *cap-to-cap* fashion appears to be characteristic of the larger metals (La to Pr) [60]. One additional $\{Ce(H_2O)_8\}$ pendant group per subunit was found to be grafted on the Wells-Dawson defect cap as “decoration”. Alkaline degradation of a plenary Wells-Dawson POM with Na_2CO_3 in hot water followed by Ce^{III} addition was also the strategy followed by Zhao et al. to prepare an arsenic analogue of these POMs [56]. They also obtained a *cap-to-belt* dimeric assembly, but in this case the cerium atoms lack the capping water molecule and show square-antiprismatic $CeO_4(O')(O_w)_3$ geometries. Monomeric subunits are similarly "decorated" with grafted $\{Ce(H_2O)_7\}$ pendant groups, and the dimers are furthermore linked in a one-dimensional assembly by additional $\{Ce(H_2O)_7\}$ bridging moieties. Two adjacent POMs in the chain are connected by two cerium atoms via interaction with two terminal oxygens each, so that the bridging ceriums show tricapped trigonal prismatic $CeOO'(O_w)_7$ distorted geometries.

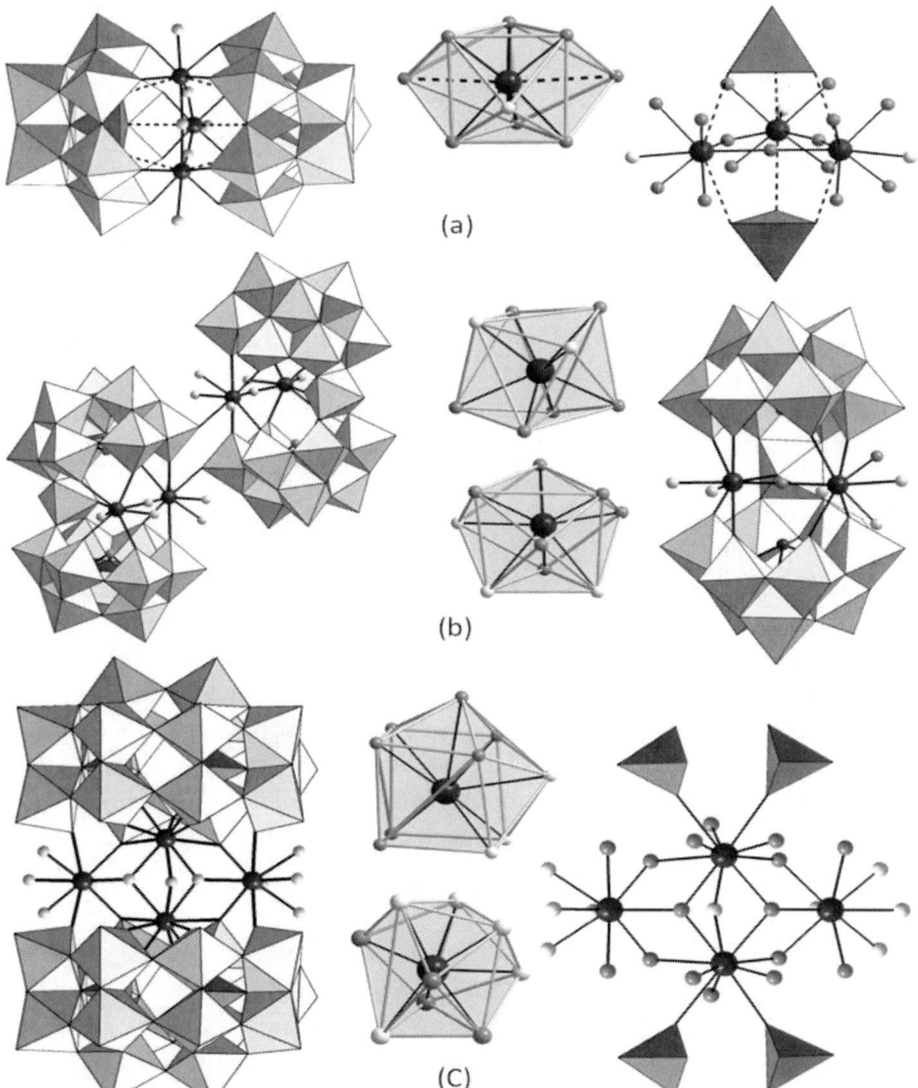

Figure 10. Structures of the $[\{Ce^{IV}_3(H_2O)_3O_3\}(A\text{-}\alpha\text{-}As^VW_9O_{34})_2]^{12-}$ (a), $[\{Ce^{IV}_2(H_2O)_5O\}\{WO(H_2O)\}(B\text{-}\alpha\text{-}As^{III}W_9O_{33})_2]_2^{16-}$ (b) and $[\{Ce^{III}_4(H_2O)_9(OH)_2\}(\alpha_1\alpha_1\text{-}P_2W_{16}O_{59})_2]^{14-}$ (c) sandwich anions with details of the central Ce^{IV}-clusters. Translucent polyhedra illustrate the different cerium coordination spheres (water molecules depicted in light grey).

Although several examples of $\{4f(L)\}_2$ species incorporating Wells-Dawson fragments are known as independent POMs, the corresponding cerium analogues based on Keggin subunits have only been observed as part of extended architectures with the exception of Kholdeeva and Hill's $[Ce^{IV}(H_2O)_3(\alpha\text{-}SiW_{11}O_{39})]_2^{8-}$ [12]. This fact is also common to other 4f metals [52a, 61]. Xu and coworkers obtained the $[Ce^{III}(H_2O)_7\{Ce^{III}(H_2O)_4(\alpha\text{-}SiW_{11}O_{39})\}]_2^{4-}$ anion by mixing the nickel-monosubstituted $[SiW_{11}O_{39}Ni(H_2O)]^{6-}$ Keggin POM with a tenfold excess of $Ce(NO_3)_3$ in water at room temperature [57]. Lowering the POM:Ce ratio resulted in the $[Ce^{III}(H_2O)_3(\alpha\text{-}SiW_{11}O_{39})]_\infty^{5-}$ chain or the Peacok-Weakley sandwich. The above anion is a monodimensional assembly in which two adjacent $[Ce^{III}(H_2O)_4(\alpha\text{-}SiW_{11}O_{39})]_2^{10-}$ dimers are linked by two $\{Ce(H_2O)_7\}$ bridging moieties via cerium coordination to two terminal oxygens of defect caps (Figure 9a). An et al. reported the $[Ce^{III}K(H_2O)_{12}][Ce^{III}(H_2O)_6]_2[Ce^{III}(H_2O)_4(\alpha\text{-}HBW_{11}O_{39})]_2\cdot 20H_2O$ compound [23] consisting in $[Ce^{III}(H_2O)_4(\alpha\text{-}HBW_{11}O_{39})]_2^{10-}$ dimers linked by additional cerium atoms in an intricate three-dimensional, all-inorganic open-framework that can be rationalized as a (3,6)-connected net showing channels filled with water molecules. This compound is synthesized from addition of $K_8[HBW_{11}O_{39}]$ to an aqueous $Ce(NO_3)_3$ solution (1:4 ratio), followed by heating after adjusting the pH to 3.5. A three-dimensional open-framework was also obtained by Pang et al. in the $[H_5\{Ag^{I}_3(H_2O)_2\}\{Ce_2^{III}(H_2O)_{12}\}\{Ce^{III}(H_2O)_4(\alpha\text{-}H_2W_{11}O_{39})\}_2]\cdot 8H_2O$ compound [58], where each $[Ce^{III}(H_2O)_4(\alpha\text{-}H_2W_{11}O_{39})]_2^{14-}$ dimer containing the monolacunary metatungstate fragment interacts with four cerium and four silver atoms to generate layers further connected by argentophilic interactions. This compound is prepared by addition of cerium and silver nitrates to an aqueous metatungstate solution at pH 5 (2:2:1 ratio), followed by acidification to pH 3.3 and heating, and it shows reversible dehydration behavior according to powder X-ray diffraction studies.

A different type of dimer was obtained by Du at al. in the hybrid inorganic-metalorganic compound $H[Cu(en)_2(H_2O)]_4[Cu(en)_2]_{1.5}[\{Ce(H_2O)(\alpha\text{-}SiW_{11}O_{39})\}_2(\mu\text{-}CH_3CO_2)_2]\cdot 11H_2O$ [59]. The $[\{Ce(H_2O)(\alpha\text{-}SiW_{11}O_{39})\}_2(\mu\text{-}CH_3CO_2)_2]^{12-}$ anion (Figure 9b) is composed of two $\{Ce(H_2O)(\alpha\text{-}SiW_{11}O_{39})\}$ subunits linked by two μ_2-bridging acetate ligands acting in a κ^3-O,O':O coordination fashion. The cerium coordination spheres are completed with one water molecule, so that they show a $CeO_4(O_{ac})_3(O_w)$ geometry that can be best described as square-antiprismatic, although highly distorted because of the acetate bite angle. Despite of being observed as part of a monodimensional compound where $\{Cu(en)_2\}$ moieties link the dimers and decorate the POM surface as pendant groups, this type of dimeric assembly is not unusual and has been described for other 4f metals and monolacunary fragments [62].

4. Ceriopolyoxotungstates Containing Polylacunary Subunits

Dilacunary polyoxotungstates with non-adjacent vacant positions act as "double" monolacunary anions, hence they can coordinate two cerium atoms in a bis(tetradentate) fashion and these will show different relative orientations of their sets of coordinative available positions. On the other hand, it is known that trilacunary POMs usually incorporate up to three electrophiles that graft on the central belt via coordination to two oxygens each. For cerium, this results in increasing the number of potential coordination sites compared to monolacunary anions. Thus, combination of cerium with polylacunary subunits can lead to a remarkable variety of architectures with very different topologies that range from relatively small sandwich-type POMs to giant assemblies among the largest polyoxotungstates known.

4.1. Sandwich-Type Ceriopolyoxotungstates

Besides Peacock-Weakley species, a few other sandwich ceriopolyoxotungstates are known. They are composed of two polylacunary subunits that encapsulate a central cluster of oxo- or hydroxo bridged Ce^{III} or Ce^{IV} atoms with nuclearities ranging from two to four.

As part of their investigations on the nature of the at that time recently isolated 9-tungstophosphate, Knoth et al. reacted this precursor with $(NH_4)_2Ce(SO_4)_3$ in water to prepare a metal-derivative that could be structurally characterized. Refluxing a yellow powder obtained from the above reaction by precipitation with KCl, they isolated the $[\{Ce^{IV}_3(H_2O)_2O_3\}(A\text{-}\alpha\text{-}PW_9O_{34})_2]^{12-}$ POM, confirming that unthermolized 9-tungstophosphate is an α-isomer and shows the A-type structure [16]. This C_{2V}-symmetric sandwich species is composed of two eclipsed $\{A\text{-}\alpha\text{-}PW_9O_{34}\}$ subunits linked by a central belt of three cerium atoms that are bridged by three μ_2-oxo ligands and further coordinated to two edge-shared $\{W_2O_{10}\}$ fragments each. Two of the ceriums show a terminal water, so that the central cluster can be seen as a ring of three corner-shared $CeO_2(O')_2(oxo)_2$ trigonal prisms, two of them being monocapped by a water molecule. Alizadeh et al. performed a similar reaction with 9-tungstoarsenate(V) and obtained the $[\{Ce^{IV}_3(H_2O)_3O_3\}(A\text{-}\alpha\text{-}As^{V}W_9O_{34})_2]^{12-}$ analogue [63]. They found some structural differences for this C_{3v}-symmetric sandwich (Figure 10a): all the cerium atoms display terminal water molecules and their distances to the oxygen atoms of the XO_4 tetrahedra are shortened to 2.86 – 2.94 Å, so that their trigonal-prismatic geometry can be considered as tricapped. They also reported a remarkable wide range of pH stability for both anions (2.5 – 9.0).

Wang and coworkers studied the interaction between $[B\text{-}\alpha\text{-}As^{III}W_9O_{33}]^{9-}$ and Ce^{IV} and isolated the $[\{Ce^{IV}_2(H_2O)_5O\}\{WO(H_2O)\}(B\text{-}\alpha\text{-}As^{III}W_9O_{33})_2]_2^{16-}$ dimer [64] from refluxing an aqueous mixture of the mentioned precursor, $Ce(SO_4)_2$ and potassium citrate at pH 3 (citrate was reported to be introduced as Ce^{IV} stabilizer to avoid instant precipitation of cerium-polyoxotungstate compounds). Each sandwich-type half of this dimer (Figure 10b) is composed of two $\{B\text{-}\alpha\text{-}AsW_9O_{33}\}$ subunits linked via one extra tungsten and a μ_2-oxo bridged cerium dimer. Each metal in the central belt coordinate to two corner-shared $\{W_2O_{11}\}$ fragments. The Ce atoms show different coordination environments: two water molecules and the terminal oxygen of the extra tungsten complete the highly-distorted $CeO_2(O')_2(O'')(oxo)(O_w)_2$ octa-coordinated sphere of one cerium, whereas three water molecules and a terminal oxygen from the adjacent sandwich subunit give a tricapped trigonal-prismatic character to the nona-coordinated $CeO_2(O')_2(O'')(oxo)(O_w)_3$ sphere of the second cerium. The observed $\{WO(H_2O)(B\text{-}\alpha\text{-}As^{III}W_9O_{33})_2\}$ fragment does actually exist as an independent POM, hence the sandwich can be rationalized as a dilacunary $[As_2^{III}W_{19}O_{67}]^{14-}$ polyoxotungstate encapsulating an oxo-bridged $\{Ce^{IV}_2(H_2O)_5O\}$ dimeric cluster. Wang reported this species to be stable in solutions with pH 2.5 – 3.5 and used it as a precursor suitable for constructing larger assemblies by aggregation with various linkers. They obtained two tetrameric species that will be commented below.

Ostuni and Pope synthesized the C_{2v}-symmetric $[\{Ce^{III}_4(H_2O)_9(OH)_2\}(\alpha_1\alpha_1\text{-}P_2W_{16}O_{59})_2]^{14-}$ sandwich (Figure 10c) by reacting the Wells-Dawson hexavacant $[P_2W_{12}O_{48}]^{14-}$ with Na_2WO_4 and $Ce(NO_3)_3$ in hot, aqueous LiCl at almost neutral pH [65]. This species was reported to be stable in solution according to ^{183}W- and ^{31}P-NMR studies and it consists in two dilacunary $\{\alpha_1\alpha_1\text{-}P_2W_{16}O_{59}\}$ subunits encapsulating a central tetracerium rhomblike cluster. The dilacunary $\{\alpha_1\alpha_1\text{-}P_2W_{16}O_{59}\}$ subunit derives from the hexavacant species by tungsten incorporation to the caps and to two adjacent vacant sites on either side of the double belt. This subunit was unprecedented at the time, illustrating the cerium ability as stabilizing agent to isolate novel POM fragments. Two different cerium types are found: the ones inserted in between the adjacent vacant sites via coordination to seven polyoxotungstate oxygens including the XO_4 groups (Ce1), and those linking the $\{P_2W_{16}O_{59}\}$ fragments via coordination to two oxygens on each vacant site (Ce2). Two μ_3-Ce1,1,2 bridging hydroxyls and one μ_2-Ce1,1 bridging water molecule complete the Ce1 coordination sphere, while each Ce2 shows four additional terminal aquo ligands. Hence, the cerium cluster can be described as two internal, face-shared $CeO_5(O_X)_2(OH)_2O_w$ bicapped cubes sharing edges with two external $CeO_2(O')_2(OH)(Ow)_4$ tricapped trigonal prisms.

4.2. Trimeric Ceriopolyoxotungstates

Two different trimeric ceriopolyoxotungstates can be found in the literature, Nogueira's $[Ce^{IV}_2(\alpha(1,4)\text{-}PW_{10}O_{38})(\alpha\text{-}PW_{11}O_{39})_2]^{17-}$ [13] and Wang's $[K\subset\{Ce^{III}(H_2O)_2(\alpha(1,4)\text{-}AsW_{10}FeO_{38})\}_3]^{14-}$ [66]. We also noticed a preliminary report on a third trimeric species, $[Ce^{III}_2(H_2O)_7(WO_2)_2(B\text{-}\alpha\text{-}As^{III}W_9O_{33})_3]^{17-}$ [67], but we were not able to find a complete report on this POM.

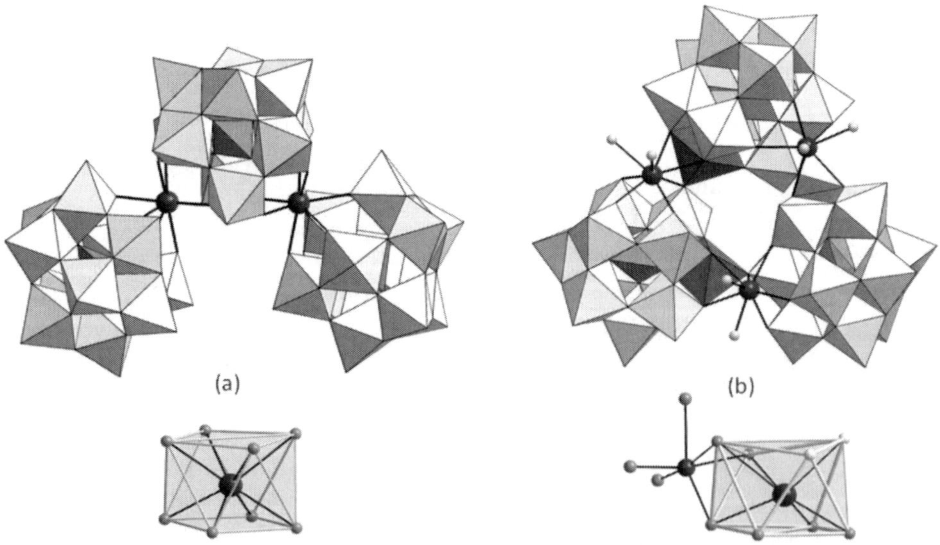

Figure 11. Structures of $[Ce^{IV}_2(\alpha(1,4)\text{-}PW_{10}O_{38})(\alpha\text{-}PW_{11}O_{39})_2]^{17-}$ (a) and $[K\subset\{Ce^{III}(H_2O)_2(\alpha\text{-}AsW_{10}FeO_{38})\}_3]^{14-}$ (b, FeO_6 octahedra represented in dark grey). Central potassium ions are omitted for clarity. Translucent polyhedra illustrate the cerium coordination spheres (water depicted in light grey).

The $[Ce^{IV}_2(\alpha(1,4)\text{-}PW_{10}O_{38})(\alpha\text{-}PW_{11}O_{39})_2]^{17-}$ trimer (Figure 11a) was prepared by alkaline degradation of phosphotungstic acid in the presence of $(NH_4)_2Ce(NO_3)_6$ with a neutral solution of potassium acetate and it was isolated from the reaction mixture as a second solid fraction after removal of $K_{10}[Ce(PW_{11}O_{39})_2]$. It contains an unusual dilacunary (α-$PW_{10}O_{38}$) Keggin subunit with vacant sites in relative addenda metal positions 1 and 4 as defined by Pope and Scully [68]. This subunit does not fulfill the principles for existing independently in solution and being isolable in the solid state. It has therefore been scarcely observed in POM chemistry [69], Nogueira's trimer thus further exemplifying how cerium coordination represents a suitable strategy for stabilizing unprecedented POM subunits. The vacant sites in ($\alpha(1,4)$-$PW_{10}O_{38}$) are not adjacent, hence this subunit coordinates two Ce^{IV} atoms in bis(tetradentate) fashion and each cerium is further capped with a monolacunary (α-$PW_{11}O_{39}$) leading to a $CeO_4(O')_4$ geometry described by the authors as bicapped trigonal-prismatic. The final C_2-symmetric assembly could be assimilated to a "double" Peacock-Weakley sandwich with angular topology further stabilized by one potassium ion in the cleft between (α-$PW_{11}O_{39}$) fragments.

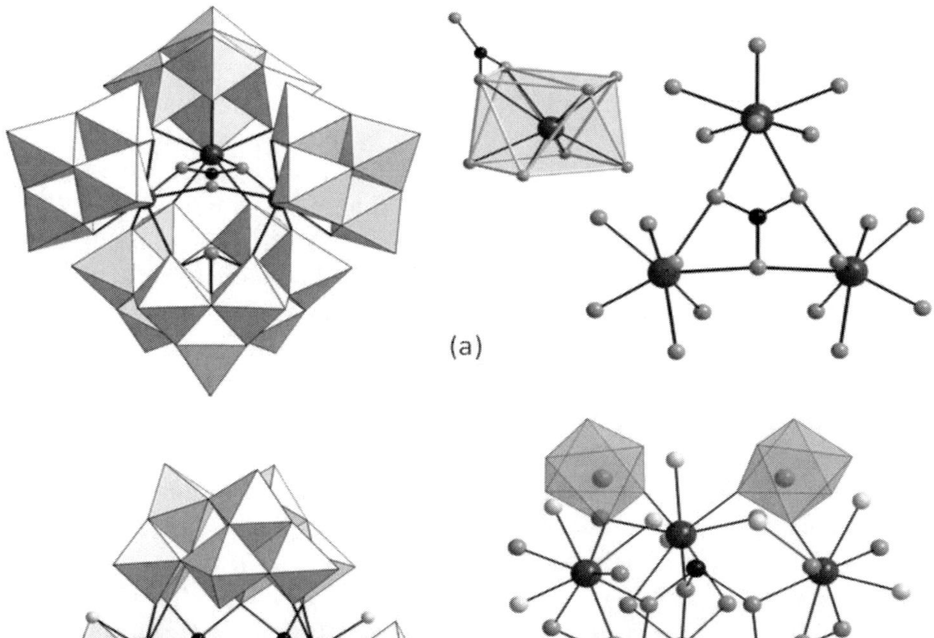

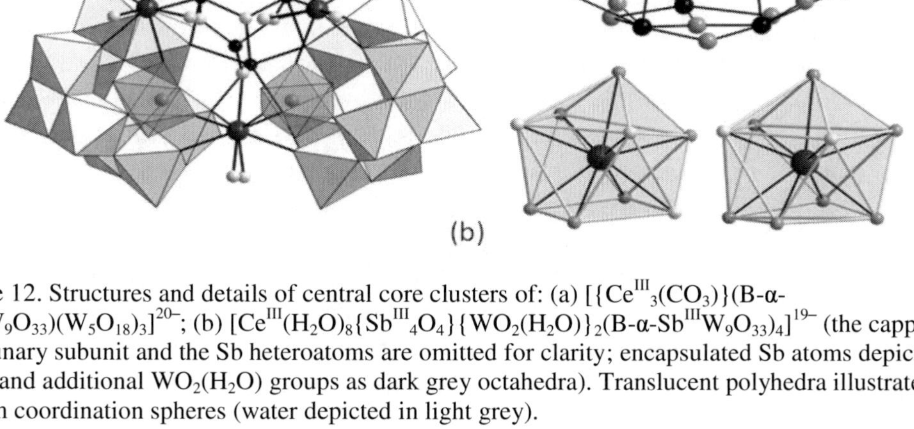

Figure 12. Structures and details of central core clusters of: (a) $[\{Ce^{III}_3(CO_3)\}(B\text{-}\alpha\text{-}Sb^{III}W_9O_{33})(W_5O_{18})_3]^{20-}$; (b) $[Ce^{III}(H_2O)_8\{Sb^{III}_4O_4\}\{WO_2(H_2O)\}_2(B\text{-}\alpha\text{-}Sb^{III}W_9O_{33})_4]^{19-}$ (the capping trilacunary subunit and the Sb heteroatoms are omitted for clarity; encapsulated Sb atoms depicted in black and additional $WO_2(H_2O)$ groups as dark grey octahedra). Translucent polyhedra illustrate the cerium coordination spheres (water depicted in light grey).

Wang's $[K\subset\{Ce(H_2O)_2(AsW_{10}FeO_{38})\}_3]^{14-}$ trimer (Figure 11b) is also based on dilacunary Keggin subunits with vacant sites in relative 1,4 positions. It was prepared by addition of $CeCl_3$, $FeCl_3$ and HTMA (acting as the appropriate cation to crystallize the title POM) to an acidified aqueous solution of Na_2HAsO_4 and Na_2WO_4, followed by basification with KOH to pH 4-5. The 10-tungstoarsenate(V) subunits fully incorporate one Fe^{III} in one vacant site, while one Ce^{III} is placed over the other. The disubstituted monomers assemble around a central potassium ion acting as template via Ce-coordination to a FeO_6 octahedral face, leading to a C_3-symmetric cyclic species that constitutes one of the few structurally characterized examples of heterometallic POMs containing O-bridged 3d-4f atoms known so far [70]. The title POM thus contains three independent $\{Fe^{III}Ce^{III}\}$ dimeric moieties where the cerium coordination sphere completed with two additional water molecules can be best described as square-antiprismatic $CeO_4(O')_2(Ow)_2$, although highly distorted because of face-sharing with the FeO_6 octahedron.

4.3. Tetrameric Ceriopolyoxotungstates

Tetrameric species represent the largest group among ceriopolyoxotungstates based on polylacunary subunits and almost all of them display trilacunary subunits with lone-pair containing heteroatoms like As^{III} or Sb^{III}. The first of these tetramers, namely $[\{Ce^{III}_3(CO_3)\}(B\text{-}\alpha\text{-}Sb^{III}W_9O_{33})(W_5O_{18})_3]^{20-}$ [71], was obtained by Naruke and Yamase in an attempted synthesis of a cerium analogue of $[Eu_3(H_2O)_3(B\text{-}\alpha\text{-}SbW_9O_{33})(W_5O_{18})_3]^{18-}$ consisting in addition of $CeCl_3$ and Sb_2O_3 dissolved in HCl to a hot, aqueous solution of WO_3, KOH and $KHCO_3$, followed by pH adjustement to 7.4 with additional $KHCO_3$. This C_{3v}-symmetric assembly is composed of three $\{Ce(W_5O_{18})\}$ subunits brought together by a central μ_3-bridging carbonate acting in a κ^2-O,O' chelating fashion toward each of the three cerium atoms (Figure 12a). This substructure is capped by an additional $(B\text{-}\alpha\text{-}SbW_9O_{33})$ subunit via coordination of each cerium to a corner-shared $\{W_2O_{11}\}$ belt-fragment. The central $\{Ce_3(CO_3)\}$ cluster shows a slightly curved triangular shape and it can be seen as three corner-shared $CeO_4(O')_2(O_{CO3})_2$ square-antiprisms with noticeable distortion because of the carbonate coordination mode.

Gouzerh and coworkers also explored one-pot approaches for the synthesis of cerium containing tungstoantimonate(III) species and isolated the $[Ce^{III}(H_2O)_8\{Sb^{III}_4O_4\}\{WO_2(H_2O)\}_2(B\text{-}\alpha\text{-}Sb^{III}W_9O_{33})_4]^{19-}$ POM from addition of Sb_2O_3 dissolved in HCl to a hot, aqueous NH_3 solution of WO_3, followed by pH adjustement to 7.5 and subsequent addition of $Ce(NO_3)_3$ [30]. It must be noticed that a similar reaction in water led to the pentatungstate-based Peacock-Weakley sandwich, illustrating the complexity of chemical equilibria involving POMs. This dissymmetric tetramer (Figure 12b) displays a complex architecture that can be rationalized as three $(B\text{-}\alpha\text{-}SbW_9O_{33})$ subunits linked by three alternate ceriums, two of them coordinated to both $\{W_2O_{10}\}$ and $\{W_2O_{11}\}$ fragments (Ce1) and the third (Ce2) exclusively to edge-shared fragments. This cyclic assembly encapsulates a (Sb_4O_4) subunit formed by a $\{Sb_3(\mu_3\text{-}O)\}$ triangle capped by a SbO_3 pyramid, which also links the three cerium atoms. A fourth $(B\text{-}\alpha\text{-}SbW_9O_{33})$ subunit containing two additional $\{WO_2(H_2O)\}$ groups grafted on the belt is attached to the cyclic assembly via corner-sharing of these groups with the subunits Ce2 links and also simultaneous coordination to Ce1 and Ce2 atoms.

Terminal water molecules complete the cerium spheres to result in highly distorted nona-coordinated geometries defined as monocapped square-antiprismatic by the authors.

In the case of tetrameric species based on tungstoarsenate(III) subunits, structural characterizations of another two compounds have been reported. Both compounds were synthesized by using the above commented $[\{Ce^{IV}_2(H_2O)_5O\}\{WO(H_2O)\}(B\text{-}\alpha\text{-}As^{III}W_9O_{33})_2]_2^{16-}$ dimer as POM precursor in reactions with organic or inorganic linkers. Wang and coworkers reacted this precursor with L-α-alanine in hot water at pH 2 and obtained the $[Ce^{IV}_4As_4W_{44}O_{151}(ala)_4(OH)_2(H_2O)_{10}]^{16-}$ POM [64]. This centrosymmetric (B-α-$As^{III}W_9O_{33}$)-based tetramer can be rationalized as follows: two trilacunary subunit are bridged by an additional $\{WO_2\}$ group via corner-sharing to $\{W_2O_{11}\}$ belt-fragments, and the so generated building-blocks are further linked by corner-shared $\{W_2O_5(\mu_2\text{-ala-}\kappa^2\text{-O:O'})\}$ dimers via analogous type of connectivity. Thus, an elongated cryptand-type assembly is generated where the third position on the belt of each trilcaunary subunit is occupied by a cerium atom, in such a way that adjacent ceriums on the long side of the cryptand are connected by a second μ_2-ala-κ^2-O:O' ligand. A corner-shared $\{W_2O_5(OH)_2\}$ dimer is encapsulated in the center and this dimer shares corners with the $\{W_2O_5(ala)\}$ linkers and simultaneously coordinates the cerium atoms, whose coordination spheres are completed by water molecules to result in $CeO_2O'O''O'''(O_{ala})(O_w)_n$ distorted square-antiprismatic (n = 2) or monocapped square-antiprismatic (n = 3) geometries. This POM (Figure 13a) represents one of the scarce examples of organic ligands directly attached to the POM-surface via tungsten coordination and the first one involving aminoacids.

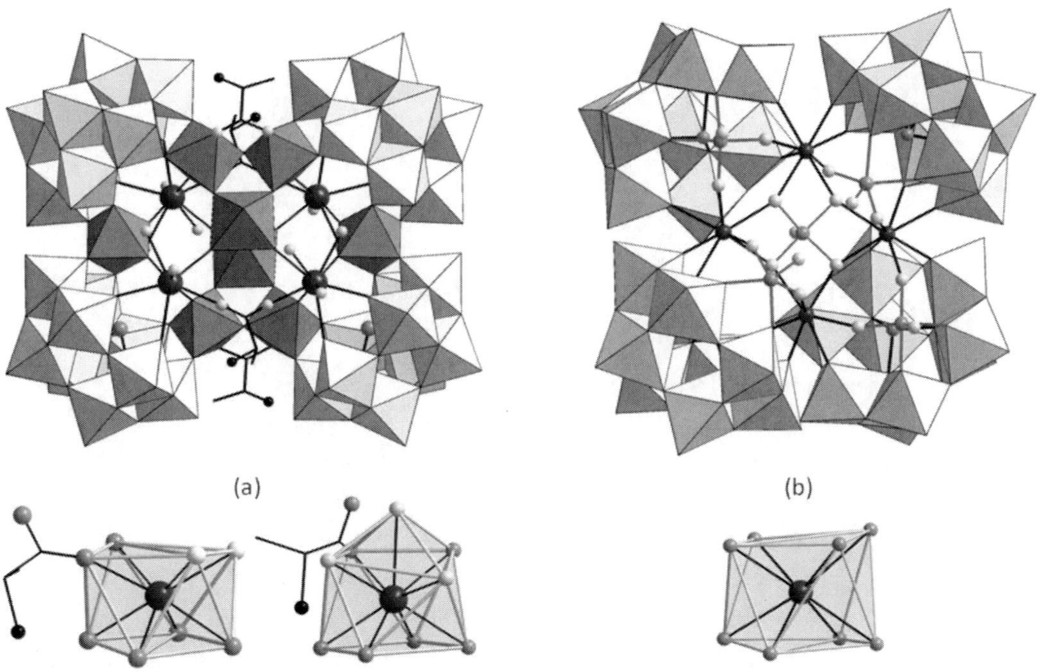

Figure 13. Structures of: (a) $[Ce^{IV}_4As_4W_{44}O_{151}(ala)_4(OH)_2(H_2O)_{10}]^{16-}$ (additional tungstens depicted as dark grey octahedra; water molecules and alanine oxygens depicted in light grey, alanine nitrogens shown in black); (b) $[Ce^{IV}_4As_4W_{41}O_{149}]^{24-}$ (additional tungstens and their bonds are shown in dark grey). Translucent polyhedra illustrate the cerium coordination spheres (water molecules shown in light grey).

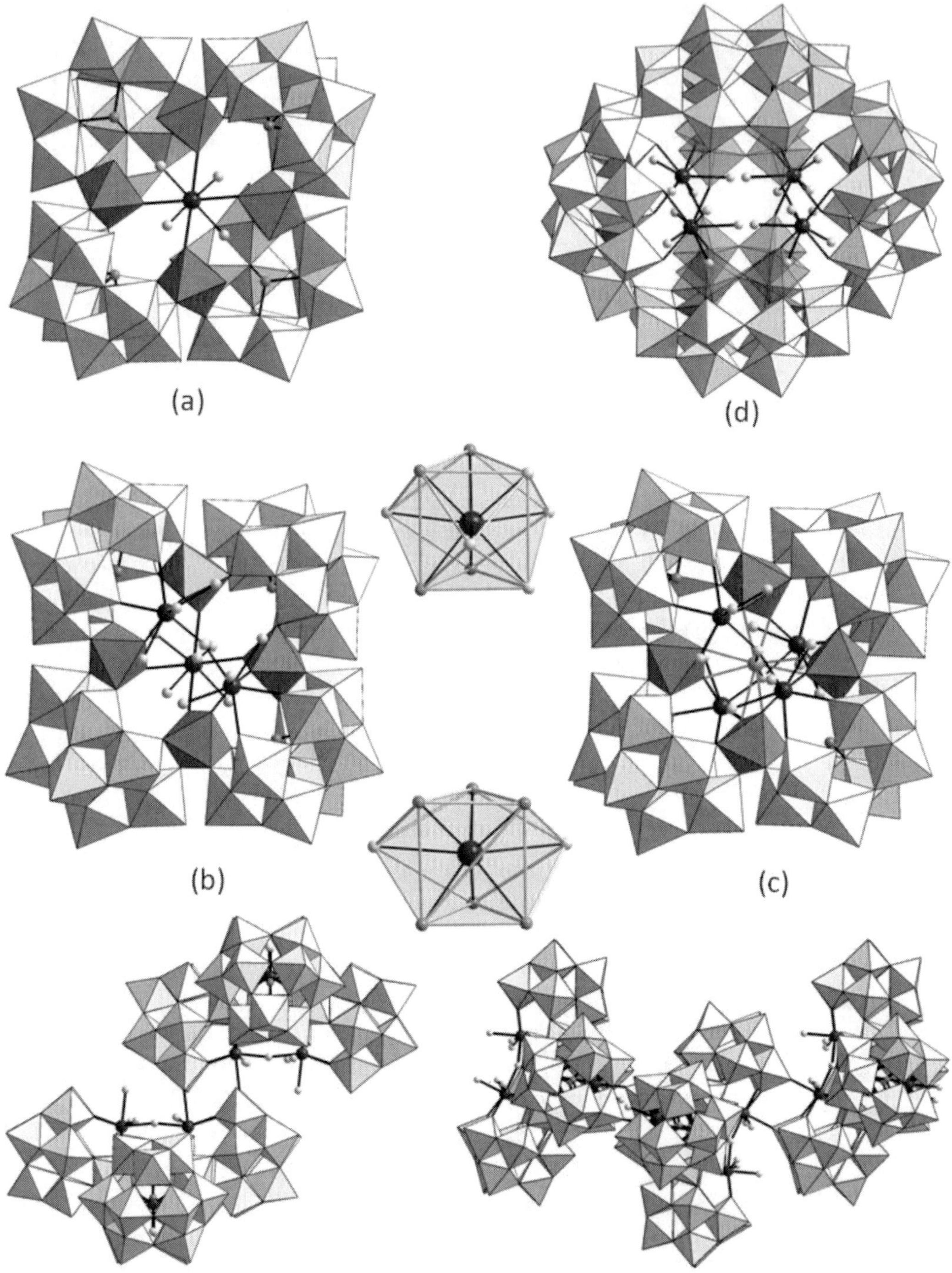

Figure 14. Structures of cerium derivatives of $[As_4W_{40}O_{140}]^{28-}$ and $[P_8W_{48}O_{184}]^{40-}$ cryptates: (a) $[\{Ce^{III}(H_2O)_5\}(As_4W_{40}O_{140})]^{25-}$. (b) $[\{Ce^{III}(H_2O)_5\}\{Ce^{III}_2(OH)(H_2O)_5\}(As_4W_{40}O_{140})]^{20-}$ and its dimeric assembly. (c) $[Ba\{Ce^{III}_2(OH)(H_2O)_5\}_2(As_4W_{40}O_{140})]^{16-}$ and its 1-dimensional assembly (Ba and its bonds depicted in grey). (d) $[K\subset Ce^{III}_2(H_2O)_{10}(H_4W_4O_{12})_2(P_8W_{48}O_{184})]^{25-}$. Additional tungstens are shown as dark grey octahedra. Translucent polyhedra illustrate the cerium coordination spheres in S1 (top) and S2 (bottom) sites (water molecules depicted in light grey).

When the $[\{Ce^{IV}_2(H_2O)_5O\}\{WO(H_2O)\}(B\text{-}\alpha\text{-}As^{III}W_9O_{33})_2]_2^{16-}$ precursor was reacted with $MnCl_2$ in hot water at pH 5, the C_{2v}-symmetric $[Ce^{IV}_4As_4W_{41}O_{149}]^{24-}$ tetramer was obtained [64]. This is also a cryptand-type assembly formed by four $(B\text{-}\alpha\text{-}As^{III}W_9O_{33})$ subunits linked by four alternate cerium atoms via coordination to $\{W_2O_{11}\}$ belt-fragments. The third position in the belt of each trilacunary subunit and the center of the puckered cycle are occupied by additional tungsten atoms, which all show square-pyramidal WO_5 geometries (Figure 13b). This type of coordination geometry is rather unusual in polyoxotungstate structures, although not unknown. Thus, adjacent cerium atoms are simultaneously connected by O–W–O bridges and one oxygen atom from the encapsulated WO_5 moiety, resulting in $CeO_2(O')_2O''O'''(O_{WO5})_2$ distorted square-antiprismatic geometries.

The $[Ce^{IV}_4As_4W_{41}O_{149}]^{24-}$ assembly can be assimilated to some extent to the cryptand-type $[As^{III}_4W_{40}O_{140}]^{28-}$ POM, which is one of the archetypal polyoxotungstoarsenates used as precursor in POM chemistry. This puckered cryptate is also formed by four $(B\text{-}\alpha\text{-}As^{III}W_9O_{33})$ subunits linked by four additional $\{WO_2\}$ groups via corner-sharing with $\{W_2O_{11}\}$ belt-fragments and it shows two different types of coordination sites: a central one delimited by the eight terminal oxygens from the additional tungstens (S1) and four peripheral sites located at the unoccupied positions on the $(B\text{-}\alpha\text{-}As^{III}W_9O_{33})$ belts (S2). The $[Ce^{IV}_4As_4W_{41}O_{149}]^{24-}$ POM can therefore be seen as a $\{As_4W_{40}O_{140}\}$-type cryptate where the linking $\{WO_2\}$ groups are replaced by cerium atoms and additional tungstens occupy the S1 and S2 coordination sites.

The $[As_4W_{40}O_{140}]^{28-}$ POM itself is indeed reactive toward 4f metals and their cerium derivatives can be seen as forming a special subclass among tetrameric ceriopolyoxotungstates. Gouzerh and coworkers obtained the $[\{Ce^{III}(H_2O)_5\}(As_4W_{40}O_{140})]^{25-}$ POM by addition of $Ce(NO_3)_3$ to a hot, aqueous solution of $NaWO_4$ and $NaAsO_2$ at pH 4.5 [30]. This monocerium-derivative shows D_{2h} symmetry and the cerium is located over the S1 site coordinated to one oxygen from each additional tungsten, its coordination sphere being completed with five water molecules to lead to a geometry defined as monocapped square-antiprismatic by the authors, although it could be best described as tricapped trigonal-prismatic (Figure 14a). The authors failed to obtain a polycerium-derivative despite of adding a twofold excess of $Ce(NO_3)_3$ to a solution of in situ generated $[As_4W_{40}O_{140}]^{28-}$, but this was achieved by Wasserman and Pope by using preformed $[As_4W_{40}O_{140}]^{28-}$ as precursor [72]. Addition of a 2M sodium acetate solution of three equivalents of the cerium source to a hot, 2M sodium chloride solution of the precursor at pH 5 resulted in the $[\{Ce^{III}(H_2O)_5\}\{Ce^{III}_2(OH)(H_2O)_5\}(As_4W_{40}O_{140})]^{20-}$ tri-cerium derivative. Aqueous NaCl was found to play an essential stabilizing role as solvent, since reactions carried out in water led to precursor decomposition. Besides one cerium atom placed on the S1 site as in Gouzerh's species, the title POM shows two other cerium atoms occupying the two adjacent S2 sites of the opposite POM face (Figure 14b). Both metals are bridged by a hydroxyl group and their octa-coordinated spheres described by the authors as distorted square-antiprismatic are completed either by three water molecules, or by two water molecules and a terminal oxygen atom from an adjacent POM. Thus, this D_{2d}-symmetric cerium-trisubstituted POM is derived as a dimeric assembly in the solid state, although ^{183}W-NMR shows that it exists as a monomeric entity in solution.

For the synthesis of a tetra-cerium derivative, the authors found blocking of the S1 site essential to obtain reproducible results. It is known that alkaline and alkaline-earth ions preferentially bind to this site, so that they used Ba^{2+} or K^+ as blocking agents. For example, the $[Ba\{Ce^{III}_2(OH)(H_2O)_5\}_2(As_4W_{40}O_{140})]^{16-}$ anion was prepared by adding a solution of a fourfold excess of the cerium source in 1M sodium acetate buffer to an aqueous solution of $BaCl_2$ and the POM precursor, followed by heating and pH adjustement to a value of 5. The obtained species shows two of the above described $\{Ce^{III}_2(OH)(H_2O)_5\}_2$ dimers occupying adjacent S2 sites on both cryptate-sides and a Ba^{2+} ion encapsulated in the S1 site via coordination to the terminal oxygens from the additional tungstens and to the bridging hydroxyl groups of both Ce-dimers. Cerium geometries are virtually identical to those of the above derivative and one cerium on each dimer shows one position in its octa-coordinated sphere occupied by a terminal oxygen from an adjacent POM. Thus, tetra-cerium derivatives assemble in infinite chains in the solid state (Figure 14c), despite of existing as monomers in solution as proved by NMR studies. All these results were also extensible to other early 4f metals from La to Gd.

Besides $[As_4W_{40}O_{140}]^{28-}$, there is also another tetrameric cryptand-type polyoxotungstate known to encapsulate 4f metals, namely $[P_8W_{48}O_{184}]^{40-}$. This POM displays a ring-like structure composed of four fused $\{P_2W_{12}O_{48}\}$ subunits and it has a large central cavity able to accommodate 3d-metal clusters with nuclearities up to twenty [73]. Zimmermann et al. reacted this cryptate with Ce^{III} (also La, Pr and Nd) in water at pH 1.5 in the presence of LiCl and obtained the $[K{\subset}Ce^{III}_2(H_2O)_{10}(H_4W_4O_{12})_2(P_8W_{48}O_{184})]^{25-}$ POM (Figure 14d) in low yield both under hydrothermal or bench, boiling conditions [74]. According to NMR studies on the reaction mother liquors, the major species present is the Wells-Dason anion, and so this POM has been attributed to an intermediate in the $\{P_8W_{48}O_{184}\}$ to $\{P_2W_{18}O_{62}\}$ transformation trapped by large 4f metals. The central cavity in the $\{P_8W_{48}O_{184}\}$ framework is filled with $(H_4W_4O_{12})$ fragments of four corner-shared WO_6 octahedra, which occupy the double-belt vacant sites on two opposite $\{P_2W_{12}O_{48}\}$ subunits. Besides a central potassium ion, the anion also contains two cerium atoms disordered over the four pockets delimited by one WO_6 octahedron from each additional $(H_4W_4O_{12})$ and two corner-shared octahedra from a $\{P_2W_{12}O_{48}\}$ belt. The cerium coordination spheres are completed with five water molecules, leading to monocapped square-antiprimatic geometries.

4.4. Hexameric Ceriopolyoxotungstates

A single cerium-containing polyoxotungstate with hexameric nature is known, namely $[(PO_4){\subset}\{Ce_3^{IV}Mn_2^{IV}O_4(OH)_2\}_3(OH)_2(H_2O)_2(\alpha\text{-}P_2W_{15}O_{56})_6]^{47-}$ [75]. This POM cannot be considered a ceriopolyoxotungstate in the strict sense, but one of the few heterometallic 3d-4f polyoxotungstates described to date, since the skeletal role played by the cerium atoms is due to being part of cerium-manganese oxo-clusters.

Fang and Kögerler obtained this phosphate-templated architecture in the course of their investigations about the influence of replacing carboxylate ligands by polyoxometalates on the magnetic properties of 3d-4f heterometallic clusters. The hexameric POM (Figure 15) was isolated by reacting the pre-formed complex $[Ce_3Mn_2O_6(OAc)_{7.5}(NO_3)_3]$ with $[P_2W_{15}O_{56}]^{12-}$ in hot water in the presence of PO_4^{3-}. The precursor shows a heterometallic core of trigonal-bipyramidal topology with the cerium and manganese atoms in equatorial and axial positions, respectively. They are bridged by six μ_3-oxo groups and six μ_2-acetate-κ^2-O:O' ligands define the cerium-manganese edges.

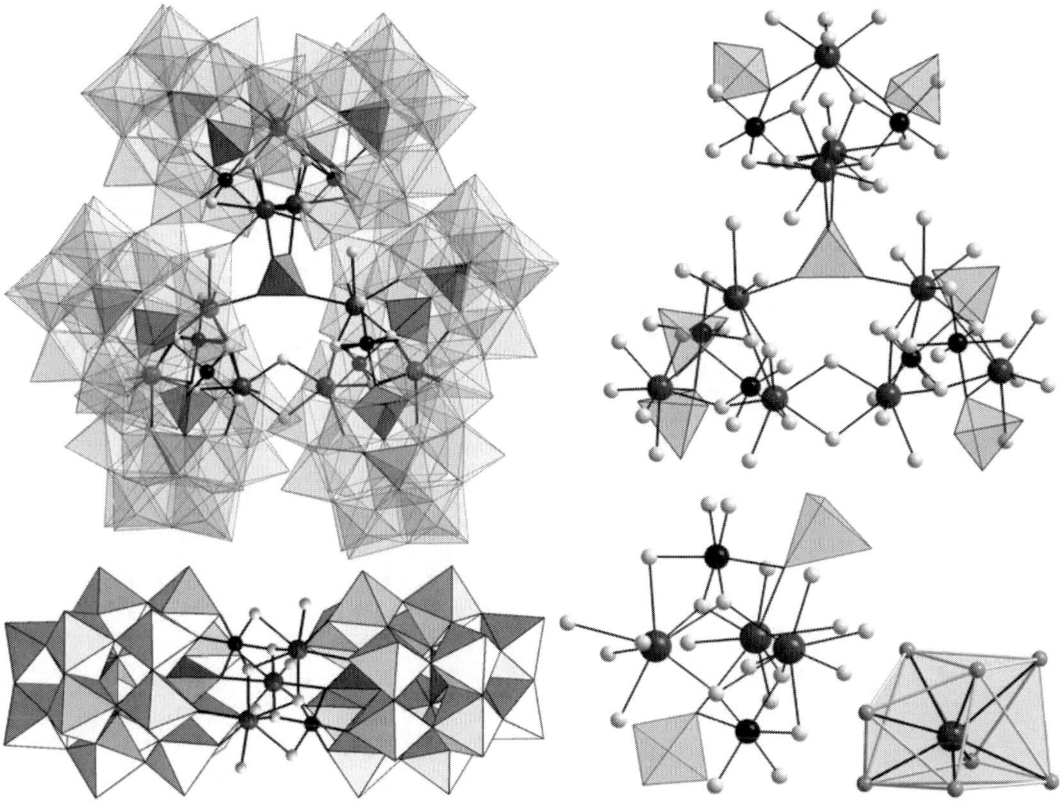

Figure 15. Structure of the hexameric $[(PO_4)\subset\{Ce_3^{IV}Mn_2^{IV}O_4(OH)_2\}_3(OH)_2(H_2O)_2(\alpha\text{-}P_2W_{15}O_{56})_6]^{47-}$ POM with a detail of the three 3d-4f clusters linked by the encapsulated phosphate (top) and representation of the $\{Ce_3^{IV}Mn_2^{IV}O_4(OH)_2(\alpha\text{-}P_2W_{15}O_{56})_2\}$ subunit with details of the sandwiched 3d-4f cluster and the coordination polyhedron of one of the cerium atoms (bottom).

These are all replaced by two trilacunary fragments upon reaction with $[P_2W_{15}O_{56}]^{12-}$ but maintaining the metal-core topology. Thus, sandwich-type $\{Ce_3^{IV}Mn_2^{IV}O_4(OH)_2(\alpha\text{-}P_2W_{15}O_{56})_2\}$ subunits are generated, where the vacant sites on each trilacunary fragment are occupied by one manganese and two cerium atoms. Four μ_3-oxygens, two μ_2-hydroxyls and two μ_2-oxygens from trilacunary-PO_4 groups bridge the metals in the sandwiched 3d-4f cluster.

Three of the above dimeric subunits are assembled in the final C_2-symmetric hexameric aggregate: two of them are linked via double μ_2-OH bridging between cerium atoms, whereas the third one is connected to these two via two Ce–O=W bonds. A central, encapsulated PO_4^{3-} group templates the formation of this macrocyclic POM by interaction with the three sandwiched 3d-4f clusters. All cerium atoms are octa-coordinated and their spheres show a highly distorted geometry in between square-antiprismatic and bicapped trigonal-prismatic. Fang and Kögerler reported this POM to be stable in solution since they did not find evidences for dissociation or central PO_4^{3-} replacement by other tetrahedral oxoanions.

4.5. Giant Ceriopolyoxotungstates

There has been a continuous interest over the past years in bringing POM chemistry to the nanosize scale by constructing multifunctional assemblies from controlled aggregation of well-defined building-blocks. This has been successfully achieved in the case of polyoxomolybdates, for which self-assembly of pentagonal {Mo(Mo$_5$)} fragments has resulted in a plethora of molecular giant species containing up to 368 molybdenum atoms that show "curved" topologies with dimensions up to 6 nm. These include icosahedral clusters known as Keplerates or circular and elyptical wheels [76]. In contrast, this strategy cannot be fully applied to polyoxotungstates because pentagonal fragments are not common in tungsten solution chemistry [77]. The vast library of lacunary polyoxotungstates available has however led to an alternative approach consisting in their aggregation via combination with various linking groups. Among these linkers, 4f metals represent the most widely used, so that five out of the six reported polyoxotungstate types containing over a hundred tungsten atoms are based on 4f metals, the other one being based on dimethyltin moieties [78]. Moreover, only two out of these five types of giant POMs contain 4f metals different from cerium [79].

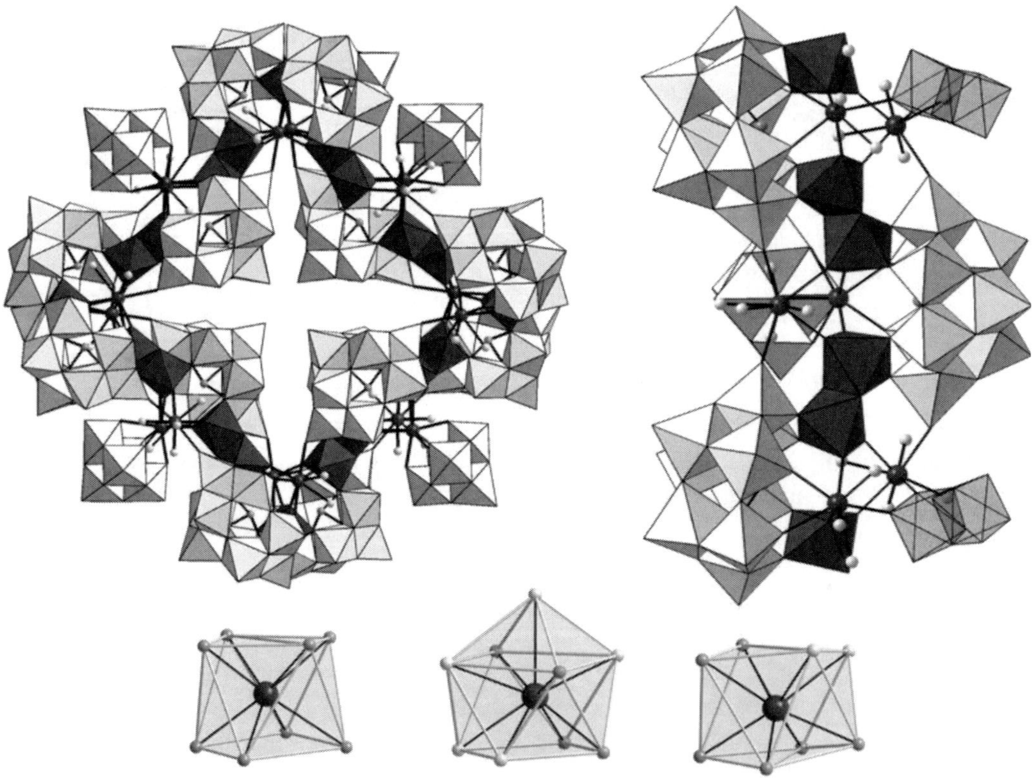

Figure 16. Structure of $[As^{III}_{12}Ce^{III}_{16}(H_2O)_{36}W_{148}O_{524}]^{76-}$ with a detail of the substructure formed by a central (B-α-$As^{III}W_9O_{33}$) subunit linked to two adjacent ones via edge-shared {W_2O_{10}} connectors (in dark grey). Additional WO_6 in external subunits are shared with adjacent substructures (one equivalent cerium atom and one {W_2O_{10}} connector are included). Translucent polyhedra illustrate the coordination spheres of cerium atoms attached to central subunits (left), located in the substructure clefts (central) and belonging to external subunits (right, water molecules depicted in light grey).

The first of such giant polyoxotungstates was reported by Pope and coworkers back in 1997 [17]. They isolated the $[As^{III}_{12}Ce^{III}_{16}(H_2O)_{36}W_{148}O_{524}]^{76-}$ cluster and its lanthanum analogue from a straightforward one-pot reaction consisting in addition of the corresponding 4f metal dissolved in aqueous acetic acid to boiling aqueous solution of Na_2WO_4 and As_2O_3 acidified to pH c.a. 7. This hexadecameric assembly constituted the only giant polyoxotungstate known until the report of the above mentioned dimethyltin-containing species in 2005. Moreover, it has remained as the largest polyoxotungstate for fifteen years, until the recent report on a family of $\{(4f)_{16}As^{III}_{16}W_{164}\}$ (4f = Eu, Gd, Tb, Dy, Ho) polyoxotungstates [79a].

Pope's disc-shaped macroanion shows a D_{2d}-symmetric structure composed of twelve (B-α-$As^{III}W_9O_{33}$) and four (W_5O_{18}) subunits linked by additional tungsten and cerium atoms (Figure 16), in such a way that each (B-α-$As^{III}W_9O_{33}$) subunit displays two additional WO_6 octahedra supported on corner-shared $\{W_2O_{11}\}$ fragments and one cerium occupying the third belt position and coordinated to these two additional WO_6. The resulting folded cyclic assembly shows a diameter of c.a. 4 nm and no large central cavity. For a better description, the title POM can be seen as formed by four angular substructures, each of them composed of a central (B-α-$As^{III}W_9O_{33}$) subunit linked to two adjacent ones via edge-sharing of additional WO_6. These substructures are further interconnected in the macrocycle by $\{WO_2\}$ groups shared by the trilacunary external subunits. Cerium atoms of the central (B-α-$As^{III}W_9O_{33}$) are capped with (W_5O_{18}) fragments, resulting in distorted square-antiprismatic geometries. Cerium atoms on external subunits are also octa-coordinated, their spheres being completed by two water molecules and two terminal oxygens from the adjacent substructure. A third type of cerium atom is located in the substructure clefts and these are coordinated to both external subunits and both edge-shared $\{W_2O_{10}\}$ connectors, their nona-coordinated spheres with monocapped square-antiprismatic geometries being completed with five water molecules. Despite of being composed of well-known building blocks, the authors found ^{183}W-NMR evidences for this supercluster to be stable in solution at pH 6.5.

Kortz and coworkers studied the interaction between 4f metals and the $[A\text{-}\alpha\text{-}GeW_9O_{34}]^{10-}$ POM, for which no 4f-metal derivative was reported at the time, and isolated the decameric $[Ce^{III}_{20}Ge_{10}W_{100}O_{376}(OH)_4(H_2O)_{30}]^{56-}$ species [80]. This giant POM with a dumbbell shape and a maximum dimension of c.a. 4.2 nm was obtained by reaction of the trilacunary tungstogermanate precursor with $CeCl_3$ (1:1) ratio in warm water at pH 5.0. It is composed of two centrosymetrically related pentameric halves linked by two Ce–O=W bridges with Ce–O bond lengths of 2.61 Å. Based on this, the authors suggested that the decameric assembly most likely dissociates in solution, although no conclusive evidences of the presence of independent pentameric halves were obtained from preliminary solution NMR studies.

Each asymmetric $\{Ce^{III}_{10}Ge_5W_{50}O_{188}(OH)_2(H_2O)_{15}\}$ pentamer (Figure 17) contains five $\{Ce_2GeW_{10}O_{38}\}$ subunits formed by a dilacunary Keggin tungstogermanate fragment coordinating two ceriums in a bis(tetradentate) fashion. These dilacunary fragments derive from the β-Keggin isomer by removal of one WO_6 octahedron from the rotated $\{W_3O_{13}\}$ cap and another one from the belt, in such a way that two enantiomeric forms with vacant sites in relative positions (4,11) and (5,10) are observed [81]. These specific dilacunary forms are unprecedented and cannot be isolated as independent clusters. They are formed in situ from the $[A\text{-}\alpha\text{-}GeW_9O_{34}]^{10-}$ precursor by tungsten gaining and subsequent skeletal rearrangement, a fact that is not unsual in polyoxotungstate chemistry. This is followed by cerium trapping, thus nicely illustrating the stabilizing ability of cerium toward novel POM fragments.

The $\{Ce^{III}{}_{10}Ge_5W_{50}O_{188}(OH)_2(H_2O)_{15}\}$ pentamer can be described on the basis of two dimeric substructures where the two $\{Ce_2GeW_{10}O_{38}\}$ subunits are joined through belt cerium atoms. The cerium atom of one subunit is capped by the $\{W_3Ce\}$ tetrameric face from the adjacent subunit via four almost coplanar bridging oxygens in a way that resembles the capping of a monovacant POM fragment. The cerium atom of the latter subunit coordinates two *cis*-related oxygens from a WO_6 octahedron adjacent to the vacant site of the former. One of the dimeric substructures contains equivalent β-(5,10) enantioneric forms, whereas both β-(5,10) and β-(4,11) enantiomers are present in the second substructure. They are assembled together through cap cerium atoms from β-(5,10) enantioners, whereas that of the β-(4,11) form constitutes the dimerization site. A fifth β-(4,11) enantiomer reinforces the assembly, being attached as a cramp to both substructures via cerium coordination to two *cis*-related oxygens from β-(5,10) and β-(4,11) subunits. All this leads to an intricate network of corner- and edge-shared WO_6 octahedra and CeO_n polyhedra, where n is either eight or nine and the coordination geometries display remarkable variety (square-antiprismatic, its monocapped derivative, tricapped trigonal-prismatic) and significant distortion degree.

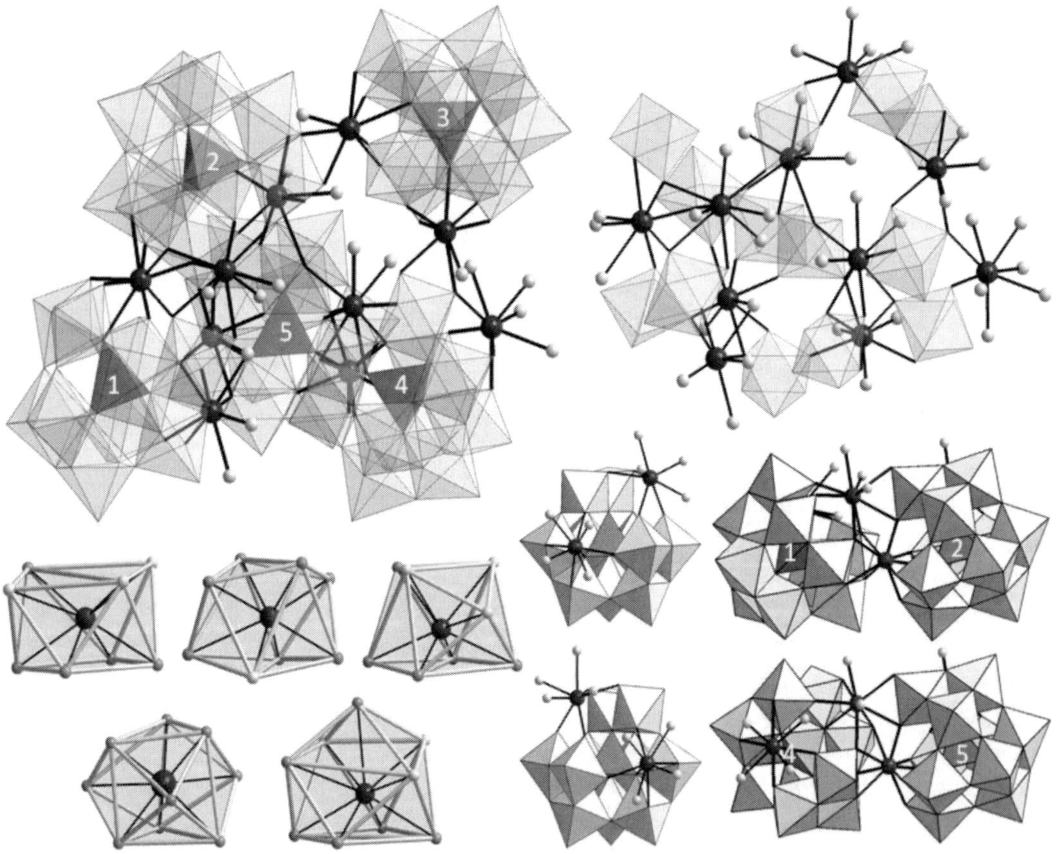

Figure 17. Top: structure of the $\{Ce^{III}{}_{10}Ge_5W_{50}O_{188}(OH)_2(H_2O)_{15}\}$ pentameric half with a detail of the WO_6-bridged cerium network. Bottom: Translucent polyhedra illustrating different cerium coordination spheres (left, water molecules depicted in light grey), cerium-stabilized β-(5,10) and β-(4,11) enantiomeric forms (middle) and dimeric substructures formed by two β-(5,10) (named as 1 and 2) or by one β-(4,11) and one β-(5,10) (named as 4 and 5) enantiomers (right).

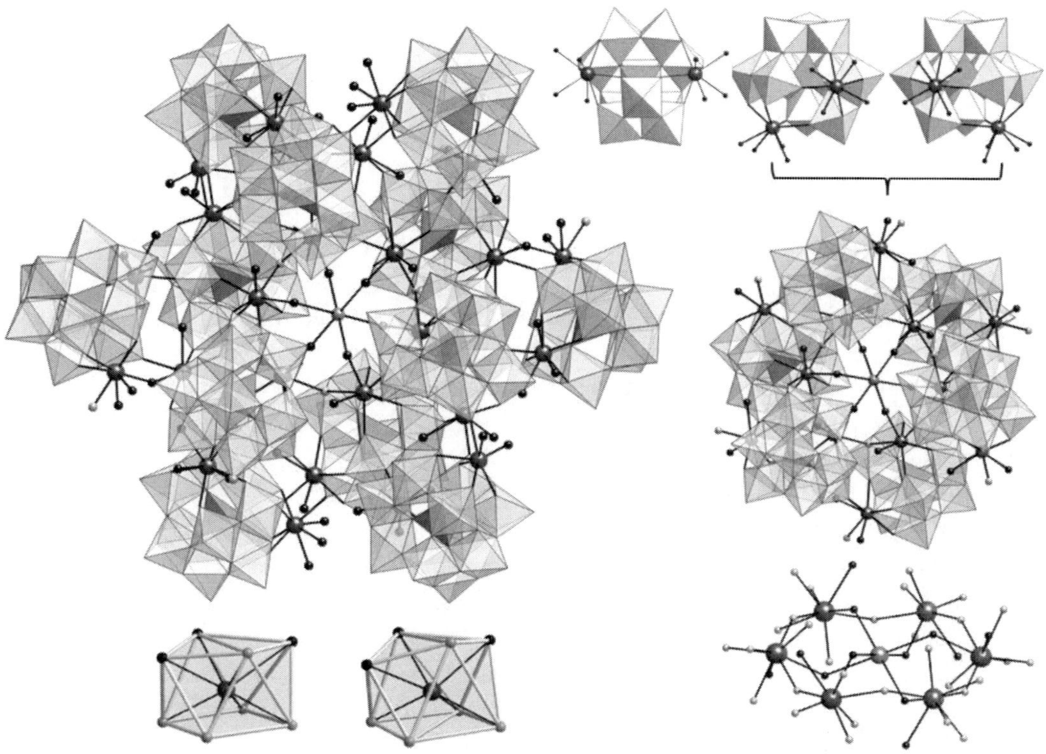

Figure 18. Left: structure of $[K{\subset}K_7Ce^{III}_{24}Ge_{12}W_{120}O_{456}(OH)_{12}(H_2O)_{64}]^{56-}$ with translucent polyhedra illustrating the less and most distorted square-antiprismatic geometries of the cerium atoms (skeletal potassiums omitted for clarity with the exception of the central one; water molecules depicted in black). Right: cerium-stabilized γ-(3,4), β-(1,8) and β-(1,5) dilacunary Keggin fragments (top), $\{K{\subset}(Ce_2GeW_{10}O_{38})_6\}$ hexameric core (middle) and central Ce_6O_6 ring interacting with the encapsulated potassium (botton).

A second giant assembly based on $\{Ce_2GeW_{10}O_{38}\}$ subunits but showing a completely different topology, namely $[K{\subset}K_7Ce^{III}_{24}Ge_{12}W_{120}O_{456}(OH)_{12}(H_2O)_{64}]^{56-}$ (Figure 18), has been recently reported by Reinoso et al. [82]. It was obtained in an attempt to synthesize a 3d-4f heterometallic Weakley-like sandwich POM from a simple one-pot procedure involving reaction of $Ce(NO_3)_3$, MCl_2 (M = Ni, Co, Mn), GeO_2 and Na_2WO_4 in a sodium acetate buffer at room temperature, followed by addition of KCl. This species constitutes the largest polyoxotungstogermanate known to date and it also contains the largest number of 4f ions observed in a polyoxometalate.

As for Kortz's decamer, the $\{Ce_2GeW_{10}O_{38}\}$ subunits forming this dodecameric assembly are also generated in situ at a similar pH. However, none of the observed dilacunary Keggin fragments are coincident with those of Kortz`s species. In this case, three different Keggin skeletons are found: the enantiomeric forms β-(1,8) and β-(1,5) with vacant sites located at the belt and the corner-shared $\{W_3O_{15}\}$ triad opposite to the rotated cap; and the γ-(3,4) fragment, which derives from the γ-isomer by removal of two WO_6 octahedra on the same POM side, each one belonging to different rotated $\{W_3O_{13}\}$ trimers. These three types of dilacunary fragments are also unprecedented and their isolation, together with Kortz's previous results, suggests that numerous dilacunary species can be present in solution at the given reaction conditions and that they can be accessed depending on the tungstogermanate source and stabilizing electrophile. This is especially relevant for γ-lacunary skeletons, for which the γ-(1,2) derivative was the only lacunary fragment known.

The crown-shaped structure of $[K\subset K_7Ce^{III}_{24}Ge_{12}W_{120}O_{456}(OH)_{12}(H_2O)_{64}]^{56-}$ comprises six outer γ-(3,4) subunits attached as cramps to a cyclic hexameric core via two Ce–O=W bridges each. This cyclic core is composed of three β-(1,8) and three β-(1,5) $\{Ce_2GeW_{10}O_{38}\}$ subunits that are linked alternately via coordination of the cerium atoms on the triad vacant sites to two *cis*-related oxygens from WO_6 octahedra adjacent to these vacant sites. A Ce_6O_6 ring is therefore generated where water molecules on the almost coplanar cerium atoms are directed to the inner part, creating a pocket were a potassium ion is located. Such a Ce_6O_6 ring and its ion encapsulation ability through internal water molecules in close analogy to the organic crown ethers are observed for the first time. Moreover, the central potassium does not appear to be a dummy captured ion, but to act as a template for the construction of the final centrosymmetric assembly, since the geometry of the $\{K(H_2O)_6\}$ moiety is imprinted in the overall POM architecture with a diameter of c.a. 4nm. In contrast to Kortz's decamer, all cerium atoms are octa-coordinated and display less distorted square-antiprismatic geometries.

The dodecameric POMs are arranged in mono-dimensional assemblies in the solid state, but according to dynamic light scattering studies, they dissociate upon dissolution to most likely result in γ-(3,4)-monomers and cyclic $\{K\subset(\{Ce_2GeW_{10}O_{38})_6\}$ hexamers, which appear to be stable in solution on the basis of IR and UV/vis spectroscopies. These studies also indicate that the latter undergo slow self-aggregation in solution to form blackberry-type supramolecular structures [83] with hydrodymanic diameter of c.a. 260 nm after about fifteen days.

5. Polyoxometalate Fragments Stabilized by Cerium

As mentioned in the sections above, cerium cannot only be considered as a linking agent to construct new polyoxotungstate architectures of increasing size and complexity through its combination with well-known lacunary building blocks, but also as a stabilizing agent of unprecedented polyoxotungstate fragments formed in situ that cannot be accessed as independent entities. Some representative examples are the $\{W_5O_{18}\}$ pentatungstate subunit in Peacock-Weakley sandwich anions, the α_1,α_1-dilacunary Wells-Dawson moiety in Ostuni and Pope's $[\{Ce^{III}_4(H_2O)_9(OH)_2\}(P_2W_{16}O_{59})_2]^{14-}$, or the five different types of dilacunary Keggin tungstogermanate fragments in Kortz's and Reinoso's giant assemblies. Some other examples of novel polyoxotungstate species trapped by cerium will be commented in this section.

Besides Peacock-Weakley decatungnstocerate anions, structural characterizations of two other cerium-containing isopolyoxotungstates have been reported to our knowledge. Li et al. obtained the $[Ce^{III}_2(H_2O)Cl(H_6W_{15}O_{54})]^{7-}$ anion from addition of $Ce(NO_3)_3$ to an aqueous solution of Na_2WO_4 (2:13 ratio) at pH around 5, followed by boiling and final addition of KCl [84]. This cluster of ideal C_3 symmetry is composed of a completely unprecedented pentadecatungstate ring enclosing a central cerium dimer (Figure 19a). The $(H_6W_{15}O_{54})$ moiety comprises three $\{W_3O_{14}\}$ fragments composed of three edge-shared WO_6 octahedra that are linked in a cyclic structure via corner-sharing with three $\{W_2O_{10}\}$ dimers. The central pocket is capped by a cerium atom on each side of the ring via coordination to eight oxygens each. Both cerium atoms show an additional terminal ligand, one water molecule and a chloride. The latter appears to be crucial for the formation of this species because the authors observed that no analogue could be obtained in the absence of chloride anions. Both cerium atoms are therefore nona-coordinated and their geometry is described as tricapped trigonal-prismatic by the authors, leading to a face-sharing dimer via three Ce–O–Ce bridges that involve the $\{W_2O_{10}\}$ connectors.

Ismail et al. also studied the reactivity between cerium and tungstate, but at significantly lower pH values. The authors obtained the $[Ce^{III}_2(H_2O)_{10}W_{22}O_{72}(OH)_2]^{8-}$ anion at room temperature by adjusting the pH of an aqueous Ce^{3+}:WO_4^{2-} mixture (1:11 ratio) to a value of 2.2 [85]. This species is not unique, but a member of a large family of 4f-metal stabilized isopolyoxotungstate analogues, for which derivatives containing Y, La and late 4f metals from Dy to Lu were also structurally characterized. The authors identified three synthetic parameters governing the formation of the title POM: temperature (heating led to isolation of the well-known metatungstate species), pH value in the 2.0 – 3.5 range (higher pH values led to the paratungstate anion) and tungstate concentration (higher than 1.5 M for isolating pure, crystalline samples). It is worth mentioning that similar reactions result in the corresponding Peacock-Weakley $[Ce(W_5O_{18})_2]^{9-}$ sandwich anion for pH values around neutrality and in the above mentioned $[Ce^{III}_2(H_2O)Cl(H_6W_{15}O_{54})]^{7-}$ for moderately acidic reaction mixtures.

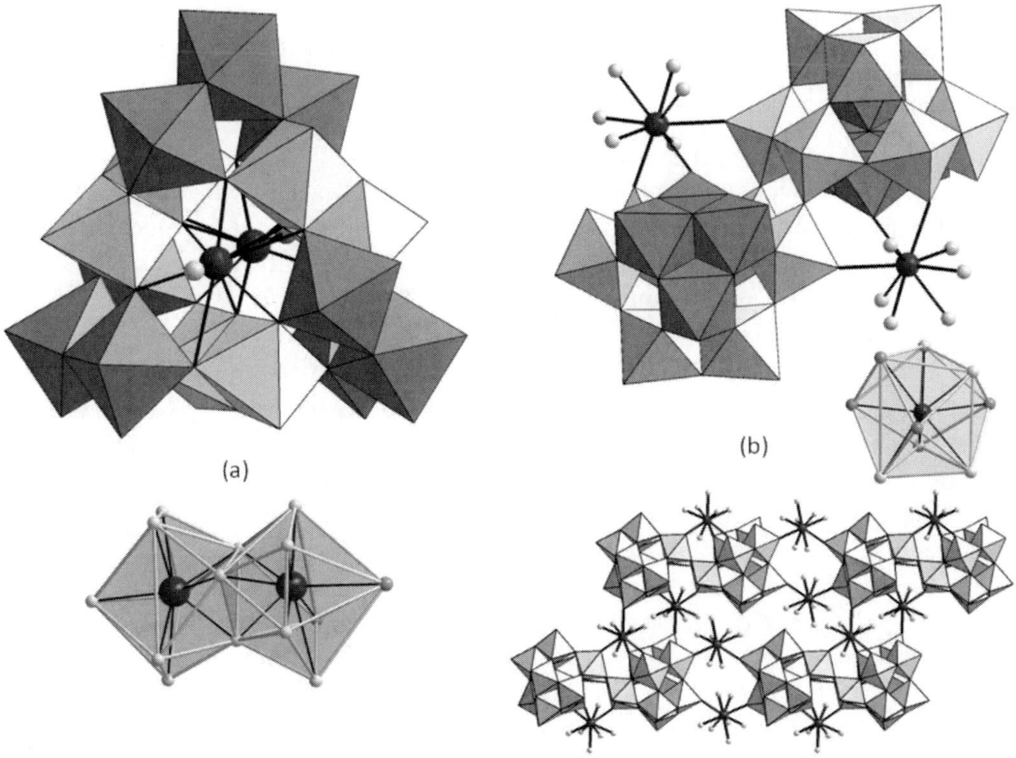

Figure 19. (a) Structure of $[Ce^{III}_2(H_2O)Cl(H_6W_{15}O_{54})]^{7-}$ with a detail of the central, face-sharing dicerium moiety (terminal chloride ligand in dark grey). (b) Structure of $[Ce^{III}_2(H_2O)_{10}W_{22}O_{72}(OH)_2]^{8-}$ and its bidimensional solid-state arrangement with the translucent polyhedron illustrating the cerium nona-coordination (water molecules depicted in light grey).

The structure of the centrosymmetric $[Ce^{III}_2(H_2O)_{10}W_{22}O_{72}(OH)_2]^{8-}$ species is composed of a $\{H_2W_{22}O_{74}\}$ cluster stabilized by coordination of two $\{Ce(H_2O)_5\}$ moieties, resulting in the first example of a isopolyoxotungstate with exposed cerium atoms displaying terminal aquo-ligands potentially accessible for ligand substitution (Figure 19b). The $\{H_2W_{22}O_{74}\}$ cluster results from the condensation of two undecatungstate subunits via corner-sharing between two edge-shared $\{W_2O_{10}\}$ fragments. These subunits can be described on the basis of two different polyoxotungstate fragments linked via corner-sharing: one tetratungstate fragment deriving from the Lindqvist-type structure by removal of two adjacent WO_6 octahedra, and one heptatungstate fragment that can be seen as a hypothetical A-α-trilacunary metatungstate species lacking one belt $\{W_2O_{10}\}$ moiety. Although the undecatungstate subunit is known to exist as an independent POM, isolation of such condensation product was attributed to stabilization by two cerium atoms coordinated to three terminal oxygens each. The nona-coordinated spheres of the cerium atoms with monocapped square-antiprismatic geometry according to the authors description are completed by a terminal oxygen belonging to an adjacent cluster, hence these clusters in the solid state arrange in chains further connected by $\{Ce(H_2O)_7\}$ linking moieties in bidimensional assemblies.

The $[Ce^{III}_2(H_2O)_{10}W_{22}O_{72}(OH)_2]^{8-}$ anion was obtained by fractional crystallization as a second batch of crystals, the first one also being a cerium-stabilized isopolyoxotungstate of formula $[\{Ce^{III}_2(H_2O)_{11}W_{28}O_{93}(OH)_2\}_2]^{28-}$, for which a preliminary communication of its cell parameters was given. The authors could not provide a complete structural elucidation of this species because of the insufficient quality of its crystals, but they did subsequently report the crystal structures of two analogous clusters containing Sm or Eu [86]. The members of the $[\{(4f)^{III}_2(H_2O)_{11}W_{28}O_{93}(OH)_2\}_2]^{28-}$ family (4f = La to Gd) represent the largest 4f-metal containing isopolyoxotungstates known to date.

Heterometallic complexes of oxo-bridged cerium and 3d metals with labile carboxylate ligands have also shown stabilizing ability toward polyoxotungstates that are generated in situ but cannot be independently isolated. In order to study the effect of replacing carboxylates by POM ligands on the structure and magnetic properties of such complexes, Fang and Kögerler selected $[CeMn_6O_9(OAc)_9(NO_3)(H_2O)_2]$ as precursor and reacted this complex with the Wells-Dawson trilacunary $[\alpha\text{-}P_2W_{15}O_{46}]^{12-}$ in a hot water/acetic acid mixture [87], resulting in the $[\{Ce^{IV}Mn^{IV}_6O_9(OAc)_8\}(\alpha_2\alpha_2\text{-}P_2W_{16}O_{57}(OH)_2)]^{8-}$ POM with ideal C_s symmetry (Figure 20a). The complex precursor is formed by a planar hexagon of Mn^{IV} atoms bridged by oxo and μ_2-acetato-κ^2-O:O' ligands and a capping Ce^{IV} atom that shows labile water and nitrato terminal ligands. The complex gets anchored to the polyoxotungstate skeleton maintaining its bowl shape upon reaction but it induces gaining of an additional tungsten atom on the Wells-Dawson species, resulting in a $(\alpha_2\alpha_2\text{-}P_2W_{16}O_{57}(OH)_2)$ dilacunary derivative. Despite of the fact that this dilacunary derivative does not fulfill the requirements for being stable in solution, ^{31}P-NMR spectroscopy indicates that the POM structure is indeed retained in solution. Grafting of the complex takes places via substitution of a bridging acetate by two oxygens from the additional WO_6 octahedron, together with replacement of the three terminal ligands on cerium by three oxygen atoms from the $(P_2W_{15}O_{56})$ fragment. The cerium atom is then nona-coordinated, showing a tricapped trigonal-prismatic geometry. The $(\alpha_2\alpha_2\text{-}P_2W_{16}O_{57}(OH)_2)$ subunit is not unprecedented and has also been observed in a few other metal derivatives [86].

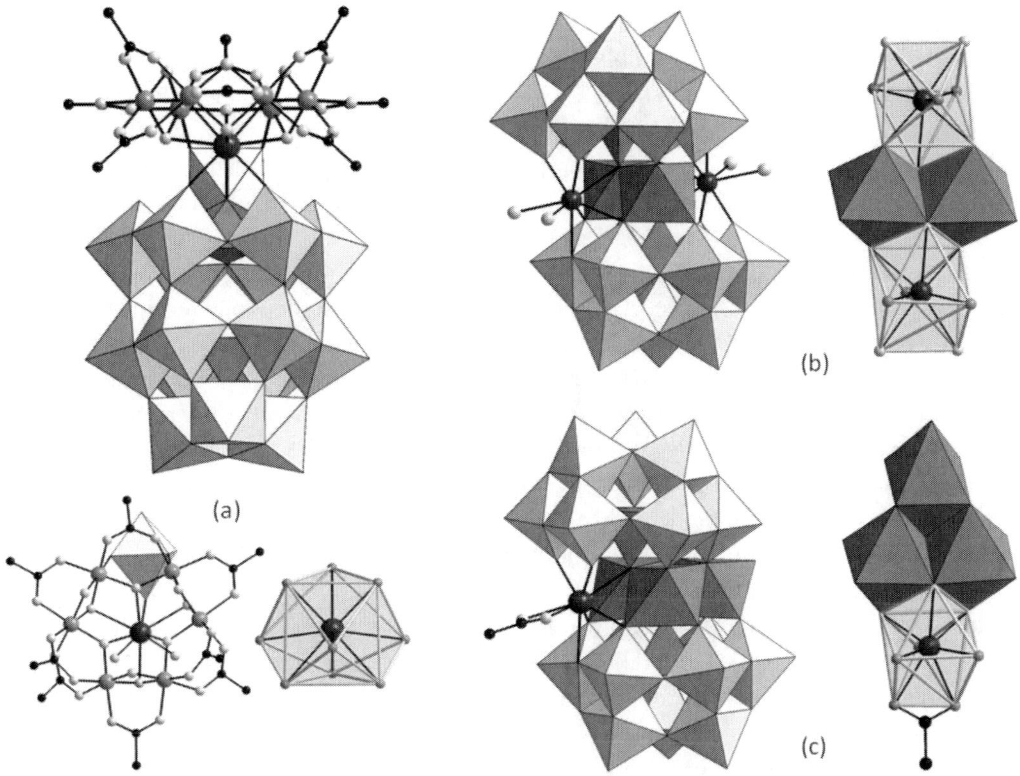

Figure 20. (a) Structure of $[\{Ce^{IV}Mn^{IV}_6O_9(OAc)_8\}(\alpha_2\alpha_2\text{-}P_2W_{16}O_{57}(OH)_2)]^{8-}$ with a detail of the grafted heterometallic complex and the cerium coordination polyhedron. (b) Structure of $[\{Ce^{III}(H_2O)_2\}_2Mn^{III}_2(B\text{-}\alpha\text{-}GeW_9O_{34})_2]^{8-}$ with a detail of its central $\{Ce_2Mn_2O_{20}\}$ rhomblike core. (c) Structure of $[Ce^{IV}(OAc)Cu^{II}_3(B\text{-}\alpha\text{-}GeW_9O_{34})_2]^{8-}$ with a detail of its central $\{CeCu_3O_{18}\}$ rhomblike core.

Reinoso and Galán-Mascarós explored the applicability of incorporating 4f metals to 3d-substituted POM precursors that can be assimilated to lacunary species as a general synthetic strategy for the preparation of heterometallic POMs containing oxo-bridged 3d-4f clusters, which constitute an emerging subclass of clusters within the vast POM area [70]. In a first stage, the authors selected Weakley-type sandwich tungstogermanates as POM precursors for reaction with Ce^{IV}. Weakley-type $[(3d)_4(H_2O)_2(B\text{-}\alpha\text{-}XW_9O_{34})_2]^{n-}$ sandwich anions comprise two trilacunary B-α-Keggin subunits encapsulating a $\{(3d)_4O_{16}\}$ rhomblike core of four edge-sharing $(3d)O_6$ octahedra with ββ-configuration of the junctions between trilacunary subunits and the central core cluster. The outer, aquated 3d metals appear to show certain lability as indicated by isolation of defect structures with sodium ions in external positions. This type of precursor showed indeed to be reactive toward Ce^{IV}, so that heterometallic derivatives incorporating one or two cerium atoms on external positions have recently been reported.

The first Weakley-like 3d-4f sandwich anion, namely $[\{Ce^{III}(H_2O)_2\}_2Mn^{III}_2(GeW_9O_{34})_2]^{8-}$, was prepared by stoichiometric reaction of $(NH_4)_2Ce(NO_3)_6$ with $[Mn^{II}_4(H_2O)_2(GeW_9O_{34})_2]^{12-}$ in water (pH = 1) at room temperature [89]. This heterometallic species with ideal C_{2h} symmetry can be described as two $\{Ce^{III}(H_2O)_2\}$ moieties incorporated to the external vacant sites of a manganese(III)-disubstituted Weakley-type tungstogermanate subunit with αα-configuration of the trilacunary/core junctions (Figure 20b). Besides two terminal aquo ligands, six terminal and bridging oxygen atoms from the sandwich subunit occupy the coordination sphere of the cerium atoms, in such a way that a completely unprecedented $\{Ce^{III}_2Mn^{III}_2O_{20}\}$ rhomblike moiety is formed via edge-sharing of two internal MnO_6 octahedra and two external highly distorted $CeO_6(O_w)_2$ square-antiprisms. Since reactions with several trivalent 4f ions were unsuccessful, the authors suggested formation of this POM taking place via release of outer $\{Mn(H_2O)\}$ groups as a result of Mn^{II} to Mn^{III} oxidation by Ce^{IV} and consequent ββ to αα isomerization, followed by cerium quenching of the resulting disubstituted sandwich species.

The authors selected the $[Cu^{II}_4(H_2O)_2(GeW_9O_{34})_2]^{12-}$ anion to explore the interaction between Ce^{IV} and a non-oxidizeable POM precursor and obtained the monocerium-containing $[\{Ce^{IV}(OAc)\}Cu^{III}_3(H_2O)(GeW_9O_{34})_2]^{11-}$ derivative from stoichiometric reaction with $(NH_4)_2Ce(NO_3)_6$ in a sodium acetate buffer at room temperature [90]. Following the above description, this C_s-symmetric species can be seen as a copper(II)-trisubstituted Weakley-type tungstogermanate subunit showing αβ-configuration of the trilacunary/core junctions with a $\{Ce^{IV}(OAc)\}$ moiety incorporated in the vacant external position (Figure 20b). Cerium coordination sphere is virtually identical to that described above but replacing the terminal water molecules by a capping κ^2-O,O'-acetate ligand. This results in a new type of rhomblike moiety, namely $\{Ce^{IV}Cu^{II}_3O_{18}\}$, formed by edge-sharing of two internal CuO_6 elongated octahedra, one external $CuO_5(O_w)$ elongated octahedron and one external highly distorted $CeO_6(O_{ac})_2$ square-antiprism. Although the Cu···Cu distances between internal and internal-external positions are inversely correlated in closely-related sandwich tungstogermanates, the authors found that the distance between internal copper atoms in the heterometallic cerium-copper species is longer than expected from the above trend and this fact was attributed to a POM skeleton breathing upon cerium incorporation. Formation of this POM was rationalized on the basis of slow evolution of the $[Cu^{II}_4(H_2O)_2(GeW_9O_{34})_2]^{12-}$ precursor to a trisubstituted derivative with consequent ββ to αβ isomerization, followed by cerium quenching.

Although defect di- and trisubstituted Weakley-type sandwich anions containing trilacunary Keggin fragments are known to exist as independent clusters for heteroatoms of the group 15 [91], the above heterometallic derivatives constitute the first examples where such type of defect structures are observed with germanium as heteroatom, and also, the first manganese(III)-disubstituted and copper(II)-trisubstituted subunits. These derivatives nicely illustrate the cerium ability to trap and stabilize different POM species generated in situ in solution, allowing them for being isolated in the solid state and structurally characterized.

CONCLUSION

Although it has been forty years now since the seminal work from Peacock and Weakley about ceriopolyoxotungstates, research on this specific group of polyoxometalates still shows a remarkable dynamism and is nowadays more active than ever as indicated by a significant increase of reports in the last decade. Combination of cerium with different lacunary polyoxotungstate building blocks acting as polydentate O-donor ligands has allowed for the construction of an outstanding variety of POM assemblies ranging from small species exemplified by $\{Ce(L)\}$ monomers or $\{Ce(L)_2\}$ sandwich anions to some of the largest polyoxotungstates known to date with more than a hundred tungsten and sixteen cerium atoms. We noticed a lack of medium to large assemblies containing five to nine lacunary subunits, but the years to come will certainly see filling of this gap. On the other hand, cerium has also been successfully applied as quenching agent of unstable polyoxotungstate skeletons, allowing for the isolation and solid-state structural characterization of several unprecedented POM fragments that cannot be directly accessed. Current research on ceriopolyoxotungstates has also strong focus on 3d-4f heterometallic POMs involving cerium and hybrid metalorganic frameworks arising from the introduction of organic ligands in cerium/polyoxotungstate systems. These two trends are still in their first stages but they will most likely push the chemistry of cerium-containing polyoxotungstates to new heights in the near future.

ACKNOWLEDGMENT

This work has been funded by Eusko Jaurlaritza/Gobierno Vasco through the Program to Support the Research Activities of Groups from the Basque University System (grant IT477-10) and the SAIOTEK Program (grant S-PE11UN062). Financial support from Universidad del País Vasco UPV/EHU is gratefully acknowledged (grant UFI11/53).

REFERENCES

[1] (a) Polyoxometalates: From Platonic Solids to Antiretroviral Activity; Pope, M. T., Müller, A., Eds.; Kluwer: Dordrecht, The Netherlands, 1994. (b) Hill, C. L., Ed. *Chem. Rev.* 1998, *98(1)*, special issue on polyoxometalates. (c) Polyoxometalate Chemistry: From Topology via self-Assembly to Applications; Pope, M. T., Müller, A., Eds.; Kluwer: Dordrecht, The Netherlands, 2001. (d) Polyoxometalate Chemistry for Nanocomposite Design; Pope, M. T., Yamase, T., Eds.; Kluwer: Dordrecht, The Netherlands, 2002. (e) *Polyoxometalate Molecular Science*; Borrás-Almenar, J. J., Coronado, E., Müller, A., Pope, M. T., Eds.; Kluwer: Dordrecht, The Netherlands, 2003. (f) Pope, M. T. in *Comprehensive Coordination Chemistry II*; McCleverty, J. A., Meyer, T. J, Eds.; Elsevier Ltd.: Oxford, UK, 2004. (g) Kortz, U., Ed. *Eur. J. Inorg. Chem.* 2009, *(34)*, special issue on polyoxometalates. (h) Long, D.-L.; Tsunashima, R.; Cronin, L. *Angew. Chem. Int. Ed.* 2010, *49*, 1736.

[2] (a) Pope, M. T. Heteropoly and Isopoly Oxometalates; Springer-Verlag: Berlin, Germany, 1983. (b) Pope, M. T.; Müller, A. *Angew. Chem. Int. Ed. Engl.* 1991, *30*, 34.

[3] Nyman, M. *Dalton Trans.* 2011, 40, 8049.

[4] See for example: Barsukova-Stuckart, M.; Izarova, N. V.; Jameson, G. B.; Ramachandran, V.; Wang, Z.; van Tol, J.; Dalal, N. S.; Ngo Biboum, R.; Keita, B.; Nadjo, L.; Kortz, U. *Angew. Chem. Int. Ed.* 2011, *50*, 2639.

[5] (a) Bassil, B. S.; Kortz, U. *Z. Anorg. Allg. Chem.* 2010, *636*, 2222. (b) Pope, M. T. in *Handbook on the Physics and Chemistry of Rare Earths, vol. 38*; Gschneider, K. A., Jr., Bunzli, J.-C. G., Pecharsky, V. K., Eds.; Elsevier: Amsterdam, The Netherlands, 2007, pp. 337-382.

[6] Peacock, R. D.; Weakley, T. J. R. *J. Chem. Soc. A* 1971, 1836.

[7] Iball, J.; Low, J. N.; Weakley, T. J. R. *J. Chem. Soc., Dalton Trans.* 1974, 2021.

[8] Belsky, A.; Hellenbrandt, M.; Karen, V. L.; Luksch, P; *Acta Crystallogr., Sect. B: Struct. Sci.* 2002, *B58*, 364.
[9] Allen, F. H. *Acta Crystallogr., Sect. B: Struct. Sci.* 2002, *B58*, 380.
[10] (a) Molchanov, V. N.; Kazanskii, L. P.; Torchenkova, E. A.; Simonov, V. I. *Kristallografiya* 1979, *24*, 167. (b) *Sov. Phys., Crystallogr.* 1979, *24*, 96.
[11] Niu J.Y.; Wang, Z.L.; Wang, J.P. *J. Coord. Chem.* 2003, *56*, 895.
[12] Kholdeeva, O. A.; Timofeeva, M. N.; Maksimov, G. M.; Maksimovskaya, R. I.; Neiwert, W. A.; Hill, C. L: *Inorg. Chem.* 2005, *44*, 666.
[13] Sousa, F. L.; Almeida Paz, F. A.; Cavaleiro, A. M. V.; Klinowski, J.; Nogueira, H. I. S. *Chem. Commun.* 2004, 2656.
[14] Sousa, F. L.; Almeida Paz, F. A.; Granadeiro, C. M. C. E.; Cavaleiro, A. M. V.; Rocha, J.; Klinowski, J.; Nogueira, H. I. S. *Inorg. Chem. Commun.* 2005, *8*, 924.
[15] Fan, L.; Xu, L.; Gao, G.; Li, F.; Li, Z.; Qiu, Y. *Inorg. Chem. Commun.* 2006, *9*, 1308.
[16] Knoth, W. H.; Domaille, P. J.; Harlow, R. L. *Inorg. Chem.* 1986, *25*, 1577.
[17] Wassermann, K.; Dickman, M. H.; Pope, M. T. *Angew. Chem. Int. Ed. Engl.* 1997, *36*, 1445.
[18] Naruke, H.; Yamase, T. *J. Alloy Compd.* 1998, *268*, 100.
[19] Rosu, C.; Weakley, T. J. R. *Acta Crystallogr., Sect. C* 1998, *C54*, IUC9800047.
[20] (a) Chen, H.; Ding, Y.; Wang, E.; Xu, X.; Qi, Y.; Chang, S.; Li, Y. *Transition Met. Chem.* 2008, *33*, 341. (b) Chen, H.; Ding, Y.; Xu, X.; Wang, E.; Chen, W.; Chang, S.; Wang, X. *J. Coord. Chem.* 2009, *62*, 347. (c) Niu, J.; Guo, D.; Zhao, J.; Wang, J. *New J. Chem.* 2004, *28*, 980.
[21] (a) Liu, C.; Luo, F.; Liu, N.; Cui, Y.; Wang, X.; Wang, E.; Chen, J. *Cryst. Growth Des.* 2006, *6*, 2658. (b) Jiang, X.-Y.; Chen, L.-J.; Huang, H.; Zhang, Q.-Z.; Yang, W.-B.; Lu, C.-Z. *Inorg. Chem. Commun.* 2006, *9*, 1103. (c) Niu, J.-Y.; Guo, D.-J.; Wang, J.-P.; Zhao, J.-W. *Cryst. Growth Des.* 2004, *4*, 241. (d) Shi, D.; Chen, Y.; Pang, H. *Z. Anorg. Allg. Chem.* 2008, *634*, 839. (e) Pang, H.-J.; Chen, Y.-G.; Meng, F.-X.; Shi, D.-M. *Z. Naturforsch.* 2008, *63b*, 16. (d) Shi, D.; Yu, H.; Pang, H.; Chen, Y. *Solid State Sci.* 2008, *10*, 847. (e) Chen, W.; Li, Y.; Wang, Y.; Wang, E. *Eur. J. Inorg. Chem.* 2007, 2216.
[22] (a) Lu, Y.; Li, Y.; Wang, E.; Xu, X.; Ma, Y. *Inorg. Chim. Acta* 2007, *360*, 2063. (b) Pang, H.; Chen, Y.; Meng, F.; Shi, D. *Inorg. Chim. Acta* 2008, *361*, 2508.
[23] An, H.; Han, Z.; Xu, T. *Inorg. Chem.* 2010, *49*, 11403.
[24] (a) Hu, M.-X.; Chen, Y.-G.; Zhang, C.-J.; Kong, Q.-J. *CrystEngComm* 2010, *12*, 1454. (b) Meng, B.-F.; You, W.-S.; Sun, X.-F.; Zhang, F.; Liu, M.-Y. *Inorg. Chem. Commun.* 2011, *14*, 35. (c) Gao, Y.; Xu, Y.; Han, Z.; Li, C.; Cui, F.; Chi, Y.; Hu, C. *J. Solid State Chem.* 2010, *183*, 1000.
[25] (a) Pang, H.-J.; Zhang, C.-J.; Chen, Y.-G.; Hu, M.-X.; Li, J. *J. Cluster Sci.* 2008, *19*, 631. (b) Li, Y.-W.; Li, Y.-G.; Wang, Y.-H.; Feng, X.-J.; Lu, Y.; Wang, E.-B. *Inorg. Chem.* 2009, 48, 6452. (c) Geletii, Y. V.; Besson, C.; Hou, Y.; Yin, Q.; Musaev, D. G.; Quiñonero, D.; Cao, R.; Hardcastle, K. I.; Proust, A.; Kögerler, P.; Hill, C. L. *J. Am. Chem. Soc.* 2009, *131*, 17360.
[26] See for example: Wong, W.-K.; Zhang, L.; Wong, W.-T.; Xue, F.; Mak, T. C. W. *J. Chem. Soc., Dalton Trans.* 1999, 615.
[27] Buchler, J. W.; De Cian, A.; Fischer, J.; Kihn-Botulinski, M.; Paulus, H.; Weiss, R. *J. Am. Chem. Soc.* 1986, *108*, 3652.

[28] AlDamen, M. A.; Cardona-Serra, S.; Clemente-Juan, J. M.; Coronado, E.; Gaita-Ariño, A.; Martí-Gastaldo, C.; Luis, F.; Montero, O. *Inorg. Chem.* 2009, *48*, 3467.

[29] (a) Ozeki, T.; Takahashi, M.; Yamase, T. *Acta Crystallogr., Sect. C* 1992, *C48*, 1370. (b) Yamase, T.; Ozeki, T.; Ueda, K. *Acta Crystallogr., Sect. C* 1993, *C49*, 1572. (c) Ozeki, T.; Yamase, T. *Acta Crystallogr., Sect. C* 1993, *C49*, 1574. (d) Yamase, T.; Ozeki, T.; Ueda, K. *Acta Crystallogr., Sect. C* 1993, *C49*, 1577. (e) Ozeki, T.; Yamase, T. *Acta Crystallogr., Sect. B* 1994, *B50*, 128. (f) Ozeki, T.; Yamase, T. *Acta Crystallogr., Sect. C* 1994, *C50*, 327. (g) Yamase, T.; Ozeki, T.; Tosaka, M. *Acta Crystallogr., Sect. C* 1994, *C50*, 1849. (h) Almeida Paz, F. A.; Balula, M. S. S.; Cavaleiro, A. M. V.; Klinowski, J.; Nogueira, H. I. S. . *Acta Crystallogr., Sect. E* 2005, *E61*, i28.

[30] Xue, G.; Vaissermann, J.; Gouzerh, P. *J. Cluster Sci.* 2002, *13*, 409.

[31] Xue, G.; Liu, B.; Wang, W.; Li, Q.; Li, H. *Hua Hsueh Hsueh Pao* 2002, *60*, 2022.

[32] Yuan, L.; Qin, C.; Wang, X.; Wang, E. *Z. Naturforsch.* 2007, *62b*, 1471.

[33] Sun, R.-Q.; Zhang, H.-H.; Zhao, S.-L.; Huang, C.-C.; Zheng, X.-L. *Jiegon Huaxue* 2001, *20*, 413 (b).

[34] Bassil, B. S.; Dickman, M. H.; von der Kammer, B.; Kortz, U. *Inorg. Chem.* 2007, *46*, 2452.

[35] Iijima, J.; Ishikawa, E.; Nakamura, Y.; Naruke, H. *Inorg. Chim. Acta* 2010, *363*, 1500.

[36] Li, B.; Zhao, J.-W.; Zheng, S.-T.; Yang, G.-Y. *J. Cluster Sci.* 2009, *20*, 503.

[37] Du, D.-Y.; Qin, J.-S.; Li, S.-L.; Wang, X.-L.; Yang, G.-S.; Li, Y.-G.; Shao, K.-Z.; Su, Z.-M. *Inorg. Chim. Acta* 2010, *363*, 3823.

[38] Du, D.-Y.; Qin, J.-S.; Yuan, G.; Lan, Y.-Q.; Wang, X.-L.; Shao, K.-Z.; Su, Z.-M. *Solid State Sci.* 2011, *13*, 1115.

[39] Niu, J.; Zhang, S.; Chen, H.; Zhao, J.; Ma, P.; Wang, J. *Cryst. Growth Des.* 2011, *11*, 3769.

[40] Chen, W.-L.; Chen, B.-W.; Li, Y.-G.; Wang, Y.-H.; Wang, E.-B. *Inorg. Chim. Acta* 2009, *362*, 5043.

[41] Ostuni, A.; Bachman, R. E.; Pope, M. T. *J. Cluster Sci.* 2003, *14*, 431.

[42] Belai, N.; Dickman, M. H.; Pope, M. T.; Contant, R.; Keita, B.; Mbomekalle, I.-M.; Nadjo, L. *Inorg. Chem.* 2005, *44*, 169.

[43] Naruke, H.; Yamase, T. *Bull. Chem. Soc. Jpn.* 2000, *73*, 375.

[44] (a) Han, Q.-X.; Wang, J.-P.; Niu, J.-Y. *Chinese J. Struct. Chem.* 2007, *26*, 234. (b) Fan, L.H.; Xu, L.; Zhang, C.-H.; Li, F.-Y.; Li, Z.-K.; Liu, X.-Z. *Struct. Chem.* 2007, *18*, 917. (c) Naruke H.; Iijima, J.; Sanji, T. *Inorg. Chem.* 2011, *50*, 7535. (d) Zhang, C.; Howell, R. C.; Scotland, K. B.; Pérez, F. G.; Todaro, L.; Francesconi, L. C. *Inorg. Chem.* 2004, *43*, 7691. (e) Jiang, N.; Xu, L.; Li, F.; Gao, G.; Fan, L. *Inorg. Chem. Commun.* 2008, *11*, 24. (f) Wang, J.-P.; Yan, Q.-X.; Du, X.-D.; Niu, J.-Y. *Chinese J. Chem.* 2008, *26*, 1239.

[45] Fedotov, M. A.; Pertsikov, B. Z.; Danovich, D. K. *Polyhedron* 1990, *9*, 1249.

[46] Bartis, J.; Sukal, S.; Dankova, M.; Kraft, E.; Kronzon, R.; Blumenstein, M.; Fracncesconi, L. C. *J. Chem. Soc., Dalton Trans.* 1997, 1937.

[47] (a) Niu, J.; Zhao, J.; Guo, D.; Wang, J. *J. Molec. Struct.* 2004, *692*, 223. (b) Iijima, J.; Naruke, H.; Sanji, T. *Chem. Lett.* 2012, *41*, 295.

[48] (a) Luo, Q.-H.; Howell, R. C.; Dankova, M.; Bartis, J.; Williams, C. W.; Horrocks, W. DeW., Jr.; Young, V. G., Jr.; Rheingold, A. L.; Francesconi, L. C.; Antonio, M. R.

Inorg. Chem. 2001, *40*, 1894. (b) Boglio, C.; Lenoble, G.; Duhayon, C.; Hasenknopf, B.; Thouvenot, R.; Zhang, C.; Howell, R. C.; Burton-Pye, B. P.; Francesconi, L. C.; Lacôte, E.; Thorimbert, S.; Malacria, M.; Afonso, C.; Tabet, J.-C. *Inorg. Chem.* 2006, *45*, 1389.

[49] Sadakane, M.; Dickman, M. H.; Pope, M. T. *Angew. Chem. Int. Ed.* 2000, *39*, 2914.

[50] Hu, M.-X.; Chen, Y.-G.; Shi, D.-M.; Pang, H.-J.; Zhang, C.-J. *Z. Naturforsch.* 2008, *63b*, 968.

[51] Zhang, C.; Ma, P.; Chen, H.; Wang, J.; Niu, J. *J. Coord. Chem.* 2011, *64*, 2178.

[52] (a) Mialane, P.; Lisnard, L.; Mallard, A.; Marrot, J.; Antic-Fidancev, E.; Aschehoug, P.; Vivien, D.; Sécheresse, F. *Inorg. Chem.* 2003, *42*, 2102. (b) Miao, J.; Zhang, S.-X.; Li, S.-J.; Gao, Y.-H.; Zhang, X.; Wang, X.-N.; Liu, S.-X. *J. Coord. Chem.* 2011, *64*, 4006. (c) Wang, J.-P.; Duan, X.-Y.; Du, X.-D.; Niu, J.-Y. *Cryst. Growth Des.* 2006, *6*, 2266. (d) Wang, J.-P.; Yan, Q.-X.; Du, X.-D.; Duan, X.-Y.; Niu, J.-Y. *Inorg. Chim. Acta* 2008, *361*, 2701.

[53] Sadakane, M.; Dickman, M. H.; Pope, M. T. *Inorg. Chem.* 2001, *40*, 2715.

[54] Sadakane, M.; Ostuni, A.; Pope, M. T. *J. Chem. Soc., Dalton Trans.* 2002, 63.

[55] Shi, D.-M.; Ma, F.-X.; Pang, H.-J.; Meng, F.-X.; Chen, Y.-G. *J. Cluster Sci.* 2007, *18*, 897.

[56] Zhao, X.-Y.; Liu, S.-X.; Ren, Y.-H.; Cao, J.-F.; Cao, R.-G.; Shao, K.-Z. *J. Solid State Chem.* 2008, *181*, 2488.

[57] Li, F.; Xu, L.; Wei, Y.; Gao, G.; Fan, L.; Li, Z. *Inorg. Chim. Acta* 2006, *359*, 3795.

[58] Pang, H.; Zhang, C.; Shi, D.; Chen, Y. *Cryst. Growth Des.* 2008, *8*, 4476.

[59] Du, D.; Qin, J.; Li, S.; Lan, Y.; Wang, X.; Su, Z. *Aust. J. Chem.* 2010, *63*, 1389.

[60] (a) Drewes, D.; Krebs, B. *Z. Naturforsch.* 2006, *61b*, 637. (b) Liu, L.; Li, F.; Xu, L.; Liu, X.; Gao, G. *J. Solid State Chem.* 2010, *183*, 350. (c) Luo, Q.; Howell, R. C.; Bartis, J.; Dankova, M.; Horrocks, W. DeW., Jr.; Rheingold, A. L.; Francesconi, L. C. *Inorg. Chem.* 2002, *41*, 6112. (d) Lu, Y.; Xu, Y.; Li, Y.; Wang, E.; Xu, X.; Ma, Y. *Inorg. Chem.* 2006, *45*, 2055. (e) Zhang, C.; Bensaid, L.; McGregor, D.; Fang, X.; Howell, R. C.; Burton-Pye, B.; Luo, Q.; Todaro, L.; Francesconi, L. C. *J. Cluster Sci.* 2006, *17*, 389.

[61] Niu, J.; Zhao, J.; Wang, J. *J. Molec. Struct.* 2004, *701*, 19.

[62] (a) Kortz, U. *J. Cluster Sci.* 2003, *14*, 205. (b) Mialane, P.; Dolbecq, A.; Rivière, E.; Marrot, J.; Sécheresse, F. *Eur. J. Inorg. Chem.* 2004, 33. (c) Hussain, F.; Sandriesser, S.; Speldrich, M.; Patzke, G. R. *J. Solid State Chem.* 2011, *184*, 214. (d) Niu, J.; Wang, K.; Chen, H.; Zhao, J.; Ma, P.; Wang, J.; Li, M.; Bai, Y.; Dang, D. *Cryst, Growth Des.* 2009, *9*, 4362.

[63] Alizadeh, M. H.; Eshtiagh-Hosseini, H.; Khoshnavazi, R. *J. Molec. Struct.* 2004, *688*, 33.

[64] Chen, W.; Li, Y.; Wang, Y.; Wang, E.; Su, Z. *Dalton Trans.* 2007, 4293.

[65] Ostuni, A.; Pope, M. T. *C. R. Acad. Sci. Paris, Sér. IIc* 2000, *3*, 199.

[66] Chen, W.; Li, Y.; Wang, Y.; Wang, E.; Zhang, Z. *Dalton Trans.* 2008, 865.

[67] Pope, M. T.; Wei, X.; Wassermann, K.; Dickman, M. H. *C. R. Acad. Sci. Paris, Sér. IIc* 1998, *1*, 297.

[68] Pope, M. T.; Scully, T. F. *Inorg. Chem.* 1975, *14*, 953.

[69] See for example: (a) Ritchie, C.; Boyd, T.; Long, D.-L.; Ditzel, E.; Cronin, L. *Dalton Trans.* 2009, 1587. (b) Pichon, C.; Dolbecq, A.; Mialane, P.; Marrot, J.; Rivière, E.; Sécheresse, F. *Dalton Trans.* 2008, 71.

[70] For a recent review, see: Reinoso, S. *Dalton Trans.* 2011, *40*, 6610.

[71] Naruke, H.; Yamase, T. *J. Alloy Compd.* 1998, *268*, 100.

[72] Wassermann, K.; Pope, M. T. *Inorg. Chem.* 2001, *40*, 2763.

[73] See for example: Mal, S. S.; Kortz, U. *Angew. Chem. Int. Ed.*, 2005, *44*, 3777.

[74] Zimmermann, M.; Belai, N.; Butcher, R. J.; Pope, M. T.; Chubarova, E. V.; Dickman, M. H.; Kortz, U. *Inorg. Chem.* 2007, *46*, 1737.

[75] Fang, X.; Kögerler, P. *Angew. Chem. Int. Ed.* 2008, *47*, 8123.

[76] See for example: (a) Müller, A.; Roy, S. *Coord. Chem. Rev.* 2003, *245*, 153. (b) Müller, A.; Roy, S. *Eur. J. Inorg. Chem.* 2005, 3561.

[77] Todea, A. M.; Merca, A.; Bögge, H.; Glaser, T.; Engelhardt, L.; Prozorov, R.; Luban, M.; Müller, A. Chem. Commun. 2009, 3351.

[78] Kortz, U.; Hussain, F.; Reicke, M. *Angew. Chem. Int. Ed.* 2005, *44*, 3773.

[79] (a) Hussain, F.; Patzke, G. R. *CrystEngComm* 2011, *13*, 530. (b) Hussain, F.; Conrad, F.; Patzke, G. R. *Angew. Chem. Int. Ed.* 2009, *48*, 9088.

[80] Bassil, B. S.; Dickman, M. H.; Römer, I.; von der Kammer, B.; Kortz, U. *Angew. Chem. Int. Ed.* 2007, *46*, 6192.

[81] The lowest numeral possible is assigned to the vacant sites. For the IUPAC nomenclature, see: Jeannin, Y. P. *Chem. Rev.* 1998, *98*, 51.

[82] Reinoso, S.; Giménez-Marqués, M.; Galán-Mascarós, J. R.; Vitoria, P.; Gutiérrez-Zorrilla, J. M. *Angew. Chem. Int. Ed.* 2010, *49*, 8384.

[83] Kistler, M. L.; Pigga, J.; Liu, T. in *Macromolecules Containing Metal and Metal-Like Elements, Volume 9: Supramolecular and Self-Assembled Metal-Containing Materials*; Abd-El Aziz, A. S., Carraher, C. E. Jr., Pittman, C. U. Jr., Zeldin, M., Eds.; John Wiley and Sons: 2009, pp. 35-59.

[84] Li, T.; Li, F.; Lu, J.; Guo, Z.; Gao, S.; Cao, R. *Inorg. Chem.* 2008, *47*, 5612.

[85] Ismail, A. H.; Dickman, M. H.; Kortz, U. *Inorg. Chem.* 2009, *48*, 1559.

[86] Ismail, A. H.; Bassil, B. S.; Suchopar, A.; Kortz. U. *Eur. J. Inorg. Chem.* 2009, 5247.

[87] Fang, X.; Kögerler, P. *Chem. Commun.* 2008, 3396.

[88] (a) Gaunt, A. J.; May, I.; Collison, D.; Travis Holman, K.; Pope, M. T. *J. Molec. Struct.* 2003, *656*, 101. (b) Fang, X.; Anderson, T. M.; Hill, C. L. *Angew. Chem. Int. Ed.* 2005, *44*, 3540. (c) Fang, X.; Anderson, T. M.; Hou, Y.; Hill, C. L. *Chem. Commun.* 2005, 5044. (d) Saku, Y.; Sakai, Y.; Nomiya, K.; *Inorg. Chem. Commun.* 2009, *12*, 650.

[89] Reinoso, S.; Galán-Mascarós, J. R. *Inorg. Chem.* 2010, *49*, 377.

[90] Reinoso, S.; Galán-Mascarós, J. R.; Lezama, L. *Inorg. Chem.* 2011, *50*, 9587.

[91] See for example: (a) Kortz, U.; Mbomekalle, I. M.; Keita, B.; Nadjo, L.; Berthet, P. *Inorg. Chem.* 2002, *41*, 6412. (b) Hou, Y.; Xu, L.; Cichon, M. J.; Lense, S.; Hardcastle, K. I.; Hill, C. L. *Inorg. Chem. 2010*, 49, 4125.

In: Cerium
Editors: Aleksey Izyumov and Gulnaz Plaksin
ISBN: 978-1-62257-670-8

Chapter 2

CERIUM-BASED CORROSION INHIBITORS

Anthony E. Hughes
CSIRO Materials Science and Engineering,
Melbourne, Australia

ABSTRACT

Cerium is recognised as a good cathodic inhibitor for a range of metals. The corrosion inhibition of cerium as a simple salt as well as in cerium-organic compounds has been the subject of extensive research since the mid 1980's. This chapter examines our current understanding of the fundamental mechanisms of corrosion inhibition using cerium salts. It then explores how the chemistry has been adapted for industrial inorganic coating applications called conversion coatings. In finishes with a review of how multifunctionality is built into cerium compounds using various ligands and how these compounds are incorporated into paint systems used for corrosion protection.

INTRODUCTION

Metals and structures made from metals are subject to degradation due to atmospheric and marine corrosion. While a number of pure metals may form passive layers comprising compact oxides that prevent ingress of corrodents such as chloride to the metal/metal oxide interface, most practical alloys have natural oxides with defects where localised attack occurs. Metal finishing is designed to provide coatings that (a) have longer lifetimes and (b) better corrosion resistance compared to the natural oxide.

The prevention of metal corrosion is a very broad field since it encompasses not only a range of metals, but also a range of application areas. Prevention of corrosion degradation is achieved by a variety of means relying mainly on cathodic protection and protective coatings. Traditionally, chromate-based compounds have been used to prevent corrosion. These compounds are oxyanions and strong oxidants and facilitate redox reactions at surfaces that help form protective layers. The very reasons why chromate is a good inhibitor means that it also presents a health risk and has been targeted for replacement for some time. Chromate is carcinogenic, teratogenic and mutagenic [1]. Of course, with these properties, release into the environment is undesirable and so chromates are also considered as environmental pollutants in addition to their other properties. Increased concerns over the hazards of chromate have led to environmental and health regulatory frameworks that have increasingly driven some major changes in the area of protective coatings in the last thirty years.

These years have been challenging times for the metal finishing and paint industries because of the extensive use of chromate. For example, important industrial metals are based on iron, aluminium, copper, zinc, titanium magnesium and tin. Application areas include automobiles, food packaging, computing and telecommunications, heavy industry, aerospace, marine, infrastructure and buildings. Nearly all metals that are used industrially are alloys. Aluminium and steel alloys alone account for some five to six hundred different types of materials each with different alloy compositions, microstructures and surface conditions. Metal finishing of these alloys is required for corrosion protection and paint adhesion. However, the surfaces of this wide range of metals and metal alloys respond differently to metal finishing treatments hence the challenge in developing cerium and other rare earth-based metal finishing treatments is in finding solution chemistries that cover this broad range of applications.

The rare earths, particularly cerium, have demonstrated some desirable properties as alternatives to chromate as an inhibitor. Research into rare earths as inhibitors, with a strong focus on cerium, has grown rapidly in the last twenty five years. There are numerous examples of patents based on rare earths and in the last decade some products based on rare earths have been released into the market. This chapter examines the field of cerium chemistry and materials science that has been investigated to develop inhibitors for a range of applications. It will also touch on some of health risks.

CERIUM AS AN INHIBITOR

In the early 1980s it was discovered that the rare earth ions conferred a degree of corrosion inhibition to metals in corrosive solutions [2-12]. The assessment of the inhibition of a compound often begins with immersion into aqueous solution containing the inhibitor and a simple salt such as sodium chloride (NaCl). NaCl is used in this type of test because chloride is one of the most aggressive anions for metal corrosion and it occurs commonly as the sodium salt in marine environments. Magnesium chloride ($MgCl_2$) is the second most common salt in sea water. For assessment of corrosion protection against these types of salts, the metal of interest is weighed prior to and after immersion for a specified period of time which is usually either a week or a month. As an example, Figure 1 shows the corrosion rate in 0.1 M NaCl for a range of cerium compounds at 1000 ppm (except where otherwise specified) and lanthanum chloride ($LaCl_3$) for the aluminium alloy 7075-T6. The corrosion rate has been determined from weight loss after immersion in the solution for a week. In the presence of $CeCl_3$ the corrosion rate was reduced by a factor of approximately twenty times.

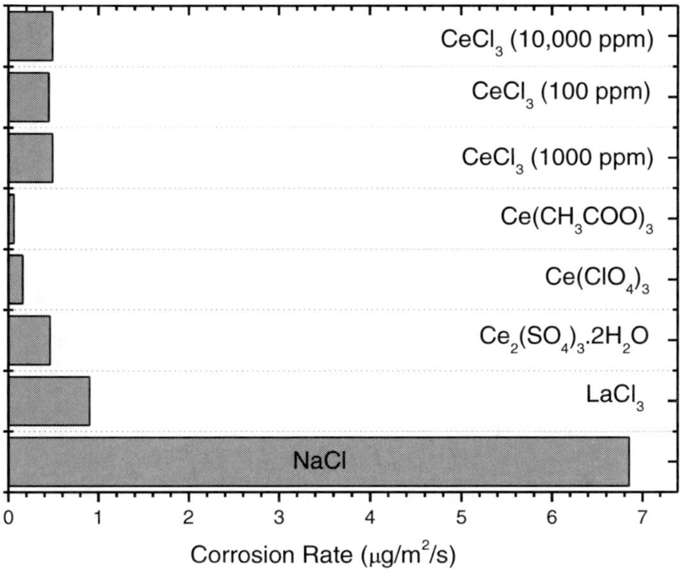

Figure 1. Corrosion rate for aluminium alloyAA7075-T6 at 1000 ppm except where otherwise specified, in 0.1M NaCl electrolyte determined after one week immersion.

The origin of the inhibition of cerium and other rare earth compounds lies in the solubility of the hydroxides as a function of pH, where insoluble compounds form under alkaline conditions. Alkaline conditions are experienced at cathodic sites and the precipitation of cerium hydroxides inhibits cathodic activity and reduces the overall corrosion rate as described below. Corrosion in salt solutions is basically an electrochemical process where the chemistry is separated into anodic and cathodic sites. At the anodic site the metal undergoes dissolution via

$$M \rightarrow M^{n+} + ne^- \quad (1)$$

The electrons generated in reaction (1) are then consumed at cathode sites, that are spatially separated from the anode. Two types of reactions occur at the cathodes including the reduction of oxygen via the so-called "4 electron pathway":

$$O_2 + 2H_2O + 4e^- \rightarrow 4OH^- \quad (2)$$

or via the "2 electron pathway":

$$O_2 + 2H_2O + 2e^- \rightarrow H_2O_2 + 2OH^- \quad (3)$$

Alternatively, hydrogen evolution can occur at the cathodes:

$$2H^+ + 2e^- \rightarrow H_2 \quad (4)$$

depending on solution conditions. Reactions (2) and (3) tends to happen in neutral solutions where the cathodic reaction can drive the pH into the alkaline region where cerium and other rare earths tend to form insoluble precipitates on the cathode thus stifling the oxygen reduction process and shutting down one half of the electrochemical reaction and reducing the overall level of corrosion [8, 11-15]. Thus the most basic explanation for the inhibition offered by cerium and the other rare earths is that the deposition proceeds in the presence of the local pH rise at the surface of active cathodes according to:

$$Ce^{n+} + nOH^- \rightarrow Ce(OH)_n \quad (5)$$

In addition, the presence of hydrogen peroxide (H_2O_2) can lead to the formation of cerium peroxo species. The cerium (III) ion in solution has nine waters of hydration [16] and the formation of the peroxo species begins with substitution of a hydrogen peroxide molecule into the hydration sphere of the Ce^{3+} ion via:

$$Ce(H_2O)_9{}^{3+} + H_2O_2 \rightarrow Ce(H_2O)_n\,(H_2O_2)^{3+} + (9\text{-}n)H_2O \quad (6)$$

Successive deprotonation and oxidation results in a cerium (IV) peroxo ion (for ease of display, the waters of hydration have been omitted):

$$Ce(H_2O_2)^{3+} + OH^- \rightarrow Ce(HO_2)^{2+} + H_2O \quad (7)$$

$$Ce(HO_2)^{2+} + OH^- \rightarrow Ce(O_2)^+ + H_2O \quad (8)$$

Excess, un-cordinated peroxide can then act as an oxidant to oxidise Ce(III) to Ce(IV):

$$Ce(O_2)^+ + \tfrac{1}{2}\,H_2O_2 \rightarrow Ce(O_2)^+ + OH^- \quad (9)$$

Cerium peroxo species have not been directly observed in experiments where metals have been immersed in aerated solutions containing cerium ions, but reactions (6) to (9) are relevant for conversion coatings described below.

Reaction (4) tends to occur in acidic solutions and more negative potentials and doesn't drive the pH to values high enough to cause precipitation.

Since cerium has the most stable tetravalent anion of the rare earths in solution then reaction (5) applies equally to $Ce(OH)_3$ as well as $Ce(OH)_4$. While reaction (5) implies the formation of specific compounds such as cerium hydroxides, a coating that provides a degree of protection to the underlying metal is more likely to involve hydrolysis and condensation reactions. Exchange of hydrolysed species on the surface may be written (adapted from Brinker and Scherrer [17]):

$$\sim M\text{-}OH + Ln(OH)(H_2O)_m{}^{n+} \rightarrow \sim M\text{-}O\text{-}Ln(OH)(H_2O)_{(m-1)}{}^{(n-1)+} + H_3O^+ \quad (10)$$

Alternatively, there may be condensation reactions defined by:

$$\sim M\text{-}OH + Ln(OH)(H_2O)_m{}^{n+} \rightarrow \sim M\text{-}O\text{-}Ln(H_2O)_m{}^{n+} + H_2O \quad (11)$$

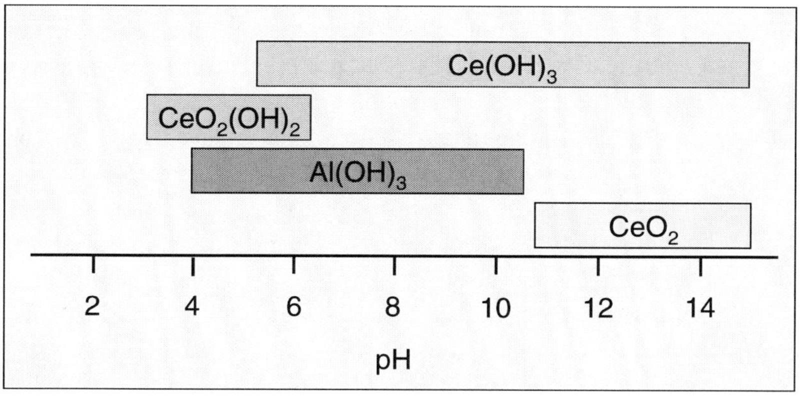

Figure 2. pH Stability regimes for various cerium compounds.

In the case of simple immersion the deposit on the surface of the metal may be a mixed metal oxy-hydroxide combining cations from both the substrate as well as cerium. The deposit may also be crystalline or amorphous depending on the ability of the precipitate to dissolve and reprecipitate to form a crystalline structure. The presence of other ligands will also modify the nature of the deposit. The presence of peroxo ligands, for example, inhibits crystallization [18]. Over long periods of time the cerium hydroxide deposits tend to form cerium dioxide (CeO_2) which has a fluorite structure. The formation of CeO_2 phase is preferred over mixed oxides which combine metal cations from the substrate due to the large Ce(IV) ion size meaning that there is little compatibility with oxides of the commodity metals (iron, zinc, aluminium, copper). This is fundamentally different from chromate conversion coatings where the Cr^{3+} cation can substitute into the Al_2O_3 structure and form a mixed compact oxide.

The degree of precipitation of cerium hydroxide depends on the extent of the pH rise at the local cathode. Yasakau et al. proposed that $Ce(OH)_3$ formed a loose deposit on the surface whereas $Ce(OH)_4$ formed a denser surface hydroxide that provided better barrier protection than $Ce(OH)_3$ [19]. According to the original Pourbaix diagram, Ce(IV) hydroxy cations precipitate as CeO_2 at a pH around 10.6 whereas $Ce(OH)_3$ precipitates at a pH of 8.1 [20]. More recent studies of Hayes et al. [21] and Hughes et al. [22] indicate that precipitation occurs at 5.3. In the study of Hughes et al. [23] they calculated the potential-pH (Pourbaix) diagram using different free energy of formation to Pourbaix and found that CeO_2 could precipitate out at pH 0.8, which was clearly wrong but emphasises the need to use these diagrams as a guide only. Cerium peroxyhydroxide, which is important for conversion coating applications (discussed below), precipitates at around pH 3.6. For reference, typical stability ranges for aluminium, cerium and lanthanum hydroxides are shown in Figure 2. This data combines information from titrations as well as thermodynamic data. It should be noted that the Pourbaix diagrams refers to stable species for which the thermodynamic data is available but transient species, which may be important in coating formation can be overlooked if only the stable species are considered.

Various cerium salts have been investigated for their inhibitive qualities. The most commonly researched ones are $CeCl_3$ and $Ce(NO_3)_3$. For aluminium and its alloys, $CeCl_3$ or $Ce(ClO_4)_3$ are preferred, whereas $Ce(NO_3)_3$ is preferred for other metals. A number of other cerium compounds have been investigated as inhibitors for paint systems and will be discussed in more detail below. In the case of the perchlorate, chloride ions are often introduced via some acidification using HCl. The preference of chloride or perchlorate for aluminium is related to the corrosion and coating process since corrosion is required to activate the surface for the electrochemical reactions that create a pH rise. For aluminium, NO_3^- ions passivate the surface thus limiting the extent of electrochemical activity where cerium deposition occurs only at the strongest electrodes. The chloride ion activates the surface thus initially increasing the electrochemical activity which raises the pH at the cathodes resulting in deposition at these sites. For metals other than aluminium, the nitrate ion activates the surface and is preferred since it does not cause any of the other corrosion effects such as pitting for which chloride is notorious.

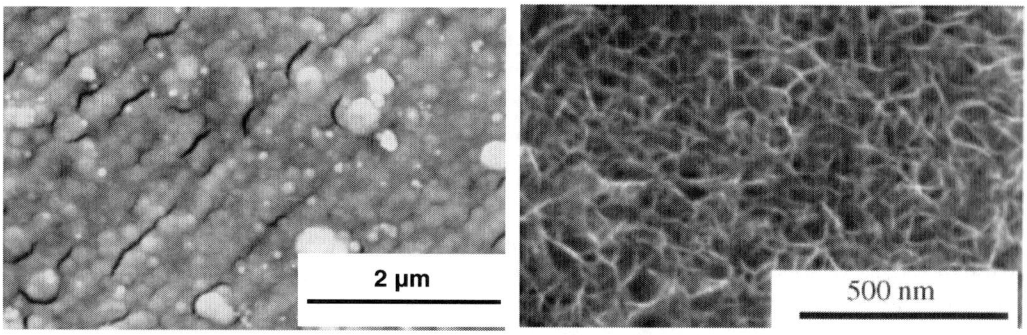

Figure 3. Left cerium-based conversion coating formed from 10mM $CeCl_3/H_2O_2$ (pH 1.9, 1%v/v) solution and Right: pseudo-boehmite coating formed in 10 mM $CeCl_3$ solution at 95°C for 60 minutes.

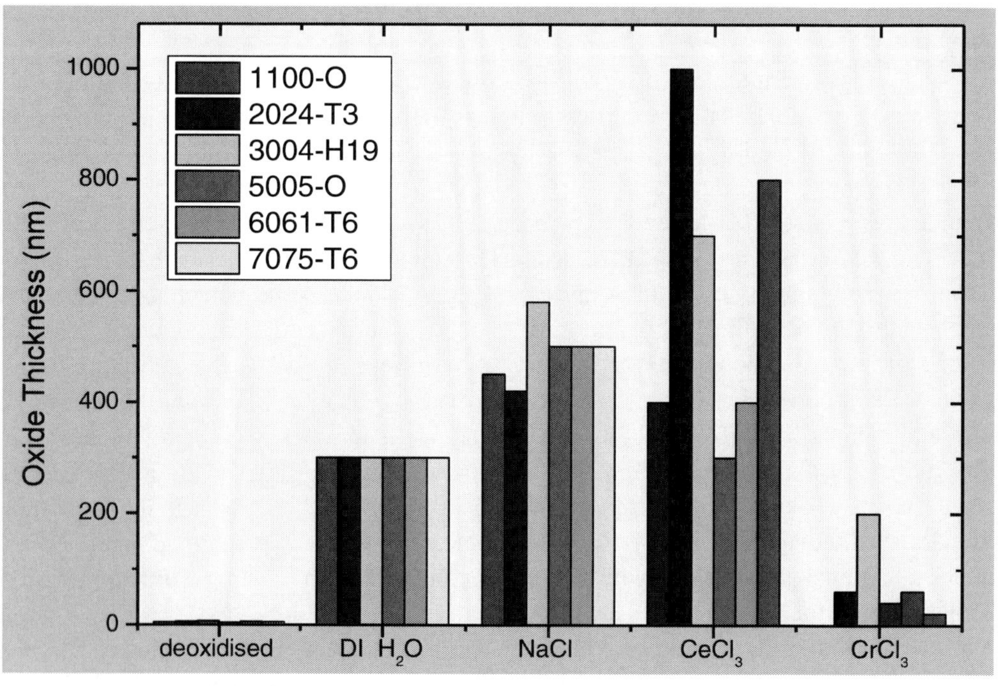

Figure 4. Example of boehmite oxide thickness on various aluminium alloys in various salt solutions after treatment of the alloy at 95°C for 60 minutes.

With respect to temperature, most of the immersion studies in simple salt solutions have been performed at room temperature. For coating work, the temperature is generally in the range of 30 to 50°C. In this region the coating deposition reactions result in more even deposition across the surface. A typical cerium-based conversion coating (described in the next section) is shown in Figure 3. As temperature increases condensation reactions occur in solution resulting in the formation of polynuclear species which are unavailable for coating formation since much of this material precipitates.

In addition, the kinetics of surface oxide formation change. In the case of Al, bayerite and then pseudo-boemite ($Al_2O_3.nH_2O$, n=2,3) and boehmite ($Al_2O_3.H_2O$) form at increasingly higher temperatures [24]. The boehmite type coatings tends to be self limiting due to the passage of reactants to the growth interface, which is probably at the barrier oxide (Al_2O_3)/boehmite interface, when formed in both pure water or electrolyte solution [24-27].

There is a considerable literature on the formation of pseudo-boehmite and boehmite coatings in the presence of cerium ions [27-42]. These coatings contain low levels of Ce incorporated into the Boehmite structure. Gorman et al. [27] proposed that the Ce modified the interface between the underlying Al alloy and the adjacent oxide. They also found that the eventual oxide thickness was different on some alloys showing dependence on the Mn plus Cu content in the presence of Ce whereas there was little difference in oxide thickness in the absence of cerium ions. For example, it can be seen in Figure 4 that in distilled (DI) H_2O the oxide is similar for all alloys. Development in NaCl results in some variation in oxide thickness, but development in $CeCl_3$ displays huge variation in thickness. By comparison, the oxide thicknesses in the presence of $CrCl_3$ are thin demonstrating the passivating nature of chromium compounds.

CERIUM-BASED INORGANIC COATINGS

There is a category of coatings called conversion coatings, which are applied to metals during manufacture for a whole range of end-use applications [43]. These coatings are an alternative to anodising in many application areas. The term "conversion" relates to converting the natural surface oxide that might develop on exposure to the environment with a coating that has better corrosion resistance and suitable morphology for paint adhesion. In reality the surface oxide is not actually "converted" from the native surface oxide, but an oxide with the desired properties is grown on the surface, usually via an immersion or spray process. The conversion coating does not usually contain the same metal cations as the parent metal.

Conversion coatings are generally inorganic in nature with the most widely used being phosphate-based coatings. In the recent past chromate conversion coatings were also widely used for many applications. Chromate conversion coatings are still used in high performance applications *i.e.*, for assured corrosion resistance, such as the aerospace industry. In many other applications chromate has been replaced by a range of other coatings technologies and cerium-based replacements are an area of active research.

Some common application areas include the auto industry, aircraft manufacture, computer casings, architectural aluminium and steel roofing and framing products. These coatings are deposited from solution and form an inorganic layer on the metal that provides corrosion protection, but more importantly a good base for keying of paint. Apart from conversion coatings, inorganic coatings are deposited electrochemically in anodising processes. Anodising involves the electrochemical growth from the parent metal and may incorporate anions that enhance corrosion resistance.

Conversion coatings are applied in a multistep process involving a number of chemical treatments as displayed in Figure 5, in this instance, for aerospace applications. The number of stages varies from one industry to another, but usually involves both alkaline and acid pretreatments designed to get the surface in a chemically reproducible state for the application of the conversion coating. It is worth describing this process in a little detail as the surface chemistry remaining after the major treatment steps has an influence on the development of the conversion coating. The first step is a detergent clean as a light treatment to remove grease and grime. This is followed by an alkaline clean that mildly etches the surface removing both remaining grease or oils as well as etching the surface oxide. This step is commonly done in a silicated sodium hydroxide bath and results in a surface that often has an accumulation of alloying elements such as Cu since their dissolution rate from the alloy is different from the aluminium in the alloy. There is also a thickened silicate layer that often contains oxide and zinc and/or magnesium oxy-hydroxide from aluminium alloys with Zn or Mg additions. After rinsing, the next step is to remove the oxide residue from the alkaline clean as well as accumulated alloying components such as Cu. This is achieved using the deoxidation step which involves immersion in mineral acids and HF, For aerospace applications, this usually involves a mixed acid bath containing chromic acid, nitric acid and hydrofluoric acid. The chromate and nitrate ions in this bath limit the etched produced by the hydrofluoric acid. As might be expected, this results in a surface oxide that contains, all components from the bath such as nitrate, chromate and fluoride ions [22, 44]. The surface oxide left after deoxidation is not ideal for the deposition of a cerium-containing conversion coating, however, the "basic" oxide left after alkaline cleaning is suitable for coating. Thus research into the use of cerium-based conversion coatings, has focused, not only on the batch chemistry of the cerium processing solution, but also on pretreatment prior to the cerium-containing conversion coating step.

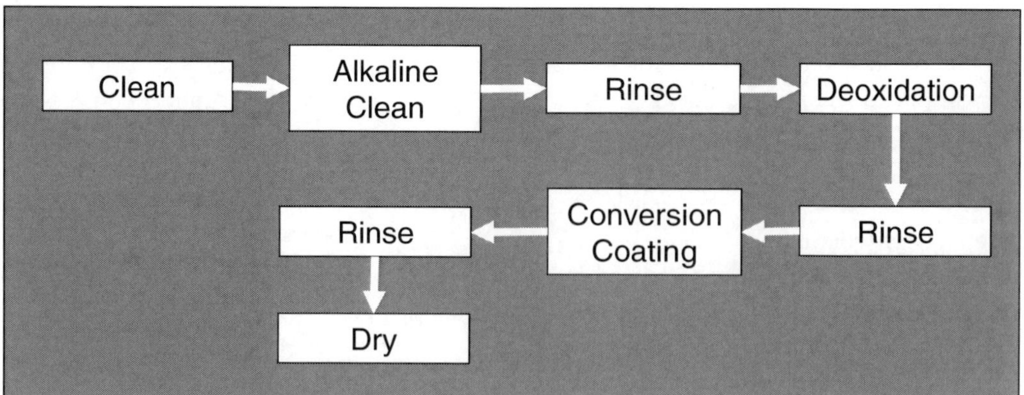

Figure 5. Treatment steps for the application of a chromate conversion coating for aerospace applications.

Cerium Bath Chemistries for Conversion Coating

The solution chemistry related to immersion experiments described above results in deposits on the surface of a number of different metals. Most coating work has focused on aluminium alloys [8, 18, 39, 45-63], but there is also many studies on other metals such as iron , zinc (for galvanising) [4, 64-70] magnesium alloys [71-73] and copper [58, 74]. For coating development a lot of early work focused on aerospace aluminium alloys (AA) such as AA2024-T3 (Al-Mg-Cu), AA7075-T6 (Al-Mg-Zn) and AA6061-T6 (Al-Mg-Si) and the discussion below relates primarily to aluminium alloys rather than other metals. These will be dealt with later.

The deposits that developed on surfaces as a result of immersion in cerium salt solutions, fall short of the process and performance requirements of a conversion coating. From a processing perspective, the time to produce these coatings could be hours or even days depending on the alloy. This figure should be compared to 1 to 2 minutes for chromate conversion coatings which the cerium conversion coating is viewed as a replacement. From a performance perspective, the coatings developed through immersion fell well short of the corrosion resistance and the paint adhesion typical of, and required for chromate conversion coatings. Their structure consisted of heavy deposits onto cathodic sites and aluminium oxide growth elsewhere. This "mixed" oxide system did not provide the barrier properties to protect the underlying aluminium alloy. Thus the challenges for developing a commercial conversion coating process were (i) acceleration of the coating process and (ii) improvement of the coating properties to achieve better corrosion resistance and paint adhesion.

The challenge of reducing the coating time was addressed by Hinton and Wilson [6] by using hydrogen peroxide as an oxidant to accelerate the coating process. Thus the coating solution was simply formed by addition of H_2O_2 to $CeCl_3$ solution acidified to pH 2 with HCl. Acidification was required to activate the surface since aluminium oxide is stable to about pH 4. Another successful coating solution contained perchloric acid with H_2O_2 [55, 59, 61, 75]. (It is assumed here that the mechanism in the presence of perchlorate is the same as in the presence of chloride.) The proposed mechanism was that hydrogen peroxide would be reduced at cathodes on the surface resulting in a local pH rise:

$$H_2O_2 + 2e^- \rightarrow + 2OH^- \quad (12)$$

According to the Pourbaix diagram [20, 21, 23], cerium solution species are insoluble at higher pH, consequently, cerium compounds would deposit onto the local cathodes via reactions like reaction (6) thereby shutting down cathodes and one half the galvanic reaction and thus reducing the overall electrochemical activity of the surface. The acidification assisted the deposition process since it destabilised the aluminium oxide and resulted in a degree of etching of the underlying alloy, albeit at anodic sites. For Cu containing aluminium alloys etching results in buildup of a very thin Cu layer at the aluminium/aluminium oxide interface that acts as a cathode [76]. Thus the etching of the surface increased the overall activity of the aluminium and facilitated deposition of the cerium-containing coating. Of course the buildup of Cu will only happen with alloys containing Cu, and a modified process was required to develop a coating more generally applicable to Al-alloys as discussed below.

Thus acidification and the addition of hydrogen peroxide reduced the coating time from times typical of coatings that developed during simple immersion (hours to days) down to as little as a few minutes. Despite the practical achievement of obtaining the coating in an expedient processing time, the proposed mechanism did not match, in detail, observations of the coating process. For example, the Pourbaix diagrams suggest that Ce(III) hydroxide should precipitate at pH 8.1 ([77]=10^{-2}M) and above, however, it was noted that precipitate started to form in the range of pH 3 to 4. This could be achieved if Ce(IV) species were present, but the coating solutions lacked the characteristic golden colour of Ce(IV) species. It was over ten years later that Hughes et al. [50] proposed that cerium peroxo species may play a role in the deposition process and also may be incorporated into the deposited coating. The role of peroxo species in the coating process was first noted, not with cerium, but when titanium addition to the coating solution was observed to accelerate deposition. Ti ions were proposed to form peroxo species faster than the cerium equivalent and transfer the protonated peroxide to the cerium (III) ion via[50]:

$$Ce^{3+}(H_2O)_9 + TiO_2^{2+} + H^+ \rightarrow Ce(HO_2)(H_2O)_n^{2+} + Ti_{aq}^{4+} \quad (13)$$

It was another six years before the role of cerium peroxo species was fully defined by Scholes et al. [18] who examined precipitates obtained from titrations of cerium-containing conversion coating solutions. They discovered that the cerium (III) peroxo species were formed over a period of 20 to 30 minutes after addition of hydrogen peroxide to the acidified cerium chloride solution (reaction (6)). During titration the deprotonation reactions (reaction (7) and (8)) and oxidation (reaction (9)) occurred in the pH range 3.5 to 4 and formed a buffer region where precipitate formed both in the titrated solution, but also during coating at local cathodes on the surface where the cathodic reactions produced a pH rise. Thus a mechanism was proposed that fitted the observations of solution behaviour and coating deposition.

A second problem with the hydrogen peroxide activation of the cerium-based coating solution was that the coating times on aluminium alloys with low copper levels could still take up to 60 minutes, which is impractical for a commercial process. This was resolved by Hughes et al. through the addition of transition metal and notably Cu to the coating solution to accelerate the coating process [50]. This was achieved through what is called in metal finishing a “displacement reaction”, where copper ions in solution are reduced on the surface of the aluminium resulting in oxidation of the aluminium via:

$$3Cu^{2+} + 2Al \rightarrow 3Cu + 2Al^{3+} \quad (14)$$

This reaction occurs readily for aluminium in solutions containing copper, so complexation was necessary to slow down the kinetics of the displacement reaction. In the end it was found that $Cu(glycinate)_2$ was the best complex and resulted in small copper deposits on the surface leading to even coating deposition. With this type of acceleration, coatings could be obtained on all classes of aluminium alloys in less than a minute. One problem with the copper in solution, even when complexed, was that it led to hydrogen peroxide decomposition. Hardin et al. [78] solved this issue by replacing the $Cu(glycinate)_2$ with $Bi(EDTA)_2$ which still accelerated the deposition of the coating but had little or no effect on hydrogen peroxide decomposition. Nevertheless H_2O_2 decomposition remains a problem with the buildup of alloying components that catalyse peroxide decomposition, in the coating bath as a function of the surface area processed through the treatment bath.

The other weakly co-ordinating anion for cerium salts that could be considered for conversion coating is the nitrate ion. This has been explored extensively for metals like Zn [67-70, 79-92] and Mg alloys [93, 94]. On aluminium, however, the nitrate ion passivates the aluminium surface and inhibits the deposition of the cerium-based conversion coating. This inhibition has led to the development of mixed coatings that have a pseudo-boehmite type coating on the aluminium matrix and cerium oxy-hydroxide deposition on the cathodic intermetallic particles [15, 34-40, 91, 95-102].

Pretreatment Steps for Cerium-based Conversion Coatings

As described above the deposition of a conversion coating is a multistep process in either immersion or spray application. The pretreatment of aluminium for chromate conversion coating is similar for most aluminium alloys except in the deoxidation step. For industries that use copper-containing high strength aluminium alloys, such as the aerospace industry, nitric acid is used to remove Cu accumulation from the surface and chromate is added to moderate the degree of etching. For other industries phosphoric acid is generally used. In the chromic acid/HF/HNO_3 pretreatment the presence of chromate and nitrate moderates the etch rate that would be experienced in HF alone since both chromate and nitrate ions passivate aluminium [26]. The result of deoxidation is a thin surface oxide comprising all components of the deoxidiser, *i.e.*, fluoride, chromate and nitrate species or reduced forms of chromate or nitrate [44, 103]. This pretreatment causes some passivation of the aluminium surface and delays the deposition of the cerium-based conversion coating [22]. Consequently, variations of the pretreatment have been investigated to find suitable chemistry or order of steps for cerium-based coatings.

Schmidt-Hansberg and Schubach investigated the order of the pretreatment steps and found that having the alkaline clean as the final step prior to cerium-based conversion coating facilitated coating deposition [104]. This treatment process was suitable for architectural applications where there was little buildup of copper. Similarly, Joshi et al. [52, 105] have found that alkaline pretreatment facilitated the deposition of the cerium-based conversion coating onto the high strength aluminium alloy 7075-T6. The reason for this facilitation is that a basic oxide is left on the surface after this type of treatment which is enriched in magnesium hydroxyoxides [44, 106] which probably facilitate the deposition via oxolation (such as reactions (10) and (11)) at the surface. Any basic oxide will facilitate this reaction with cerium ions in solutions For aluminium processing basic surface oxides that are commonly encountered include magnesium and zinc oxides (from alloying elements) and silicate containing oxides (from pretreatments).

Cerium-based Deoxidiser

As described above the chromate/HF/HNO_3 deoxidation leaves a thin chromate-containing layer on the surface after treatment. Hughes et al. [23, 48, 108] considered the possibility of developing a cerium-based deoxidiser that left a cerium containing thin layer on the surface of the aluminium in order to facilitate the deposition of a cerium-based conversion coating. Similarly Smith et al. [109] considered the addition of cerium salts to standard mineral acid pretreatment solutions to facilitate coating deposition.

A range of deoxidiser compositions were explored spanning etch rates that were low compared to the standard chromate-based deoxidiser to those that were similar to it. It was found that cerric anions in the presence of HF resulted in stable solution compositions for achieving a high etchrate. The cerric anions were $Ce(IV)(SO_4)_4^{2-}$ and $Ce(IV)(NO_3)_4^{2-}$ in the form of their ammonium salts. Both these anions were stable in the presence of HF and either HNO_3 or H_2SO_4 at around pH 0.5. While cerium fluorides have a low solubility, this did not appear to be an issue for these cerium-based deoxidiser formulations, and it is possible that either the solution complex is stable to ligand exchange in the presence of fluoride, excess sulphate or nitrate anions stabilise the complex or the hexafluorocerate anion $(Ce(F_6)^{2-})$ [110, 111] is formed in solution. None of these options was confirmed. Even in the presence of nitrate anions, additional oxidants were necessary to remove copper deposits that remained on the surface. The preferred oxidant was potassium persulphate ($K_2S_2O_8$).

Unlike the chromate system, the cerium-based system appeared to perform best where a low etch rate deoxidiser was used. This deoxidiser comprised $(NH_4)_2Ce(IV)(SO_4)_4$ in the presence of H_2SO_4. A combination of transmission electron microscopy, Rutherford backscattering spectroscopy and X-ray photoelectron spectroscopy (XPS) showed that the aluminium oxide thickened to around 50 nm from 7nm after the alkaline clean, during treatment in the low etch rate, cerium-based deoxidizer [23]. XPS revealed that some Ce(III) and Ce(IV) were incorporated into the surface oxide suggesting both adsorption and, in the case of Ce(IV), oxidation. It was suggested that the Ce(III) was incorporated into the aluminium oxide and that cerium (IV) resulted from oxidation of surface cerium hydroxide deposits to CeO_2.

Performance

The standard performance tests in many metal finishing industries usually involves test in corrosive test environments for the fully formed conversion coating as well as the painted (primed and topcoated). Much of the research literature includes either testing using techniques that are not part of standard performance testing (e.g. electrochemical procedures) or are not performed as according to the standard (e.g. testing area too small, different testing conditions). This section only focuses on testing done strictly in accordance with testing standards. In this respect there are not many research reports that conform to these requirements.

Table 1 compares the performance of several different conversion coating processes as well as a boehmite coatings grown in the presence of cerium ions (Ce-boehmite). The tests include neutral salt spray (NSS) and paint adhesion using a dolley that is glued to the surface and then the pull-off force is measured. The coatings are chromate conversion coating (CrCC), Ce-based conversion coating (CeCC), Permanganate-based process (MnCC)[112-114] and a cobalt-based conversion coating (CoCC) using cobalt amine complexes [115, 116]. Apart from the CrCC all the chromate-replacement coatings required a sealing step to improve the NSS performance. For MnCC it was an organic seal in an oil-based system, CoCC used a vanadate based system and both the cerium processes used silicate sealing [43, 49] although phosphate also works well [54, 56].

With respect to the NSS performance there are a number of slightly different performance requirements for passing neutral salt spray and the one that will be focused on here is 336 hours in NSS. There are two standards that refer to the 336 h in NSS: MIL-C-81706 requires five 10" x 3" panels to be tested with a pass being no visible pits and test panels must be 6° from the normal. It is used to qualify process lines. Mil_C-5541E refers uses the same number and area of panels but refers to ASTM B117 where panels can be tested from 5 to 20° and accepts a small number of pits (5) for specified test area. Testing at 6° from the normal is milder than testing at higher angles [117].

The performance data is presented in Table 1 and Figure 6 where the 336h pass line is also indicated. The data for three different aluminium alloys is presented including AA2024-T3 (Al-Cu-Mg), AA7076-T6 (Al-Zn-Mg) and AA7475-T7651 (Al-Zn-Mg) all of which are high strength aluminium alloys used in aircraft manufacture. The most obvious feature of Figure 6 is that the performance of all the conversion coatings varies with the alloy. For the CrCC the NSS performance easily passes the 336h requirement on AA2024-T3 and AA7075-T6 whereas it fails on AA7475-T7651. This result is interesting in that AA7475 is a "cleaner" form of AA7075-T6, *i.e.* it has lower levels of tramp (impurity) elements. However, it is not the improved cleanliness of the alloy, but its surface condition that has led to the difference in performance.

Figure 7 shows EELS elemental maps from TEM studies of the surface oxide of AA2024-T3 and AA7475-T7651 in the "as-received" condition. The surface oxide for AA2024 T3 is slightly less than 100 nm and can easily be removed by the pretreatment process described above, however, the oxide on AA7475-T7651 is much thicker and not so easily removed by the pretreatment. Consequently the conversion coating has not formed properly on the alloy surface. Hence its poor performance.

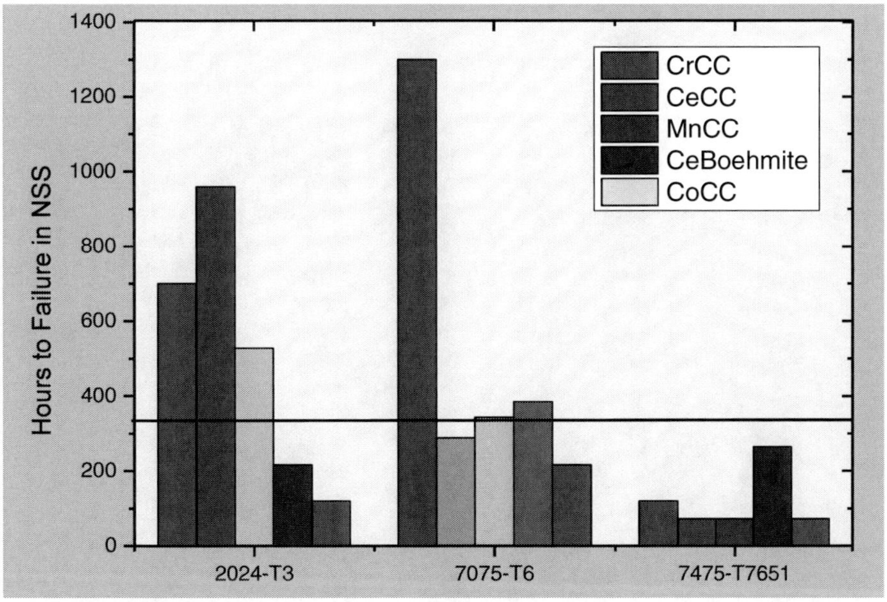

Figure 6. Comparison of neutral salt spray performance for several alternatives to the chromate conversion coating (CrCC). Alternatives are described in the text.

Table 1. Summary of NSS/Adhesion Performance of Various Conversion Coated Processes

Coating Process	Corrosion Resistance Unpainted NSS Time to failure (hrs)	Paint AdhesionPATTI (% of Cr value)	
		Uns	Sld
AA2024-T3			
CrCC	700	100%	-
MnCC	528	186%	52%
CoCC	120	-	217%
CeCC	960	119%†	124%†
Ce Boehmite	216	64%	119%
AA7075-T6			
CrCC	1300	100%	-
MnCC	342	80%	73%
CoCC	216	44%	58%
CeCC	288	96%	93%
Ce-Boehmite	384	31%	133%
AA7475-T7561			
CrCC	120	100%	-
MnCC	72	182%	81%
CoCC	72	-	217%
CeCC	72	85%	97%
Ce-Boehmite	72-264	85%	128%

Uns =Unsealed conversion coated panel.

Sld =Sealed conversion coated panel.

† = Tested panel had been both primed and top-coated.

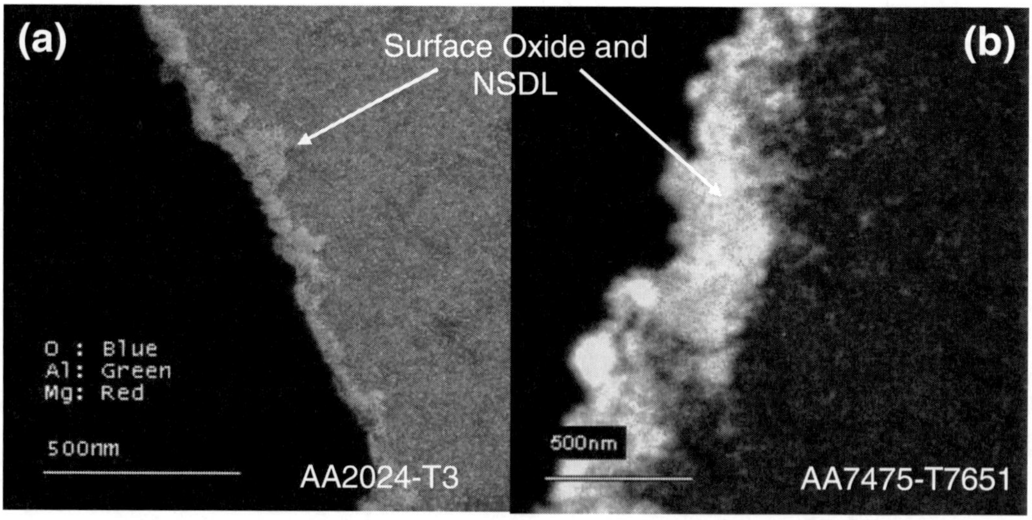

Figure 7. Comparison of the surface oxides on (a) AA2024-T3 to (b) AA7475-T7651. The surface oxide and near surface deformed layer (NSDL) is indicated the width of the grey regions.

From the perspective of cerium-based processes replacing the chromate process, it can be seen that the CeCC outperforms the CrCC on AA2024-T3, but does not pass the 336 NSS hour performance requirement on AA7075-T6. On the other hand the Ce-based boehmite process does pass the 336h requirement on AA7075-T6, but does not out-perform the CrCC. A similar mixture of performance results for the other types of coatings is also observed. These types of mixed results have made the replacement of the CrCC a complicated process, but cerium-based processes continue to be of interest.

The other type of test performed on conversion coatings are the corrosion and paint adhesion performance under paint. In Table 1 adhesion tests are presented as a percentage of the paint adhesion of the chromate conversion coating that they are intended to replace. Table 1 includes data for the unsealed as well as the sealed versions. The unsealed is included here since metals are often coated with inhibiting primers as well as topcoats. If the principal task of the conversion coating is to improve the adhesion then the seal step used for additional corrosion resistance may not be required under a paint system. This shortens the process and reduces the costs. As with the NSS results there are a range of performances with some outstanding individual results for some alloys but no overall consistency. The CeCC shows similar results to the CrCC with the best performance on AA2024-T3. In the unsealed state the Ce-boehmite coating shows poorer performance to the CrCC which is probably attributable to a high level of surface hydroxyl groups which as not suitable for bonding to the solvent-based polymer used in this particular test. On the other hand both the silicate sealed CcCC and Ce-boehmite generally show better adhesion than the CrCC. This is somewhat surprising in that silicate finishes are not favoured in the metal finishing industry due to the possibility of residual NaOH or KOH which might adversely effect topcoat adhesion. These results show that proper preparation can result in very good adhesion. One other test worth mentioning here is a filiform test and cerium-based conversion coatings have been demonstrated to perform well in this test [104, 118, 119].

Similar to the results presented here several round robin studies have found a range of performances for chromate replacements [120] which complicates the choice of replacement of chromate. The varied performance is due to a range of factors from pretreatments, solution formulation, ageing of coatings and baths and many other factors that differ from the research laboratory to a metal finishing process line.

Other Coating Developments

A number of other avenues of investigation for the use of rare earths in the coatings area have also been investigated. These will be covered briefly here since they form an important part of the scientific literature, but have not been developed into commercial coating processes.

Sol-Gel

Sol-gel science is an obvious path to the development of either an inorganic coating or a hybrid organic/inorganic coating system [121-130]. Sol-gel processes for aerospace applications have seen good success with Si-Zr-based systems for adhesive bonding applications [131, 132]. A review of sol-gel since is beyond the scope of this chapter, but a brief summary of the approach follows.

Sol-gel approaches generally involve the use and reaction of metal alkoxides. The most widely explored are silicon alkoxides followed by titanium and zirconium alkoxides. The prevelance of research into these alkoxides is largely due to their commercial availability as fine chemicals. Examples of the use of cerium compounds as inhibitors have involved the dissolution of $Ce(NO_3)_3$ into alkoxide precursors [131, 133], the incorporation of Ce-oxide nanoparticles [134] and other cerium compounds such as organometallics [135, 136] and cerium molybdate as particles or nanowires [137]. These types of incorporations often depend specifically on the chemistry of the compound and the metal-alkoxide system. It can be generally said that the incorporation of the cerium compound has a positive influence on corrosion resistance, but the magnitude of the influence is varied.

Sealing

Porous coatings can be developed on a number of metals. Phosphating, for example, results in an open structure that can benefit from sealing with active inhibitor on a range of metals. For aluminium, Boehmite (AlOOH or $\frac{1}{2}Al_2O_3.H_2O$) and pseudoboehmite ($\frac{1}{2}Al_2O_3.nH_2O$ where $1 > n \leq 3$) coatings also develop open porous structures that swell and close upon sealing [24]. These types of coatings have been discussed above in the context of boehmite growth in the presence of cerium ions, but some work has looked at sealing these coatings with cerium salts particularly $Ce(NO_3)_3$ [138]. Anodising produces a better structured aluminium oxide with properties that can be better manipulated [139]. Post anodising steps include impregnation of inhibitors and colouring agents prior to or during "sealing" in a hot (close to boiling) water which swells the pores and closes then due to increased hydration. The performance of anodised coatings is improved if a post anodising step involving sealing in cerium salts is used [140].

CERIUM IN ORGANIC COATINGS

The incorporation of cerium inhibitors into a polymer matrix that forms the basis of a paint system is not so straight forward. For cerium compounds, the two key problems are (i) the interaction of the inhibitor with groups within the polymeric matrix and (ii) the Ce(III)/Ce(IV) redox pair. Generally Ce(III) compounds are selected as the additive, however if there is dissolution of the inhibitor then the cerium (IV) cation may be created by reaction with water during inhibitor preparation or when water and oxygen enter the polymer matrix via:

$$2Ce^{3+} + 2(H_2O) + O_2 \rightarrow 2Ce^{4+} + 4OH^- \qquad (14)$$

Ce(IV) is a strong oxidant and used as an oxidising agent in organic reactions [16]. The reaction of cerium species with the epoxy is feasible and Schuman and Stoffer [141] reported an exothermic complexation of cerium species with amines and a colour change, from transparent to brown, in a number of epoxy systems. These reactions with the polymer resin are not unexpected since a number of studies have investigated the use of cerium (Ce(IV)) as a catalyst in the opening of epoxy rings [142, 143] and as cure initiators for epoxys [144]. Further, cerium compounds are widely used for oxidation of alcohols to aldehydes and ketones [145], and rare earth complexation with glycols and their subsequent oxidation have also been reported [146]. The issue of cerium (IV) being a strong oxidant is probably related to the pursuit of Pr-based inhibitors for paint systems. Nevertheless various Ce-inhibitors have been incorporated into a range of coating matrices and demonstrated good performance. Data on properly formulated coatings, is, however, lacking with the focus being more on sol-gel and conversion coating systems.

A number of different cerium-based inhibitors that have been tested in laboratory-based, "model" primers are listed in Table 2. Paint inhibitors are chosen on the basis of low solubility unlike aqueous inhibitors which have a high solubility. The desirable properties of paint inhibitors is that they have good inhibition at low levels (typically (10^{-3}M), but the low solubility provides long term release for ongoing inhibition. The cerium-based inhibitors that have been studied can be broadly divided into inorganic and organic cerium inhibitors.

The number of cerium-based inorganic inhibitors is small compared to the potentially huge number of combinations of cerium complexes that can be formed with organic ligands. Consequently, exploring this space has been left to high throughput techniques based on electrochemistry [163-169] and high throughput corrosion tests using microfluidics approaches [170].

Table 2. Cerium-based inhibitors under development for paint coatings

Inorganic	*Organophosphates*	*General Organic*
Molybdate [41, 42, 137] Phosphate Sulphate Iodate[109, 147]	Di-butyl phosphate[148-154] Di-phenyl Phosphate[149, 155]	Saliscylate [156] Glycollate[157] Cinnemate and deriveratives [158-160] Formate [157, 161] Tartrate[157, 161] Oxalates[162] Acetates[162]

Figure 8 shows a section of an epoxy containing a $Ce(dbp)_3$ inhibitor. This type of coating showed improved filiform corrosion resistance compared to the epoxy without the inhibitor as exemplified by the corrosion number (Figure 10). Filiform corrosion is a type of corrosion where corrosion filaments develop underneath paint from a defect in the paint (e.g. scratch) that is open to the air and corrodents. The corrosion number is derived from the average filament length multiplied by the filiform site density. Figure 10 shows that even small additions of $Ce(dbp)_3$ result in some improvement (lowering) of the corrosion number. With 5.2 wt % $Ce(dbp)_3$ the corrosion number is lowered by two orders of magnitude.

From a practical research point of view, Figure 8 shows some of the complexities associated with the development of new inhibitors for paint systems as there is considerable layering in this laboratory manufactured paint film. In this instance the film was prepared using a drawbar. At the air/polymer surface there is a layer that is apparently devoid of inhibitor; these are called "skin" or "varnish" layers and are thought to occur due to solvent evaporation effects during the initial stages of drying and are common even in commercial paint systems. Beneath the skin layer are two more layers that have different densities of particles. Mulitlayers are often observed in these types of coatings and can be due to settling effects or multiple pass coating, particularly during spray processing. The kinetics of leaching of inhibitors is important for the best performance from an inhibited primer but poorly understood. The influence of the distribution of the inhibitor in the film on leaching is not well researched, but it is assumed that a good distribution is important for maximum performance. So one issue with new inhibitors, including cerium-based ones, is how to properly disperse the inhibitor. In this instance the nanowires of $Ce(dbp)_3$ are reasonably well dispersed as displayed in TEM sections of the paint in Figure 11.

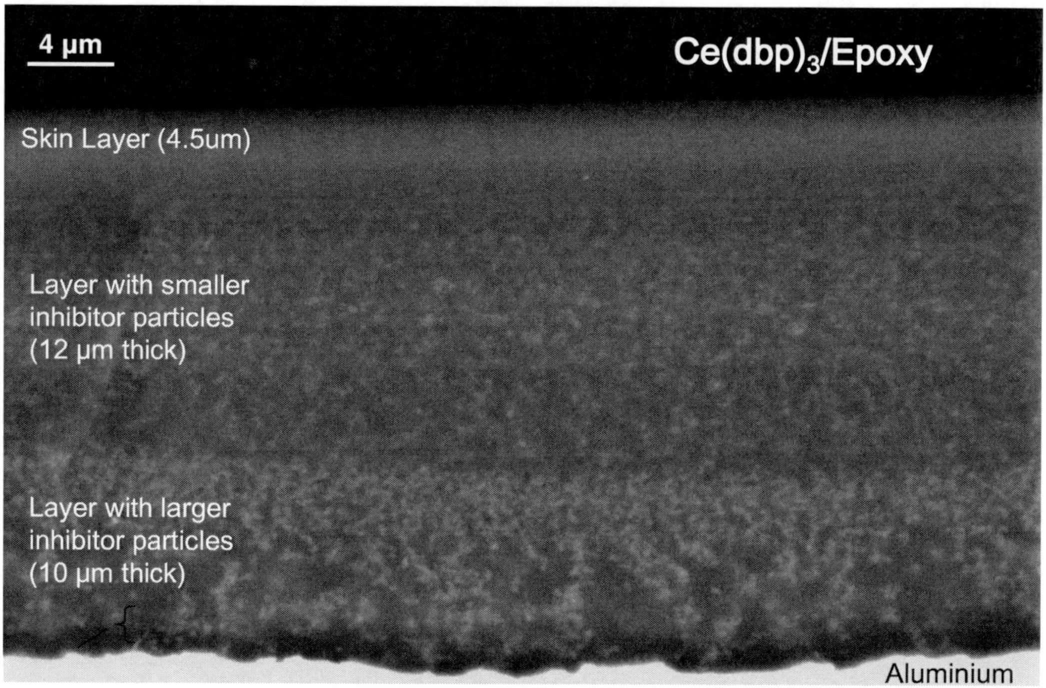

Figure 8. Cross section of an epoxy containing nanofilaments of $Ce(dbp)_3$ inhibitor. Image collected at the University of Manchester as part of a TU-Delft-CSIRO.

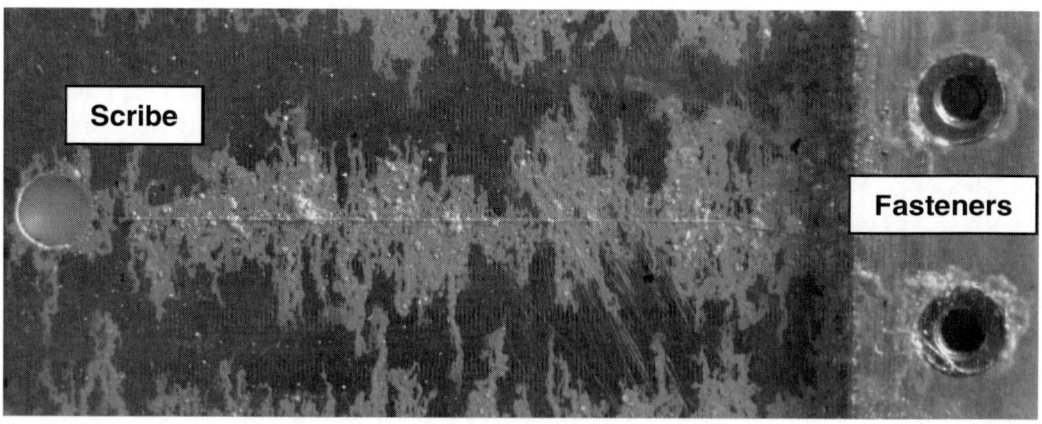

Figure 9. Example of severe filiform corrosion from a scribe and around two fasteners. System: Alkaline cleaned aluminium AA29024-T3 with a polyurethane coating [119].

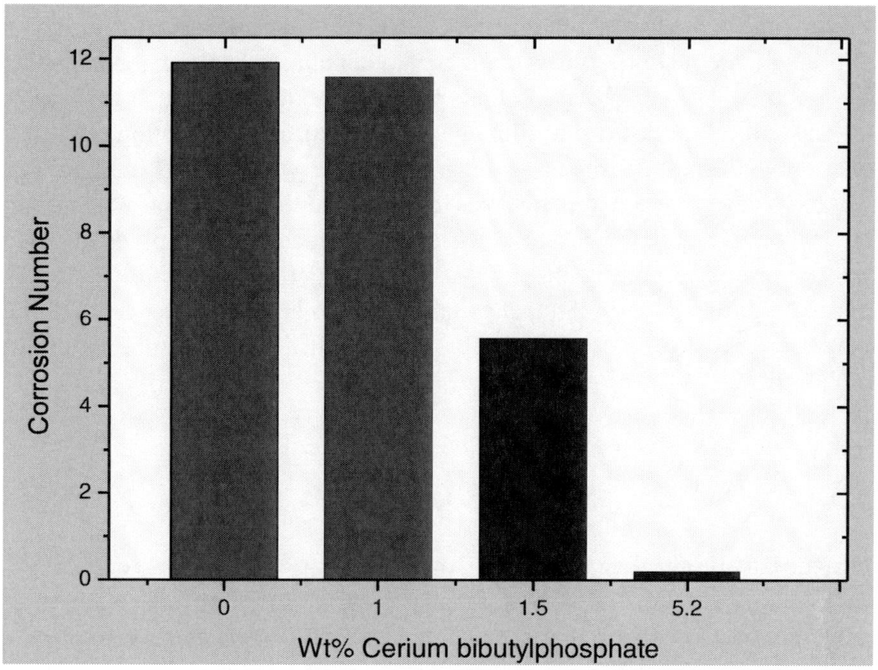

Figure 10. Filament number versus the loading of $Ce(dbp)_3$ inhibitor in an epoxy matrix. Image collected at the University of Manchester as part of a TU-Delft-CSIRO collaboration.

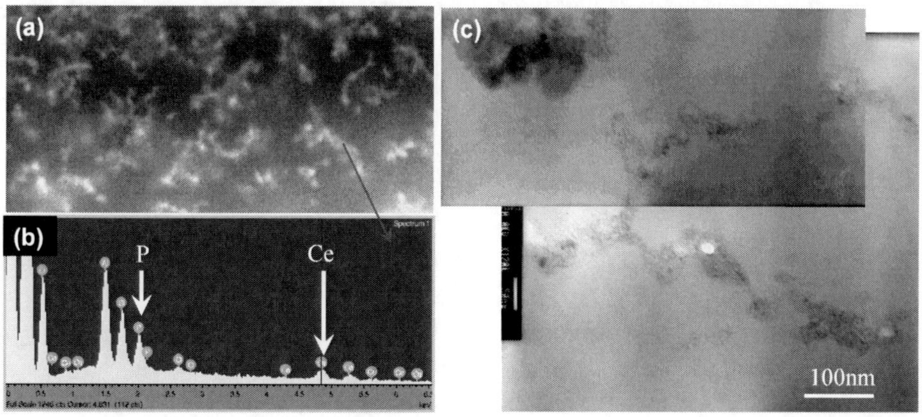

Figure 11. (a) SEM image of a cross-section of the $Ce(dbp)_3$ inhibited epoxy film shows the $Ce(dbp)_3$ nanowires. (b) X-ray analysis of one of the nanowires and (c) TEM montage showing the $Ce(dbp)_3$ nanowires. The data source is the same as for Figure 8.

Figure 11 (a) shows the good dispersion of the nanowires which are confirmed as Ce-organophosphate in Figure 11(b). The EDS result also shows the presence of Al, Si and Cl. The Al and Cl are probably related to the corrosion process since these data were obtained after filiform testing. Si was also detected in the EDS and highlights the interaction of the inhibitor with components of the paint system since a silane is added as a dispersing agent [153] highlighting the interaction of the inhibitor with chemicals in the paint formulation. These interactions may cause considerable changes to the paint curing process and degree of crosslink.

Apart from interaction with other components in the paint system, for commercially formulated primers, the resin (polymer matrix) also contains other compounds with different functionality such as UV absorbers, colour pigments, extenders and others, all of which change the transport properties of the films without necessarily interacting with the inhibitor *i.e.* they may only change diffusion path length. They may also act as inhibitor sinks if the inhibitor strongly absorbs onto the surfaces of these additives. From a paint formulation perspective, all these materials need to be below the critical volume pigment concentration (CVPC) where these is enough resin to coat all the particles in the primer. This is an area where there is very little research, if any, available in the literature for cerium-based inhibitors and much work remains to be done to design systems that are compatible with cerium and other new inhibitors.

CONCLUSION

This chapter has summarised the current state of the art with respect to the use of cerium compounds as corrosion inhibitors in a range of metal finishing and paint applications. Cerium has demonstrated corrosion resistance that makes it a contender to replace chromate in some applications. Indeed the cerium-based conversion coating has already undergone extensive pre-commercial trials and was available for a short period for treatment of architectural aluminium applications. Future research is likely to focus on cerium organometallic complexes given the large number of possible cerium complexes that might be developed that draw on the inhibitive qualities of both the cerium cation and the organic anion.

REFERENCES

[1] N. I. Sax, *Dangerous Properties of Industrial Materials*, 5 th ed. New York: Van Nostrand Reinhold Company, 1979.

[2] B. R. W. Hinton, "Corrosion inhibition with rare-earth-metal salts," *Journal of Alloys and Compounds,* vol. 180, pp. 15-25, Mar 1992.

[3] H. S. Isaacs, *et al.*, "The Electrochemical Response Of Steel To The Presence Of Dissolved Cerium," *Journal Of The Electrochemical Society,* vol. 138, pp. 390-393, Feb 1991.

[4] B. R. W. Hinton and L. Wilson, "The corrosion inhibition of zinc with cerous chloridE," *Corrosion Science,* vol. 29, pp. 967-&, 1989.

[5] D. R. Arnott, *et al.*, "Cationic-film-forming inhibitors for the protection of the AA7075 aluminum-alloy against corrosion in aqueous chloride solution," *Corrosion,* vol. 45, pp. 12-18, Jan 1989.

[6] L. Wilson, Hinton, B.R.W., "A Method of Forming a Corrosion Resistant Coating," Australia Patent, 1988.

[7] D. R. Arnott, *et al.*, "Cationic film-forming inhibitors for the corrosion protection of AA 7075 aluminum-alloy in chloride solutions," *Materials Performance,* vol. 26, pp. 42-47, Aug 1987.

[8] D. R. Arnott, *et al.*, "Auger and XPS studies of cerium corrosion inhibition on 7075-aluminium alloy," *Applied Surface Science,* vol. 22-3, pp. 236-251, 1985.

[9] D. R. Arnott, N. E. Ryan and B.R.W. Hinton, "Cerium Conversion coatings for Corrosion Protection of Alumnium Alloys," *Materials Forum,* vol. 9, p. 12, 1985.

[10] D. R. Arnott, *et al.*, "Auger and XPS Studies of Cerium Corrosion Inhibition on 7075 Aluminium Alloy," *Applied Surface Science,* vol. 22-23, pp. 236-251, 1985.

[11] B. R. W. Hinton, Ryan, N.E., Arnott, D.R., Trathen, P.N., Wilson, L., Williams, B.E., "The Inhibition of Aluminium Alloy Corrosion by rare Earth metal Cations," *Corrosion Australasia,* vol. , pp. 12 - 17, 1985.

[12] B. R. W. Hinton, *et al.*, "The inhibition of aluminum-alloy corrosion by cerous cations," *Metals Forum,* vol. 7, pp. 211-217, 1984.

[13] N. C. Rosero-Navarro, *et al.*, "Electrochemical techniques for practical evaluation of corrosion inhibitor effectiveness. Performance of cerium nitrate as corrosion inhibitor for AA2024T3 alloy," *Corrosion Science,* vol. 52, pp. 3356-3366, Oct 2010.

[14] A. J. Aldykiewicz, *et al.*, "The investigation of cerium as a cathodic inhibitor for aluminum-copper alloys," *Journal of the Electrochemical Society,* vol. 142, pp. 3342-3350, Oct 1995.

[15] A. Aballe, *et al.*, "Inhibition of the corrosion process of alloy AA5083 (Al-Mg) in seawater by cerium cations. An EIS study," *Materials and Corrosion-Werkstoffe Und Korrosion,* vol. 52, pp. 344-350, May 2001.

[16] F. A. C. a. G. Wilkinson, *Advanced Inorganic Chemistry*, 5th ed. New York: John Wiley & Sons, 1988.

[17] C. J. Brinker, Scherer, G.W., *Sol-Gel Science*, !st Edition ed. San Diego: Academic Press, 1990.

[18] F. H. Scholes, *et al.*, "The role of hydrogen peroxide in the deposition of cerium-based conversion coatings," *Applied Surface Science,* vol. 253, pp. 1770-1780, Dec 2006.

[19] K. A. Yasakau, *et al.*, "Lanthanide salts as corrosion inhibitors for AA5083. Mechanism and efficiency of corrosion inhibition," *Journal of the Electrochemical Society,* vol. 155, pp. C169-C177, 2008.

[20] M. Pourbaix, *Atlas of Electrochemical Equilibria in Aqueous Solutions*. Houston: NACE and Cebelcor, 1974.

[21] S. A. Hayes, *et al.*, "The phase stability of cerium species in aqueous systems - I. E-pH diagram for the Ce-$HClO_4$-H_2O system," *Journal of the Electrochemical Society,* vol. 149, pp. C623-C630, Dec 2002.

[22] A. E. Hughes, *et al.*, "XPS and SEM characterization of hydrated cerium oxide conversion coatings," *Surface and Interface Analysis,* vol. 23, pp. 540-550, Jul 1995.

[23] A. E. Hughes, *et al.*, "Development of cerium-based conversion coatings on 2024-T3 Al alloy after rare-earth desmutting," *Surface and Interface Analysis,* vol. 36, pp. 290-303, Apr 2004.

[24] W. Vedder and Da. Vermilyea, "Aluminum+water reaction," *Transactions of the Faraday Society,* vol. 65, pp. 561-&, 1969.

[25] R. S. Alwitt, "Growth of hydrous oxide-films on aluminum," *Journal of the Electrochemical Society,* vol. 121, pp. 1322-1328, 1974.

[26] Da. Vermilyea and W. Vedder, "Inhibition of aluminum+water reaction," *Transactions of the Faraday Society,* vol. 66, pp. 2644-&, 1970.

[27] J. D. Gorman, *et al.*, "Oxide formation on aluminium alloys in boiling deionised water and NaCl, $CeCl_3$ and $CrCl_3$ solutions," *Corrosion Science,* vol. 45, pp. 1103-1124, Jun 2003.

[28] F. Mansfeld, "Use of rare earth metal salt solutions for corrosion protection of aluminum alloys and mild steel," *Russian Journal Of Electrochemistry,* vol. 36, pp. 1063-1071, Oct 2000.

[29] F. Mansfeld, "Corrosion protection of aluminium alloys and mild steel provided by rare earth metal salt (REMS) solutions," in *Passivity and Localized Corrosion - an International Symposium in Honor of Professor Norio Sato.* vol. 99, M. Seo, *et al.*, Eds., ed, 1999, pp. 259-268.

[30] F. Mansfeld and Y. Wang, "Development of stainless aluminum-alloys by surface modification," *Materials Science and Engineering a-Structural Materials Properties Microstructure and Processing,* vol. 198, pp. 51-61, Jul 1995.

[31] H. W. White, *et al.*, *Characterization of corrosion protective films on steel and aluminum alloys*, in Nondestructive Characterization of Materials, 757-764 from the proceedings of the 6th International Symposium on Nondestructive Characterization of Materials Jun 07-11, 1993, Oahu, Hawaii, (1994), pp 757-764.

[32] F. Mansfeld, Wang, F., "Corrosion Protection of High Copper Aluminium Alloys by Surface Modification," *British Corrosion Journal,* vol. 29, pp. 194 - 200, 1994.

[33] F. Mansfeld, *et al.*, "The ce-mo process for the development of a stainless aluminum," *Electrochimica Acta,* vol. 37, pp. 2277-2282, Sep 1992.

[34] M. Bethencourt, *et al.*, "Using EIS to analyse samples of Al-Mg alloy AA5083 treated by thermal activation in cerium salt baths," *Corrosion Science,* vol. 50, pp. 1376-1384, May 2008.

[35] B. Davo, *et al.*, "Inhibition of stress corrosion cracking of alloy AA8090 T-8171 by addition of rare earth salts," *Corrosion Science,* vol. 47, pp. 1227-1237, May 2005.

[36] M. Bethencourt, *et al.*, "Advanced generation of green conversion coatings for aluminium alloys," *Applied Surface Science,* vol. 238, pp. 278-281, Nov 2004.

[37] M. Bethencourt, *et al.*, "Combination of thermal activation and addition of H2O2 to improve cerium-based immersion treatment of alloy AA5083," *Materials and Corrosion-Werkstoffe Und Korrosion,* vol. 54, pp. 77-83, Feb 2003.

[38] M. Bethencourt, *et al.*, "High protective, environmental friendly and short-time developed conversion coatings for aluminium alloys," *Applied Surface Science,* vol. 189, pp. 162-173, Apr 2002.

[39] A. Aballe, *et al.*, "On the mixed nature of cerium conversion coatings," *Materials and Corrosion-Werkstoffe Und Korrosion,* vol. 53, pp. 176-184, Mar 2002.

[40] A. Aballe, *et al.*, "Electrochemical noise applied to the study of the inhibition effect of $CeCl_3$ on the corrosion behaviour of Al-Mg alloy AA5083 in seawater," *Electrochimica Acta,* vol. 47, pp. 1415-1422, Feb 2002.

[41] A. E. Hughes, *et al.*, "The characterisation of Ce-Mo-based conversion coatings on Al-alloys .1," *Corrosion Science,* vol. 38, pp. 1957-1976, Nov 1996.

[42] J. D. Gorman, *et al.*, "The characterisation of Ce-Mo-based conversion coatings on Al-alloys .2," *Corrosion Science,* vol. 38, pp. 1977-1990, Nov 1996.

[43] R. G. Buchheit, Hughes, A.E., "Chromate and Chromate-Free Coatings," in *Corrosion: Fundamentals, Testing and Protection.* vol. 13A, C. Moosbrugger, Ed., Materials Park, Oh, USA: ASM International, 2003, pp. 720 -735.

[44] A. E. Hughes, *et al.*, "Characterisation of surface preparation of 2024 aluminium alloy for conversion coating," *Materials Science and Technology,* vol. 12, pp. 928-936, Nov 1996.

[45] A. E. Hughes, *et al.*, "Factors influencing the deposition of Ce-based conversion coatings, part I: The role of Al^{3+} ions," *Surface & Coatings Technology,* vol. 203, pp. 2927-2936, Jun 2009.

[46] D. Lau, *et al.*, "Factors influencing the deposition of Ce-based conversion coatings, Part II: The role of localised reactions," *Surface & Coatings Technology,* vol. 203, pp. 2937-2945, Jun 2009.

[47] A. E. Hughes, *et al.*, "Towards replacement of chromate inhibitors by rare earth systems," *Corrosion Reviews,* vol. 25, pp. 591-605, 2007.

[48] S. K. Toh, *et al.*, "Characterization of non-Cr-based deoxidizers on Al alloy 7475-T7651," *Surface and Interface Analysis,* vol. 36, pp. 1523-1532, Dec 2004.

[49] A. E. Hughes, Hardin, S.G., Harvey, T.G. Nikpour, T., Hinton, B.R.W. Hinton, Galassi, A., McAdam, Stonham, A., Harris, S.J., Church, S., Figgures, C., Dixon, D., Bowden, C., Morgan, P., Toh, S.K., McCulloch, D., DuPlessis, J., "Characterisation of Various Conversion Coatings on AA2024-T3," *ATB Metallurgie,* vol. 43, p. 9, 2003.

[50] A. E. Hughes, Hardin, S.G., Wittel, K.W., Miller, P.R., "Accelerated Cerium-Based Conversion Coatings," presented at the NACE 2000, Research Topics Symposium, Orlando, USA, 2000.

[51] A. E. Hughes, Nelson, K.J.H., "Rare Earth Conversion Coatings - A Status Report," presented at the Interfinish '96, Birmingham, UK, 1996.

[52] S. Joshi, *et al.*, "Alkaline activation of Al 7075-T6 for deposition of cerium-based conversion coatings," *Surface & Coatings Technology,* vol. 205, pp. 4312-4319, May 2011.

[53] S. Joshi, *et al.*, "Characterization of Cerium-Based Conversion Coatings on Al 7075-T6 Deposited from Chloride and Nitrate Salt Solutions," *Journal of the Electrochemical Society,* vol. 158, pp. C88-C93, 2011.

[54] D. K. Heller, *et al.*, "The effect of post-treatment time and temperature on cerium-based conversion coatings on Al 2024-T3," *Corrosion Science,* vol. 52, pp. 360-368, Feb 2010.

[55] B. L. Treu, *et al.*, "Characterization of Localized Surface States of Al 7075-T6 during Deposition of Cerium-Based Conversion Coatings," *Journal of the Electrochemical Society,* vol. 157, pp. C282-C287, 2010.

[56] D. K. Heller, *et al.*, "Effect of Phosphate Source on Post-Treatment of Cerium-Based Conversion Coatings on Al 2024-T3," *Journal of the Electrochemical Society,* vol. 156, pp. C400-C406, 2009.

[57] P. S. Jones, *et al.*, "Spray deposition of cerium oxide-based conversion coatings on Al 2024-T3," *International Journal of Applied Ceramic Technology,* vol. 5, pp. 63-73, 2008.

[58] S. You, *et al.*, "Response of nanocrystalline cerium-based conversion coatings on Al 2024-T3 to chloride environments," *Materials Letters,* vol. 61, pp. 3778-3782, Jul 2007.

[59] B. E. Rivera, *et al.*, "Deposition and characterization of cerium oxide conversion coatings on aluminum alloy 7075-T6," *Surface & Coatings Technology,* vol. 176, pp. 349-356, Jan 2004.

[60] B. Y. Johnson, *et al.*, "Effect of coating parameters on the microstructure of cerium oxide conversion coatings," *Materials Science and Engineering a-Structural Materials Properties Microstructure and Processing,* vol. 361, pp. 225-231, Nov 2003.

[61] W. G. Fahrenholtz, *et al.*, "Characterization of cerium-based conversion coatings for corrosion protection of aluminum alloys," *Surface & Coatings Technology,* vol. 155, pp. 208-213, Jun 2002.

[62] P. Campestrini, *et al.*, "Formation of a cerium-based conversion coating on AA2024: relationship with the microstructure," *Surface & Coatings Technology,* vol. 176, pp. 365-381, Jan 2004.

[63] B. Davo and J. J. de Damborenea, "Corrosion and inhibition of medium-strength aluminum alloys," *Revista De Metalurgia,* vol. 40, pp. 442-446, Nov-Dec 2004.

[64] K. Aramaki, "Prevention of zinc corrosion in oxygenated 0.5 M NaCl by treatment in a cerium(III) nitrate solution and modification with sodium hexadecanoate," *Corrosion Science,* vol. 48, pp. 3298-3308, Oct 2006.

[65] K. Aramaki, "The effect of modification with hydrogen peroxide on a hydrated cerium(III) oxide layer for protection of zinc against corrosion in 0.5 M NaCl," *Corrosion Science,* vol. 48, pp. 766-782, Mar 2006.

[66] K. Aramaki, "A self-healing protective film prepared on zinc by treatment in a $Ce(NO_3)_3$ solution and modification with $Ce(NO_3)_3$," *Corrosion Science,* vol. 47, pp. 1285-1298, May 2005.

[67] M. G. S. Ferreira, *et al.*, "Silanes and rare earth salts as chromate replacers for pre-treatments on galvanised steel," *Electrochimica Acta,* vol. 49, pp. 2927-2935, Jul 30 2004.

[68] K. Aramaki, "Self-healing protective films prepared on zinc by treatments with cerium(III) nitrate and sodium phosphate," *Corrosion Science,* vol. 44, pp. 2621-2634, Nov 2002.

[69] M. F. Montemor, *et al.*, "Composition and corrosion behaviour of galvanised steel treated with rare-earth salts: the effect of the cation," *Progress In Organic Coatings,* vol. 44, pp. 111-120, Apr 2002.

[70] K. Aramaki, "Self-healing mechanism of an organosiloxane polymer film containing sodium silicate and cerium(III) nitrate for corrosion of scratched zinc surface in 0.5 M NaCl," *Corrosion Science,* vol. 44, pp. 1621-1632, Jul 2002.

[71] A. L. Rudd, *et al.*, "The corrosion protection afforded by rare earth conversion coatings applied to magnesium," *Corrosion Science,* vol. 42, pp. 275-288, Feb 2000.

[72] K. Brunelli, *et al.*, "Effect of HCl pre-treatment on corrosion resistance of cerium-based conversion coatings on magnesium and magnesium alloys," *Corrosion Science,* vol. 47, pp. 989-1000, Apr 2005.

[73] M. Dabala, *et al.*, "Cerium-based chemical conversion coating on AZ63 magnesium alloy," *Surface & Coatings Technology,* vol. 172, pp. 227-232, Jul 2003.

[74] J. Edington, *et al.*, "Development of a spontaneous immersion process for deposition of cerium oxide coatings on copper substrates," *Surface & Coatings Technology,* vol. 200, pp. 5733-5737, May 2006.

[75] B. Y. Johnson, *et al.*, "Microstructural characteristics of cerium oxide conversion coatings obtained by various aqueous deposition methods," *Materials Characterization,* vol. 54, pp. 41-48, Jan 2005.

[76] A. E. Hughes*, et al.*, "Study of deoxidation of 2024-T3 with various acids," *Materials Science and Technology,* vol. 17, pp. 1642-1652, Dec 2001.

[77] C. Luo, Hughes, A.E., Zhou X., Thompson G.E., *Corrosion Science,* vol 61, pp 35-44, 2012.

[78] S. G. Hardin, Wittell, K.W., Hughes, A.E., Nelson, K.J.H., "Process and Solution for Providing A Conversion Coating on a Metallic Surface I," United States Patent, 2004.

[79] K. Aramaki, "Preparation of self-healing protective films on a zinc electrode treated in a cerium(III) nitrate solution and modified with sodium phosphate and cerium(III) nitrate," *Corrosion Science,* vol. 46, pp. 1565-1579, Jun 2004.

[80] K. Aramaki, "Self-healing mechanism of a protective film prepared on a $Ce(NO_3)_3$-pretreated zinc electrode by modification with $Zn(NO_3)_2$ and Na_3PO_4," *Corrosion Science,* vol. 45, pp. 1085-1101, 2003.

[81] K. Aramaki, "Self-healing protective films prepared on zinc electrodes by treatment in a cerium(III) nitrate solution and modification with sodium phosphate and calcium or magnesium nitrate," *Corrosion Science,* vol. 45, pp. 2361-2376, Oct 2003.

[82] K. Aramaki, "Improvement in the self-healing ability of a protective film consisting of hydrated cerium(III) oxide and sodium phosphate layers on zinc," *Corrosion Science,* vol. 45, pp. 451-464, Feb 2003.

[83] K. Aramaki, "XPS and EPMA studies on self-healing mechanism of a protective film composed of hydrated cerium(III) oxide and sodium phosphate on zinc," *Corrosion Science,* vol. 45, pp. 199-210, Jan 2003.

[84] K. Aramaki, "Cerium(III) chloride and sodium octylthiopropionate as an effective inhibitor mixture for zinc corrosion in 0.5M NaCl," *Corrosion Science,* vol. 44, pp. 1361-1374, 2002.

[85] K. Aramaki, "Preparation of chromate-free, self-healing polymer films containing sodium silicate on zinc pretreated in a cerium(III) nitrate solution for preventing zinc corrosion at scratches in 0.5M NaCl," *Corrosion Science,* vol. 44, pp. 1375-1389, 2002.

[86] K. Aramaki, "Self-healing mechanism of an organosiloxane polymer film containing sodium silicate and cerium(III) nitrate for corrosion of scratched zinc surface in 0.5M NaCL," *Corrosion Science,* vol. 44, pp. 1621-1632, 2002.

[87] K. Aramaki, "Synergistic inhibition of zinc corrosion in 0.5M NaCl by combination of cerium(III) chloride and sodium silicate," *Corrosion Science,* vol. 44, pp. 871-886, 2002.

[88] K. Aramaki, "Cerium(III) chloride and sodium octylthiopropionate as an effective inhibitor mixture for zinc corrosion in 0.5 M NaCl," *Corrosion Science,* vol. 44, pp. 1361-1374, Jun 2002.

[89] K. Aramaki, "Preparation of chromate-free, self-healing polymer films containing sodium silicate on zinc pretreated in a cerium(III) nitrate solution for preventing zinc corrosion at scratches in 0.5 M NaCl," *Corrosion Science,* vol. 44, pp. 1375-1389, Jun 2002.

[90] K. Aramaki, "Synergistic inhibition of zinc corrosion in 0.5 M NaCl by combination of cerium(III) chloride and sodium silicate," *Corrosion Science,* vol. 44, pp. 871-886, Apr 2002.

[91] K. Aramaki, "Treatment of zinc surface with cerium(III) nitrate to prevent zinc corrosion in aerated 0.5 M NaCl," *Corrosion Science,* vol. 43, pp. 2201-2215, Nov 2001.

[92] K. Aramaki, "The inhibition effects of cation inhibitors on corrosion of zinc in aerated 0.5 M NaCl," *Corrosion Science,* vol. 43, pp. 1573-1588, Aug 2001.

[93] S. Maddela, *et al.*, "Influence of Surface Pretreatment on Coating Morphology and Corrosion Performance of Cerium-Based Conversion Coatings on AZ91D Alloy," *Corrosion,* vol. 66, Nov 2010.

[94] O. V. Karavai, *et al.*, "Localized electrochemical study of corrosion inhibition in microdefects on coated AZ31 magnesium alloy," *Electrochimica Acta,* vol. 55, pp. 5401-5406, Jul 2010.

[95] B. Davo and J. J. de Damborenea, "Use of rare earth salts as electrochemical corrosion inhibitors for an Al-Li-Cu (8090) alloy in 3.56% NaCl," *Electrochimica Acta,* vol. 49, pp. 4957-4965, Oct 30 2004.

[96] A. Aballe, *et al.*, "Influence of the cathodic intermetallics distribution on the reproducibility of the electrochemical measurements on AA5083 alloy in NaCl solutions," *Corrosion Science,* vol. 45, pp. 161-180, Jan 2003.

[97] P. Traverso, *et al.*, "Corrosion inhibition of Al 6061 T6/Al_2O_3 10% (v/v) composite in 3.5% NaCl solution with addition of cerium (III) chloride," *Surface And Interface Analysis,* vol. 34, pp. 185-188, Aug 2002.

[98] A. Aballe, *et al.*, "$CeCl_3$ and $LaCl_3$ binary solutions as environment-friendly corrosion inhibitors of AA5083 Al-Mg alloy in NaCl solutions," *Journal of Alloys and Compounds,* vol. 323, pp. 855-858, Jul 2001.

[99] S. Bohm, *et al.*, "Kinetic and mechanistic studies of rare earth-rich protective film formation using in situ ellipsometry," *Journal Of The Electrochemical Society,* vol. 147, pp. 3286-3293, Sep 2000.

[100] A. Aballe, *et al.*, "Electrochemical impedance spectroscopy study of the behaviour of lanthanide-rich films formed on aluminium-magnesium alloys," in *Electrochemical Methods in Corrosion Research Vi, Pts 1 and 2.* vol. 289-2, P. L. Bonora and F. Deflorian, Eds., ed, 1998, pp. 557-566.

[101] M. Bethencourt, *et al.*, "Lanthanide compounds as environmentally-friendly corrosion inhibitors of aluminium alloys: A review," *Corrosion Science,* vol. 40, pp. 1803-1819, Nov 1998.

[102] M. Bethencourt, *et al.*, "Protection against corrosion in marine environments of AA5083 Al-Mg alloy by lanthanide chlorides," *Journal of Alloys and Compounds,* vol. 250, pp. 455-460, Mar 1997.

[103] C. E. Moffitt, *et al.*, "An XPS study of the elemental enrichment on aluminum alloy surfaces from chemical cleaning," *Surface & Coatings Technology,* vol. 137, pp. 188-196, Mar 2001.

[104] T. Schmidt-Hansberg, Schubach, P., "A Comparative Study of Innovative Aluminium Pretreatments," *ATB Metallurgie,* vol. 43, pp. 9 - 14, 2003.

[105] S. Joshi, *et al.*, "Effect of alkaline cleaning and activation on aluminum alloy 7075-T6," *Applied Surface Science,* vol. 257, pp. 1859-1863, Jan 2011.

[106] K. J. H. Nelson, *et al.*, "Characterisation of aluminium alloys after HNO_3/HF-NaOH-HNO_3/HF pretreatment," *Materials Science and Technology,* vol. 17, pp. 1211-1221, Oct 2001.

[107] A. de Frutos, *et al.*, "Influence of pre-treatments in cerium conversion treatment of AA2024-T3 and 7075-T6 alloys," *Surface & Coatings Technology,* vol. 202, pp. 3797-3807, May 2008.

[108] A. E. Hughes, *et al.*, "Desmutting of aluminium alloy 2024-T3 using rare earth electrolyte," *Materials Science and Technology,* vol. 15, pp. 1124-1132, Oct 1999.
[109] K. R. B. C.J.E. Smith, V.M. Evans, S.A. Garrett and K.S. Smith, "Research into Chromate-free Treatments for the Protection of Alumnium Alloys," in *83rd Meeting of hte AGARD SMP on Environmentally Compliant Surface Treatments of MAterials for Aerospace Applications*, Florence Italy, 1996, pp. 8-1 to 8-10.
[110] A. K. Sengupta and U. Bhattacharyya, "Fluorocerrates(Iv) Of Some Organic-Base Cations And Hexamminecobalt(Iii) Cation," *Journal of the Indian Chemical Society,* vol. 60, pp. 519-520, 1983.
[111] R. A. Penneman and Rosenzwe.A, "Compounds In Cerium Tetrafluoride-Ammonium Fluoride-Water System," *Inorganic Chemistry,* vol. 8, pp. 627-&, 1969.
[112] J. W. Bibber, "An overview of nonhexavalent chromium conversion coatings--Part I: aluminum and its alloys," *Metal Finishing,* vol. 99, pp. 15-22, 2001.
[113] J. Bibber, "An update on chrome-free pretreatments for aluminum," *Metal Finishing,* vol. 96, pp. 28-31, 1998.
[114] A. E. Hughes, *et al.*, "Development of permanganate-based coatings on aluminum alloy 2024-T3," *Corrosion,* vol. 62, pp. 773-780, Sep 2006.
[115] M. P. Schriever, "Non-Chromated Oxide Coating for Aluminum Substrates," United States Patent, 1999.
[116] A. E. Hughes, *et al.*, "SEM and RBS characterization of a cobalt-based conversion coating process on AA2024-T3 and AA7075-T6," *Surface and Interface Analysis,* vol. 36, pp. 1585-1591, Dec 2004.
[117] A. E. Hughes and T. Nikpour, "Influence of salt fog fallout on the failure of chromated aluminum alloy 2024-T3 panels in neutral salt fog," *Corrosion,* vol. 61, pp. 285-292, Mar 2005.
[118] A. E. Hughes, *et al.*, "A morphological study of filiform corrosive attack on cerated AA2024-T351 aluminium alloy," *Corrosion Science,* vol. 47, pp. 107-124, Jan 2005.
[119] S. Intem, *et al.*, "Underfilm corrosion on polyurethane-coated aluminum alloy 2024-T3 containing dissimilar metal fasteners," *Jct Research,* vol. 3, pp. 313-322, Oct 2006.
[120] Natonal Defence for Evnironmental Excellance (NDCEE), "Alternatives fo Chrome Conversion Coatings on aluminium Alloys 2024, 6061, 7075, and Ion Vapour deposited Aluminium on Steel.," Engineering and Technical Services for Joint Group on Acquisution Pollution Prevention (JG-APP) Pilot Projects
1998.
[121] S. Y. Zhang, *et al.*, "Preparation and corrosion resistance studies of nanometric sol-gel-based CeO_2 film with a chromium-free pretreatment on AZ91D magnesium alloy," *Electrochimica Acta,* vol. 55, pp. 870-877, Jan 2010.
[122] X. K. Zhong, *et al.*, "Effect of cerium concentration on microstructure, morphology and corrosion resistance of cerium-silica hybrid coatings on magnesium alloy AZ91D," *Progress in Organic Coatings,* vol. 69, pp. 52-56, Sep 2010.
[123] M. Schem, *et al.*, "CeO2-filled sol-gel coatings for corrosion protection of AA2024-T3 aluminium alloy," *Corrosion Science,* vol. 51, pp. 2304-2315, Oct 2009.
[124] D. Raps, *et al.*, "Electrochemical study of inhibitor-containing organic-inorganic hybrid coatings on AA2024," *Corrosion Science,* vol. 51, pp. 1012-1021, May 2009.

[125] P. C. R. Varma, *et al.*, "Influence of magnesium nitrate on the corrosion performance of sol-gel coated AA2024-T3 aluminium alloy," *Surface & Coatings Technology,* vol. 204, pp. 277-284, Oct 2009.

[126] N. N. Voevodin, *et al.*, "Corrosion protection for aerospace aluminum alloys by Modified Self-assembled Nanophase Particle (MSNAP) sol-gel," *Surface & Coatings Technology,* vol. 201, pp. 1080-1084, Oct 2006.

[127] M. L. Zheludkevich, *et al.*, "Nanostructured sol-gel coatings doped with cerium nitrate as pre-treatments for AA2024-T3 - Corrosion protection performance," *Electrochimica Acta,* vol. 51, pp. 208-217, Oct 2005.

[128] J. H. Osborne, "Observations on chromate conversion coatings from a sol-gel perspective," *Progress in Organic Coatings,* vol. 41, pp. 280-286, May 2001.

[129] N. P. Tavandashti and S. Sanjabi, "Corrosion study of hybrid sol-gel coatings containing boehmite nanoparticles loaded with cerium nitrate corrosion inhibitor," *Progress in Organic Coatings,* vol. In Press, Corrected Proof.

[130] M. L. Zheludkevich, *et al.*, "Sol-gel coatings for corrosion protection of metals," *Journal of Materials Chemistry,* vol. 15, pp. 5099-5111, 2005.

[131] J. H. Osborne, *et al.*, "Testing and evaluation of nonchromated coating systems for aerospace applications," *Progress in Organic Coatings,* vol. 41, pp. 217-225, May 2001.

[132] Y. J. Du, *et al.*, "Inorganic/organic hybrid coatings for aircraft aluminum alloy substrates," *Progress in Organic Coatings,* vol. 41, pp. 226-232, May 2001.

[133] K. A. Yasakau, *et al.*, "Influence of inhibitor addition on the corrosion protection performance of sol-gel coatings on AA2024," *Progress in Organic Coatings,* vol. 63, pp. 352-361, Oct 2008.

[134] M. Garcia-Heras, Jimenez, A. Casal, B., Galvan, J.C., Radzki, S., Villegas, M.A., *Jpournal of Alloys and Compounds,* vol. 380, pp. 219-224, 2004.

[135] Y. J. Chung, Jeanjaquet, S.L., Kendig, M.W. , "Corrosion Inhibiting Sol-Gel Coatings for Metal Alloys," United States Patent, 2003.

[136] R. S. Koefod, "Corrosion-Inhibiting Salt Deicers," United States Patent, 1996.

[137] K. A. Yasakau, Tedim, J., Zheludkevich, M.L., Drumm, R., Shem M., Wittmar, Vieth, M., A., Ferreira, M.G.S., "cerium molybdate nanowires for active corrosion protection of aluminium alloys," *Corrosion Science,* vol. 58, pp. 41-51, 2012.

[138] R. Rungta, "Rare Earth Coating Process for Aluminium Alloys," United States Patent, 1994.

[139] S. Wernick, Pinner, R., Sheasby, P.G., *The Surface Treatment and Finishing of Aluminium and its Alloys*, 5 th ed. Teddington: Finishing Publications and ASM International, 1987.

[140] M. Curioni, *et al.*, "Reducing the energy cost of protective anodizing," *Journal of Applied Electrochemistry,* vol. 41, pp. 773-785, Jul 2011.

[141] T. P. Schuman, *et al.*, "Cerium-based inhibitors of aluminium alloy corrosion," presented at the International Waterbourne, High-Solids and Powder Coatings Symposium, New Orleans, LA, USA, 2002.

[142] L. R. Reddy, Reddy, M.A., Bhanumathi, N., Rao, K.R., "Cerium chloride-catalysed cleavage of epoxides with aromatic amines," *Synthesis-Stuttgart* vol. 6, pp. 831–832, 2001.

[143] G. Sabitha, Babu, R.S., Rajkumar, M., Reddy, Ch.S., Yadav, J.S., *Tetrahedron Letters,* vol. 42, 2001.

[144] P. Castell, Galià, M., Serra, A., Salla, J.M., Ramis, X., *Polymer,* vol. 41, 2000.

[145] Z. Khan, Raju, D., Kabir-ud-Din, *Colloids Surf. A: Physicochem. Eng. Aspects* vol. 225, 2003.

[146] F. R. Duke, Bremer, R.F., *Journal of the American Chemical Society,* vol. 73, 1951.

[147] S. L. E. C.J.E. Smith, V.M. Evans, H.J. Kimptoon, K.S. Smith and M. Gibson, "Methods of Assessment of Corrosion Inhibitors in Aerospace Paints," in *Proceedings of the 9th European Symposium on Corrosion Inhibitors (9SEIC)* ed. London: Her Britannic Majesty's Stationary Office, 2000, pp. 415 - 426.

[148] F. H. Scholes, *et al.*, "Interaction of $Ce(dbp)_3$ with surface of aluminium alloy 2024-T3 using macroscopic models of intermetallic phases," *Corrosion Engineering Science and Technology,* vol. 44, pp. 416-424, Dec 2009.

[149] M. Forsyth, *et al.*, "Inhibition of corrosion on AA2024-T3 by new environmentally friendly rare earth organophosphate compounds," *Corrosion,* vol. 64, pp. 191-197, Mar 2008.

[150] D. Ho, *et al.*, "Cerium dibutylphosphate as a corrosion inhibitor for AA2024-T3 aluminum alloys," *Journal of the Electrochemical Society,* vol. 153, pp. B392-B401, 2006.

[151] B. R. W. Hinton, *et al.*, "Raman, EDS and SEM studies of the interaction of corrosion inhibitor $Ce(dbp)_3$ with AA2024-T3," presented at the 4th International Symposium on Aluminium Surface Science and Technology, Beaune, France, 2006.

[152] N. Birbilis, *et al.*, "Inhibition of AA2024-T3 on a phase-by-phase basis using an environmentally benign inhibitor, cerium dibutyl phosphate," *Electrochemical and Solid State Letters,* vol. 8, pp. C180-C183, 2005.

[153] J. Mardel, *et al.*, "The characterisation and performance of $Ce(dbp)_3$-inhibited epoxy coatings," *Progress in Organic Coatings,* vol. 70, pp. 91-101, 2011.

[154] J. Mardel, *et al.*, "The characterisation and performance of $Ce(dbp)_3$-inhibited epoxy coatings," *Progress in Organic Coatings,* vol. 70, pp. 91-101, Feb-Mar 2011.

[155] T. A. Markley, *et al.*, "Corrosion protection of AA2024-T3 using rare earth diphenyl phosphates," *Electrochimica Acta,* vol. 52, pp. 4024-4031, Mar 2007.

[156] M. Forsyth, *et al.*, "ATR characterisation of synergistic corrosion inhibition of mild steel surfaces by cerium salicylate," *Corrosion Science,* vol. 44, pp. 2651-2656, 2002.

[157] C.J.E. Smith, K. R. Baldwin and M. A. Hewins, "A Study into the Corrosion Inhibition of an Aluminium Alloy by Cerium Salts," in *Progress in the Understadning and Prevention of Corrosion* 1993
ed: Journal of Metals, 1993, pp. 1652 - 1663.

[158] G. B. Deacon, *et al.*, "Synthesis and Characterisation of Rare Earth Complexes Supported by para-Substituted Cinnamate Ligands," *Zeitschrift Fur Anorganische Und Allgemeine Chemie,* vol. 635, pp. 833-839, 2009.

[159] F. Blin, *et al.*, "The corrosion inhibition mechanism of new rare earth cinnamate compounds - Electrochemical studies," *Electrochimica Acta,* vol. 52, pp. 6212-6220, Jun 2007.

[160] F. Blin, *et al.*, "Corrosion mitigation of mild steel by new rare earth cinnamate compounds," *Journal of Applied Electrochemistry,* vol. 34, pp. 591-599, Jun 2004.

[161] M. C. G. K.R. Baldwin, P.L. Lane and C.J.E. Smith, "The Development of Alterantives to Chromate Inhibitors for the Protection of Aerospace Aluminium Alloys," in *7th European Sumposium on Corrosion Inhibitors*, University of Ferrara, 1990, pp. 771-785.

[162] H. E. Hager, Johnson, C.J., Blohowiak, K.Y., Wong., C.M., Jones, J.H., Cook Jr, R.L., Taylor, S.R., "Chromate Free Protective Coatings," United States Patent, 2000.

[163] S. R. Taylor and B. D. Chambers, "Identification and characterization of nonchromate corrosion inhibitor synergies using high-throughput methods," *Corrosion,* vol. 64, pp. 255-270, Mar 2008.

[164] S. R. Taylor and B. D. Chambers, "The discovery of non-chromate corrosion inhibitors for aerospace alloys using high-throughput screening methods," *Corrosion Reviews,* vol. 25, pp. 571-590, 2007.

[165] B. D. Chambers and S. R. Taylor, "The high throughput assessment of aluminium alloy corrosion using fluorometric methods. Part II - A combinatorial study of corrosion inhibitors and synergistic combinations," *Corrosion Science,* vol. 49, pp. 1597-1609, Mar 2007.

[166] B. D. Chambers and S. R. Taylor, "The high throughput assessment of aluminium alloy corrosion using fluorometric methods. Part I - Development of a fluorometric method to quantify aluminium ion concentration," *Corrosion Science,* vol. 49, pp. 1584-1596, Mar 2007.

[167] S. J. Garcia, *et al.*, "Validation of a fast scanning technique for corrosion inhibitor selection: influence of cross-contamination on AA2024-T3," *Surface and Interface Analysis,* vol. 42, pp. 205-210, Apr 2010.

[168] S. J. Garcia, *et al.*, "The influence of pH on corrosion inhibitor selection for 2024-T3 aluminium alloy assessed by high-throughput multielectrode and potentiodynamic testing," *Electrochimica Acta,* vol. 55, pp. 2457-2465, Feb 2010.

[169] T. H. Muster, *et al.*, "A rapid screening multi-electrode method for the evaluation of corrosion inhibitors," *Electrochimica Acta,* vol. 54, pp. 3402-3411, Apr 2009.

[170] P. A. White, *et al.*, "High-throughput channel arrays for inhibitor testing: Proof of concept for AA2024-T3," *Corrosion Science,* vol. 51, pp. 2279-2290, Oct 2009.

In: Cerium
Editors: Aleksey Izyumov and Gulnaz Plaksin

ISBN: 978-1-62257-670-8

Chapter 3

TETRAVALENT CERIUM CHEMISTRY

Farid M. A. Sroor and Frank T. Edelmann
Chemisches Institut der Otto-von-Guericke-Universität Magdeburg, Magdeburg, Germany

ABSTRACT

This chapter gives an overview on the chemistry of tetravalent lanthanide compounds, especially those of tetravalent cerium. Following a brief Introduction it covers the tetrahalides, dioxides and other lanthanides(IV) salts. Coordination compounds of cerium in the oxidation state +4 include halogeno complexes and complexes of oxo acids, β-diketonates and related Schiff-base complexes as well as porphyrinates and related complexes. The chapter also covers the chemistry of organocerium(IV) complexes as well as "pseudo-organometallics" of tetravalent cerium. The latter comprise mainly the alkoxides and siloxides as well as various amido and amidinato complexes of cerium in the oxidation state +4. True organocerium(IV) complexes are thus far restricted to a handful of cyclopentadienyl derivatives and some cerocene sandwich complexes. Cerium(IV) complexes comprising specially designed *N*-heterocyclic ligands comprise the latest addition to the field.

Keywords: Cerium; Halides; Cerium dioxide; Alkoxides; Siloxides; Amides; Amidinates; Cyclopentadienyl complexes; Carbene complexes; Cyclooctatetraenyl complexes

ABBREVIATIONS AND ACRONYMS

acac	Acetylacetonate
9-BBN	9-Borabicyclo[3.3.1]nonane
CAN	Ceric ammonium nitrate, $(NH_4)_2[Ce(NO_3)_6]$
Cerocene	Bis(η^8-cyclooctatetraenyl)cerium
COT	Cyclooctatetraene or η^8-cyclooctatetraenyl ligand
COT"	1,4-Bis(trimethylsilyl)cyclooctatetraenyl
COT^{BIG}	1,4-Bis(triphenylsilyl)cyclooctatetraenyl

Cp	Cyclopentadienyl anion
CVD	Chemical vapor deposition
Cy	Cyclohexyl
Dipp	2,6-Diisopropylphenyl
DMAP	4-(Dimethylamino)pyridine
DME	1,2-Dimethoxyethane
$E_{1/2}$	Half-wave potential
HTTA	Thenoyltrifluoroacetone
NN'_3	$N(CH_2CH_2NSiMe_2{}^tBu)_3$
NR'_2	$NC(Me)_2(CH_2)_3CMe_2$
OTf	Triflate anion, $CF_3OSO_2^-$
py	Pyridine
SCE	Silver / silver chloride-electrode
SOFC	Solid oxide fuel cell
Uranocene	Bis(η^8-cyclooctatetraenyl)uranium

1. Introduction

Besides the ubiquitous oxidation state Ln^{3+}, the higher oxidation state Ln^{4+} is also encountered with some lanthanoids, for example in the case of the ions Ce^{4+} (f^0, orange-yellow), Pr^{4+} (f^1, colorless), Nd^{4+} (f^2, blue-violet), Tb^{4+} (f^7, colorless), and Dy^{4+} (f^8, orange-yellow) (Table 1). However, all three states $Ln^{2+,3+,4+}$ are never encountered for the same element. Thus the highly important mechanistic steps of oxidative addition and reductive elimination typical for the *d*-block metals cannot occur with the *f*-block metals as they would involve M^{2+} or M^{4+} transformations, respectively [1].

Among the tetravalent lanthanide ions, only Ce^{4+} is readily available in aqueous solution (E Ce^{3+}/Ce^{4+} = +1.44 V in 2M H_2SO_4, 1.61 V in 1M HNO_3, 1.70 V in 1M $HClO_4$). The different values for the reduction potentials indicate that stable complexes are formed in these acidic media. Thermodynamically, the oxidation of water by the Ce^{4+} aqua ion should be possible, but the system is kinetically stable. In contrast, the much more positive normal potentials of the other tetravalent lanthanide ions Pr^{4+}, Nd^{4+}, Tb^{4+}, and Dy^{4+} (*e.g.* Pr: +2.86 V) make them very strong oxidizing agents which readily decompose water under evolution of O_2.

Table 1. Possible oxidation states for rare earth metals

Ce	Pr	Nd	Pm	Sm	Eu	Gd	Tb	Dy	Ho	Er	Tm	Yb	Lu
				+2	+2						+2	+2	
+3	+3	+3	+3	+3	+3	+3	+3	+3	+3	+3	+3	+3	+3
+4	+4	+4					+4	+4					

This is also the main reason why no organometallic compounds have ever been reported for lanthanide(IV) ions other than Ce^{4+}. Thus later in this Chapter (Sections 4 and 5) we will discuss not only "real" organocerium(IV) compounds, *i.e.* those containing direct Ce-C bonds, but also several classes of "*pseudo*-organometallics" which resemble true organolanthanide complexes in their solubility in organic solvents and/or enhanced volatility. In particular, these include the alkoxides, siloxides, amides and amidinates of tetravalent cerium.

Thus organocerium(IV) complexes can be divided into three classes, of compounds, *i.e.* cyclopentadienyl complexes, complexes containing *N*-heterocyclic carbene ligands, and cerocenes, *i.e.* bis(cyclooctatetraenyl) sandwich complexes.

The oxidation of Ce^{3+} to Ce^{4+} with the use of strong oxidizing agents like MnO_4^- or $S_2O_8^{2-}$ (Scheme 1) enables the selective separation of cerium from lanthanide mixtures. The resulting Ce^{4+} can be precipitated from aqueous nitric acid solution in the form of ceric ammonium nitrate, $(NH_4)_2[Ce(NO_3)_6]$ [1].

$$2\,Ce^{3+} + S_2O_8^{2-} \longrightarrow 2\,Ce^{4+} + 2\,SO_4^{2-}$$

Scheme 1.

The readily occurring transition from colorless Ce^{3+} to bright yellow or orange Ce^{4+} forms the basis for the use of cerium(IV) sulfate solutions in redox titrations ("cerimetric" analysis). The ease of access to various tetravalent cerium compounds makes cerium(IV) most valuable in research as well as in various practical applications. Important fields of application for cerium(IV) compounds include organic syntheses, bioinorganic chemistry, materials science, and industrial catalysis (*e.g.* vehicle emissions control, oxygen storage *etc.*) [1].

2. Inorganic Cerium(IV) Compounds

2.1. Cerium(IV)-Tetrahalides

Only the tetrafluorides of Ce, Pr, and Tb exist, the three lanthanides with the most stable +4 oxidation state. Fluorine is most likely to support a high oxidation state, and even though salts of ions like $[CeCl_6]^{2-}$ are known, no binary tetrachlorides have ever been isolated as pure materials. Anhydrous LnF_4 (Ln = Ce, Pr, Nd, Tb, Dy) can be made by fluorination of the trifluorides or, in the case of Ce, by fluorination of metallic Ce, $CeCl_3$, or CeO_2. The method employing cerium dioxide appears to be the most straightforward one. CeF_4 can be crystallized from aqueous solution as a monohydrate. In the solid state, all tetrafluorides adopt the MF_4 structure with dodecahedral eight-coordination and are thus isomorphous with UF_4.

Factors that favor formation of the tetrafluorides include a low value of I_4 for the metal and a high lattice energy. This is most likely to be found with the smallest halide, *i.e.* fluoride. The low bond energy of F_2 is an additional supporting factor [1].

Other tetrahalides do not exist. Thus, although both $CeCl_3$ and salts of the $[CeCl_6]^{2-}$ ion are quite stable, $CeCl_4$ cannot be isolated. The reasons for this are those that enable fluorine to support high oxidation states. Similar factors indicate that tetrabromides and tetraiodides are much less likely to be isolated [1].

2.2. Cerium(IV)-Dioxide

Rare earth metals generally react with dioxygen under formation of the lanthanide(III) oxides, Ln_2O_3, with the exception of cerium, praseodymium, and terbium. In these cases, CeO_2, Pr_6O_{11}, and Tb_4O_7 are formed, respectively. Lanthanide dioxides, LnO_2 (Ln = Ce, Pr, Tb) crystallize in the fluorite (CaF_2) structure.

Defects in the oxygen positions lead to various mixed Ln^{3+}/Ln^{4+} oxides of the type (for praseodymium) $Pr_{12}O_{22}$, $Pr_{11}O_{20}$, $Pr_{10}O_{18}$, Pr_9O_{16}, Pr_8O_{14}, Pr_7O_{12}, and Pr_6O_{10} before the row ends with Pr_2O_3. Under forcing conditions, *e.g.* heating in pure oxygen under pressure, these will eventually yield PrO_2 (or TbO_2 in the case of terbium) [1].

Among these materials, *cerium dioxide* (ceria) is of particular importance. Very pure ceria forms a white powder, but more often it appears pale-yellow, and less pure samples can even be brownish. A brownish coloration could be indicative for the presence of impurities such as praseodymium and neodymium.

Nevertheless, impure ceria can be used for applications where purity is not critical, *e.g.* for polishing glass or stones. Ceria can be prepared by calcination of suitable precursors such as cerium nitrate, $Ce(NO_3)_3$, cerium oxalate, $Ce_2(C_2O_4)_3$, or cerium hydroxide, $Ce(OH)_3$, in air. At room temperature and under atmospheric pressure it is more stable than the Ce_2O_3 phase. Cerium dioxide is basic and can be dissolved in acids (although with some difficulty) [1].

Cerium dioxide has found numerous practical applications, *e.g.* as oxidation catalyst and as catalyst support (*e.g.* for gold nanoparticles), in ceramics, self-cleaning ovens, and catalytic converters, for sensitizing photosensitive glasses and for polishing glass and stones. Ceria is also being used in infrared filters and as a replacement for radioactive thorium dioxide in incandescent mantles.

While cerium dioxide is transparent for visible light, it is a strong ultraviolet light absorber. Thus it has been envisaged as a prospective replacement for ZnO and TiO_2 in sunscreens, although its photocatalytic activity is somewhat lower. Most of the practical uses of cerium dioxide in organic synthesis and catalysis are based on its oxidizing properties. For example, its use in the walls of so-called self-cleaning ovens make use of the fact that it assists oxidation of sticky hydrocarbon deposits during the high-temperature cleaning process. Ceria is also of great current interest as a material for solid oxide fuel cells (SOFC's) because of its relatively high oxygen ion conductivity [2].

Another very important application of ceria is its use in catalytic converters in automotive applications. Cerium dioxide is able to release or store oxygen in the exhaust stream of an automotive engine because the material is able to become non-stoichiometric in its oxygen-content. The catalytic activity of ceria has been found to depend directly on the number of oxygen.

$$2\,CeO_2 \longrightarrow Ce_2O_3 + 1/2\,O_2$$

$$Ce_2O_3 + H_2O \longrightarrow 2\,CeO_2 + H_2$$

Scheme 2.

It is able to effectively reduce the NO_x emissions and also to oxidize toxic carbon monoxide to non-toxic CO_2. The use of inexpensive ceria in such catalysts also presents an economic advantage as it reduces the amount of platinum needed for reducing NO_x emissions and achieving complete combustion of harmful exhaust gases.

In addition, ceria has been found to be a useful co-catalyst in a variety of industrially important reactions including various oxidation reactions, the Fischer-Tropsch reaction, as well as the water-gas shift reaction and steam-reforming of diesel fuel to give hydrogen gas and carbon dioxide (in combination with various transition metal or metal oxide catalysts) [1,2].

Furthermore a laboratory demonstration of thermochemical water splitting cycles based on the CeO_2/Ce_2O_3 pair (Scheme 2) has been reported. Thermal reduction of Ce^{4+} to Ce^{3+} (endothermic step) has been performed in a solar reactor featuring a controlled inert atmosphere. The feasibility of this first step has been demonstrated and the operating conditions have been defined (T = 2000 °C, P = 100–200 mbar).

The subsequent hydrogen generation step (water-splitting with Ce_2O_3) carried out in a fixed bed reactor was complete with a fast kinetic in the studied temperature range 400–600 °C. The recovered CeO_2 was then recycled in the first step. In this process, water is the only material input and heat is the only energy input. The only outputs are hydrogen and oxygen, and these two gases are obtained in different steps avoiding a high temperature energy consuming gas-phase separation. Furthermore, pure hydrogen is produced which can be used directly in fuel cells. These results have shown that the cerium oxide two-step thermochemical cycle is a promising process for hydrogen production [3].

Ceria nanoparticles are among the most important nanomaterials for a wide range of applications in catalysts [4,5], fuel cells [6], chemical-mechanical polishing for microelectronics, phosphor/luminescence, as well as metallurgical and glass/ceramic applications [7].

These applications are mainly based on the high thermodynamic affinity of cerium for oxygen, its characteristic redox chemistry involving Ce(3+)/Ce(4+), and absorption/excitation energy bands associated with its electronic structure. The sizes and size distributions of ceria nanoparticles are important for determining their properties [8]. The most important route for the synthesis of CeO_2 nanoparticles is the chemical approach [9]. It is well established that nanoparticles usually have specific shapes, especially when they are small, because a single crystal nanoparticle is enclosed by crystallographic facets that have lower energy. Spherical particles usually have a higher surface energy due to the presence of high index crystallographic facets. These structural characteristics are likely to affect the nanoparticle performance in technological applications.

A detailed study of the surface structure of ceria nanoparticles has been carried out. For particles in the size range of 3-10 nm, the particles were found to be dominated by truncated octahedral shapes [10]. A simple and economical route based on an ethyleneglycol-mediated process has been developed to synthesize three-dimensional (3D) flowerlike ceria micro/nanocomposite structures using cerium trichloride as a reactant. Scanning electron microscopy (SEM) and X-ray diffraction (XRD) were used to investigate the evolution process of the ceria precursor, and a two-stage growth process has been identified during the morphology evolution. Ceria with the same flowerlike micro/nanocomposite structure was readily obtained by calcination of the ceria precursor.

This novel micro/nanocomposite structure combined the advantages of both microstructure and nanostructure. It was shown that the as-obtained ceria could be used as not only an effective sorbent for the removal of pollutants such as of As(V) and Cr(VI) in water treatment but also as an excellent support for gold nanoparticles to remove CO by catalytic oxidation, thus demonstrating a promising potential in environmental remediation [11]. A general synthesis of dispersible rare-earth oxide nanocrystals by thermolysis of their benzoylacetonate complexes in oleic acid/oleylamine has been reported which is also applicable to the preparation of ceria nanoparticles. Due to the selective adsorption of the capping ligands on certain cubic faces during crystal growth, nanocrystals with different morphologies, such as nanopolyhedra, nanoplates, and nanodisks, were created, which exhibited a striking ability to self-assemble into large-area nanoarrays.

These novel nanometric architectures with unusual optical properties resulting from surface-modification effects can be expected to enrich the field of lanthanide chemistry, benefit the development of a shape-control theory for nanocrystals, and promise the fabrication of advanced rare-earth functional materials with new applications [12].

Moreover, uniform-sized ceria nanocrystals with quasispherical, wire, and tadpole shapes have been synthesized from a nonhydrolytic sol–gel reaction of cerium(III) nitrate and diphenyl ether in the presence of appropriate surfactants. This synthetic procedure offers several important advantages for the production of ceria nanocrystals.

First, uniform-sized nanocrystals can be obtained on a scale of tens of grams in a single reaction without the need for a further size-sorting process. When reactors are set up in parallel, even several kilograms of monodisperse nanocrystals can be readily obtained. Second, ceria nanocrystals with various shapes can be produced simply by changing the experimental conditions.

In particular, novel tadpole-shaped nanowires with an extremely small diameter of 1.2 nm have been produced [13]. A facile aqueous solution-based chemical protocol for the synthesis of ceria and samarium-doped ceria (Sm-CeO_2) nanoparticles at low temperature, as well as at room temperature, has also been developed.

The process uses clear aqueous solution of a simple cerium salt and ammonium carbonate as a precursor solution. By this method, crystalline CeO_2 nanocrystals with particle sizes in the range of 3-15 nm were synthesized without using organic substrates.

The particle sizes of CeO_2 could be controlled by varying the reaction parameters. The synthesized CeO_2 nanoparticles possessed high contents of oxygen vacancies (i.e. high concentration of Ce^{3+}) and a high specific surface area, which enhances the catalyst activity of CeO_2. The synthesized nanoparticles exhibited excellent catalytic activity toward the conversion of allylic compounds to the corresponding en-one (allylic oxidation) as compared to bulk CeO_2.

The catalytic activity was found to be highly dependent on the surface area, crystallinity, and extent of Ce^{3+} present in the catalyst [14]. The synthesis, structure-sensitive properties, and promising applications have recently been compiled in a review article [15].

Finally, a surprising application of cerium dioxide nanoparticles in nanomedicine has been reported. It was discovered that cerium dioxide nanoparticles can scavenge reactive molecules in the eye and prevent degenerative retinal disorders in rats. The results suggested that nanoceria particles could be used to treat a variety of problems that cause blindness [16].

2.3. Other Cerium(IV) Salts

Pale yellow *cerium(IV) hydroxide* can be prepared by addition of bases, such as aqueous ammonia, to solutions of cerium(IV) salts, *e.g.* cerium(IV) nitrate or ceric ammonium nitrate. Nanocrystalline *cerium(IV) hydroxide* (NCs-$Ce(OH)_4$) is an intermediate in the production of cerium dioxide, which has been synthesized successfully using a novel and simple wet chemical route at ambient temperature for the preparation of nanocrystalline CeO_2 powder and films on mass scale for various purposes. The average crystallite size of NCs-$Ce(OH)_4$ has been estimated by the Scherrer equation to be 3–4 nm. SEM examinations showed that the surface texture was uniformly agglomerated and homogeneous. Thermal analysis suggested that the cerium(IV) ion is in the tetrahydrated form. Absorption and luminescence spectroscopic studies have been examined for future application in the development of optical devices [17]. *Cerium(IV) nitrate* can be crystallized in the form of its pentahydrate, $Ce(NO_3)_4{\cdot}5H_2O$, which presumably contains 11-coordinate $Ce(NO_3)_4(H_2O)_3$ molecules with all four nitrato units acting as chelating ligands as in the corresponding thorium nitrate complex. *Cerium(IV) sulfate*, also called ceric sulfate, is a yellow to yellow-orange commercially available Ce^{4+} compound. It can be prepared by heating of cerium dioxide with concentrated sulfuric acid (Scheme 3) [1]. Cerium(IV) sulfate exists as the anhydrous salt $Ce(SO_4)_2$, but a few hydrated forms are also known: $Ce(SO_4)_2{\cdot}nH_2O$ (n = 4, 8, or 12). It is moderately soluble in water and dilute acids. Its neutral solutions slowly decompose, depositing light yellow CeO_2. Solutions of ceric sulfate have an intense yellow color. The tetrahydrate will lose the water when heated to 180-200 °C. The Ce^{4+} ion is a strong oxidizer, especially under acidic conditions. If ceric sulfate is added to dilute hydrochloric acid, then elemental chlorine is formed, albeit slowly. With stronger reducing agents it reacts much faster. For example, with sulfite in acidic solutions it reacts quickly and completely. Ceric sulfate is frequently used in analytical chemistry for redox titrations, often together with a redox indicator. Cerium(IV) sulfate is also one of the reagents in the oscillating Belousov–Zhabotinsky reaction. In this reaction mixture consisting of potassium bromate, cerium(IV) sulfate, malonic acid and citric acid in dilute sulfuric acid, the concentration ratio of the Ce^{4+} and Ce^{3+} ions oscillates, causing the color of the solution to oscillate between yellow and colorless. This is due to the Ce^{4+} ions being reduced by malonic acid to Ce^{3+} ions, which are then oxidized back to Ce^{4+} ions by bromate(V) ions [1]. *Cerium(IV) acetate*, $Ce(OAc)_4$, has been synthesized by first heating a solution of $Ce(OAc)_3$ and anhydrous $Ce(NO_3)_3$ in a mixture of glacial acetic acid and acetic anhydride until NO_2 evolution was complete. The resulting solution of $Ce(OAc)_3$ was ozonized at 70 °C to form $Ce(OAc)_4$ in quantitative yield and excellent purity. The presence of nitrate was essential for obtaining such good yield and purity. IR and X-ray diffraction measurements showed that $Ce(OAc)_4$ is isomorphous with $Th(OAc)_4$ and $U(OAc)_4$ [18,19]. *Cerium(IV) trifluoromethanesulfonate* has been prepared by the reaction of cerium(IV) carbonate with trifluoromethanesulfonic acid. The powerful oxidizing ability of this compound was observed in the oxidation of benzyl alcohols and alkylbenzenes [20].

$$CeO_2 + 2\,H_2SO_4 \longrightarrow Ce(SO_4)_2 + 2\,H_2O$$

Scheme 3.

Further cerium(IV) salts include, among others, the hydrated *cerium(IV) chromates* $Ce(CrO_4)_2{\cdot}H_2O$ and $Ce(CrO_4)_2{\cdot}2H_2O$ [21]. *Cerium(IV) perchlorate* is readily formed by the rection of cerium(IV) hydroxide with $HClO_4$, but this process is complicated by hydrolysis and partial reduction of Ce^{4+}. In addition, a considerable amount of water of crystallization was found to be retained by the perchlorate salt. Even with the use of the monohydrate of $HClO_4$ (84.4% $HClO_4$) instead of the dihydrate (72.6% $HClO_4$), considerable hydrolysis and reduction of Ce(IV) perchlorate took place. The reaction of anhydrous $HClO_4$ with cerium(IV) hydroxide resulted in the formation of perchloratoceric acids, the compositions of which are very complicated [22].

3. Coordination Compounds of Tetravalent Cerium

With only a few exceptions, the coordination chemistry of rare earth elements in the oxidation state is basically the coordination chemistry of tetravalent cerium. Even then well-characterized coordination compounds are limited to only a few classes of compounds. Notable are *e.g.* halogeno complexes and complexes of oxo acids, β-diketonates and related Schiff-base complexes as well as porphyrinates and related complexes [1]. Two other important classes of cerium(IV) compoinds, the alkoxides and amides of Ce^{4+}, can be regarded as "*pseudo*-organometallics" and will be discussed together with the organocerium(IV) complexes in Section 4.

3.1. Halogeno Complexes

Several halogeno complexes of tetravalent lanthanides are known. For example, the tetrafluorides form three series of fluoro complexes, $[LnF_6]^{2-}$ (*e.g.* in $K_2[PrF_6]$), $[LnF_7]^{3-}$ (*e.g.* in $Cs_3[NdF_7]$), and $[LnF_8]^{4-}$ (*e.g.* in $(NH_4)_4[CeF_8]$). In the case of the fluorocerates(IV), ammonium salts like $(NH_4)_4[CeF_8]$ (square antiprismatic coordination) and $(NH_4)_3[CeF_7(H_2O)]$ can be isolated by crystallization from aqueous solution. The synthesis of alkali metal derivatives of the types M_2CeF_6 and M_3CeF_7 (M = Na, K, Rb, Cs) requires the use of solid state methods such as the the reaction of CeO_2/MCl mixtures with elemental fluorine [23]. The same is true for the fluoro metallates of other lanthanides in the oxidation state +4, for which the alkali metal fluoro complexes M_2LnF_6 (M = Na, K, Rb, Cs; Ln = Tb, Pr) and Cs_3LnF_7 (Ln = Pr, Nd, Tb, Dy) are known. All these materials can only be prepared by anhydrous solid-state methods such as the fluorination of Ln_2O_3/MCl mixtures using elemental fluorine or fluorination of Cs_3LnCl_6 with XeF_2 [1].

Hexachlorocerates(IV) stabilized by bulky Group 15 cations are easily prepared as shown in Scheme 4 starting with hydrated cerium dioxide [1]. Among these, especially the bright yellow pyridinium salt $(pyH)_2[CeCl_6]$ has frequently been employed as alternative Ce(IV) precursor [24]. Its preparation is quite straightforward and involves precipitation of cerium(IV) hydroxide by adding sodium hydroxide to ceric ammonium sulfate, followed by suspension of the freshly prepared $Ce(OH)_4$ in ethanol and subsequent treatment with HCl gas and pyridine. Other derivatives with bulky cations can be prepared in an analogous manner.

Although cerium tetrachloride is unstable, the dipyridinium hexachlorocerate(IV) is so stable that it may be dried at 120 °C *in vacuo* [25].

$$Ce(OH)_4 \xrightarrow[\text{EtOH, } [R_4E]Cl]{\text{HCl (g)}} [R_4E]_2[CeCl_6]$$

$$R_4E = Et_4N, \text{Pyridinium}, Ph_3PH, Ph_4As$$

Scheme 4.

The corresponding hexachloroceric acid has been isolated in the form of its adduct with diglyme (diglyme = diethyleneglycol dimethylether) by treatment of hydrated ceria with thionyl chloride in the presence of diglyme according to Scheme 5 [1]:

$$CeO_2 \cdot x\,H_2O \xrightarrow[\text{diglyme}]{SOCl_2} [H_2(\text{diglyme})_3][CeCl_6]$$

Scheme 5.

$$[H_2(\text{diglyme})_3][CeCl_6] \xrightarrow[\text{EtOAc}]{2\,L} CeCl_4L_2 \quad [L = Ph_3AsO;\ {}^tBu_2SO;\ (Me_2N)_3PO;\ \textit{trans}]$$

Scheme 6.

$$(NH_4)_2[Ce(NO_3)_6] \xrightarrow[Ph_3PO]{HCl_{(g)}} CeCl_4(Ph_3PO)_2 \quad (\textit{cis})$$

Scheme 7.

The diglyme-stabilized hexachloroceric acid also served as starting material for a series of stable adducts of cerium(IV) tetrachloride with *O*-donor ligands (Scheme 6). Like their uranium(IV) congeners, these complexes adopt an octahedral coordination geometry with the *O*-donor ligands arranged in the *trans*-positions [1]. In contrast, the closely related bis(triphenylphosphine oxide) adduct comprises the *cis*-configuration. This compound was prepared according to Scheme 7 [1]. While pure cerium(IV) tetrahalides other than CeF_4 have never been isolated, it has been possible to convert hexachlorocerates (*cf.* Scheme 4) into the homologous salts containing the $[CeBr_6]^{2-}$ anion. These purple hexabromocerates(IV) can be prepared by treatment of the hexachloro precursors with gaseous HBr. It was not possible to make the corresponding hexaiodocerates(IV), most likely due to internal redox reactions [1].

3.2. Complexes of Oxo Acids

Ceric ammonium nitrate, $(NH_4)_2[Ce(NO_3)_6]$ (= CAN), is perhaps the most common Ce^{4+} precursor.

$$(NH_4)_2[Ce(NO_3)_6] \xrightarrow[Me_2CO]{Ph_3PO} Ce(NO_3)_4(Ph_3PO)_2 \ (\textit{trans})$$

Scheme 6.

This bright orange, water-soluble salt is widely used as a standard oxidant in quantitative analysis as well an oxidizing agent in organic synthesis. The hexanitratocerate anion forms directly upon dissolution of cerium(III) oxide in nitric acid. Thus CAN can be prepared by dissolving Ce_2O_3 in hot concentrated nitric acid, followed by precipitation with ammonia [1]. In the hexanitratocerate(IV) anion, the bite angle of the nitrato ligands is so small that they can all act as bidentate ligands, giving rise to an icosahedral coordination geometry (coordination number 12) around cerium [26]. Starting from CAN, some 10-coordinate neutral nitrato complexes containing additional *O*-donor ligands have been prepared, such as the bis(triphenylphosphine oxide) derivative shown in Scheme 6 [1].

CAN also served as a useful precursor for the synthesis of cerium(IV) alkoxides and mixed-ligand alkoxide/nitrato complexes [27]. Practical applications include its use as a component of chrome etchant, a material that is used in the production of liquid crystal displays and photomasks. The most important and versatile applications of CAN, however, are those in organic synthesis, where it is used as a strong oxidizing agent, as catalyst in the synthesis of heterocycles (*e.g.* quinolines, quinoxalines) and as a deprotection agent.

Ceric ammonium sulfate, $(NH_4)_2[Ce(SO_4)_3]$, is frequently used as a safe and non-toxic oxidizing agent in redox titrations (color change from orange Ce^{4+} to colorless Ce^{3+}). Monohydrated *ceric ammonium phosphate*, $(NH_4)_2[Ce(PO_4)_2]\cdot H_2O$, was hydrothermally synthesized from the system $NH_3/CeO_2/H_3PO_4$ and characterized by powder X-ray diffraction and various other analytical techniques [28].

Several interesting complexes containing the bis-μ-peroxo-hexacarbonatodicerate(IV) anion, $[(CO_3)_3Ce(\mu\text{-}O_2)_2Ce(CO_3)_3]^{8-}$, have been isolated and structurally characterized. Examples include $K_2Na_6[(CO_3)_3Ce(\mu\text{-}O_2)_2Ce(CO_3)_3]\cdot 13H_2O$, $Rb_8[(CO_3)_3Ce(\mu\text{-}O_2)_2Ce(CO_3)_3]\cdot 12H_2O$, and $Cs_8[(CO_3)_3Ce(\mu\text{-}O_2)_2Ce(CO_3)_3]\cdot 10H_2O$ [29,30]. A $Ce(\mu\text{-}O_2)_2Ce$ unit with bridging, doubly bidentate peroxy groups has also been found in the *N,N,N',N'*-ethylenediamine-tetraacetato (= EDTA) derivative $Na_2K_2[(EDTA)Ce(\mu\text{-}O_2)_2Ce(EDTA)]$ [29].

Cerium(IV) β-Diketonates and Related Complexes

Homoleptic β-diketonate complexes of tetravalent cerium, $Ce[R^1C(O)CHC(O)R^2]_4$, are easily accessible and form a well-investigated class of cerium(IV) complexes [31,32]. Scheme 7 lists a small selection of known complexes of this type.

Most remarkably, the corresponding homoleptic β-diketonate complexes of trivalent cerium, $Ce[R^1C(O)CHC(O)R^2]_3$, tend to be readily oxidized in the presence of air. Thus a typical synthetic procedure for preparing the more stable Ce^{4+} derivatives involves the reaction of the Na salt of the diketonate with $Ce(NO_3)_3\cdot 6H_2O$ in aqueous EtOH solution. When the reaction is carried out in air, oxidation will easily take place and the cerium(IV) product is formed directly. Alternatively, the cerium(IV) precursor $(NH_4)_2[Ce(NO_3)_6]$ (= CAN) can be treated with the sodium β-diketonate in ethanol. The homoleptic $Ce[R^1C(O)CHC(O)R^2]_4$ complexes form dark red or purple crystalline solids, which are soluble in organic solvents such as toluene or CH_2Cl_2. The coordination geometry around cerium is in most cases square antiprismatic, but distorted dodecahedral and distorted bicapped trigonal prismatic geometries have also occasionally been found [31,32].

Mixed-ligand complexes containing chelating *β*-diketonate ligands are also known. For example, $Ce(OPPh_3)_2(TTA)_2(NO_3)_2$ (HTTA = thenoyltrifluoroacetone) has been prepared from $Ce(OPPh_3)_2(NO_3)_4$ (*cf.* Scheme 6) and HTTA. Due to their relatively high volatility, homoleptic cerium(IV) tetrakis(*β*-diketonates) have been successfully employed as molecular precursors for the MOCVD (= metal-organic chemical vapor deposition) production of pure CeO_2 and yttria-doped CeO_2 thin films [33,34].

Recently, the very bulky tripodal Schiff-base ligand $N[CH_2CH_2N{=}CH(C_6H_2{}^tBu_2\text{-}3,5\text{-}OH\text{-}2)_3$ (= $H_3TRENDSAL$) has also been found to form very stable complexes with tetravalent cerium. As with the *β*-diketonates, it was found that the cerium(III) derivative Ce(TRENDSAL) is quite air-sensitive. Thus a straightforward preparation of the cerium(IV) chloro derivative involves treatment of hydrated cerium trichloride with the *in situ*-prepared ligand in the presence of oxygen as illustrated in Scheme 8. The use of ceric ammonium nitrate (= CAN) as precursor allowed the synthesis of the corresponding nitrato complex $(TRENDSAL)CeNO_3$ [35].

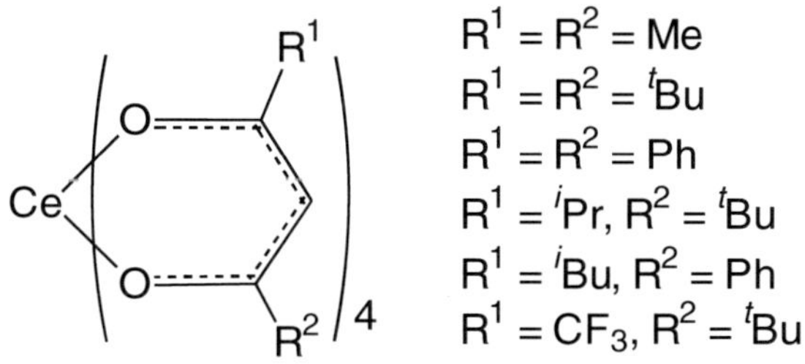

Scheme 7.

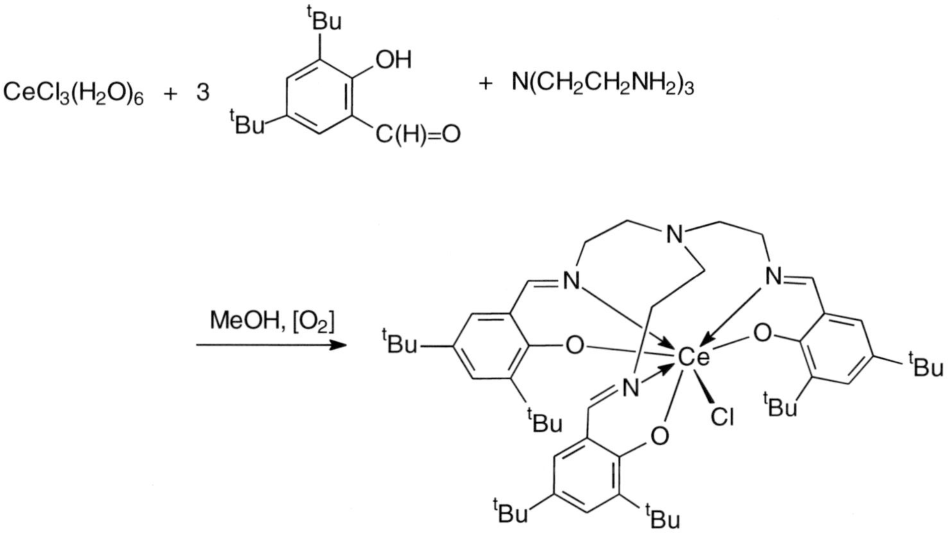

Scheme 8.

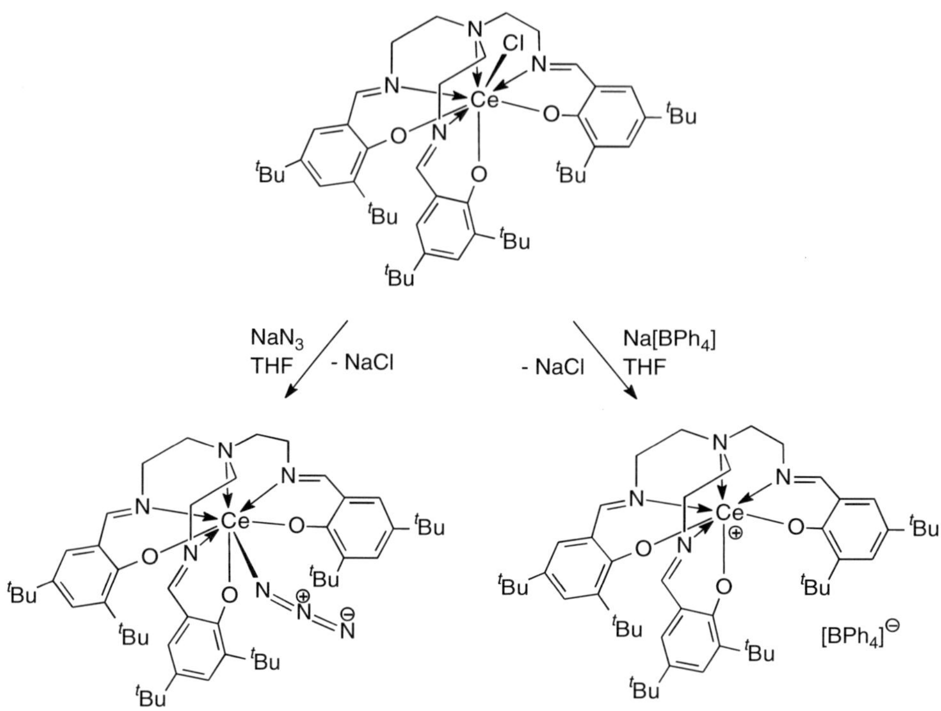

Scheme 9.

As shown in Scheme 9, the first cerium(IV) azide complex, (TRENDSAL)CeN_3, was readily obtained by treatment of (TRENDSAL)CeCl with a large excess of NaN_3 in THF (black, block-like crystals, 53% yield). Halide abstraction from (TRENDSAL)CeCl upon reaction with Na[BPh_4] in acetonitrile afforded the first cationic complex of tetravalent cerium, [Ce(TRENDSAL)][BPh_4], which was isolated as a black, crystalline solid [36].

3.3. Cerium(IV) Sandwich Complexes with Tetrapyrrole-Type Ligands

Formally tetravalent cerium is long known to form stable sandwich complexes with tetrapyrrole-type macrocyclic ligands such as porphyrines, phthalocyanines, and 2,3-naphthalocyanines (Scheme 10) [37-39]. For the rare earth sandwich complex series, virtually all the neutral complexes can be formulated as Ln^{III}(ring-1^{2-})(ring-$2^{\cdot-}$), in which a trivalent rare earth metal center is sandwiched between a dianionic macrocycle and a radical anionic ligand, having different extent of electron delocalization. Among the entire lanthanide series, cerium is the only exception. Having an electronic configuration of [Xe]$4f^1 5d^1 6s^2$, cerium may also utilize the electron in the extended 4f orbital, leading to predominantly tetravalent neutral sandwich complexes.

Several synthetic methods have been established for the preparation of cerium(IV) sandwich complexes. Scheme 11 illustrates some typical preparations. For example, the homoleptic bis(phthalocyaninato) complex Ce[Pc($OC_{12}H_{25}$)$_8$]$_2$ was prepared by treatment of Ce(acac)$_3$·$n$$H_2O$ with 4,5-bis(dodecyloxy)phthalonitrile in the presence of 1,8-diazabicyclo[5.4.0]undec-7-ene (DBU) (Scheme 11a). This represents a general procedure which has been employed previously to prepare the pentyloxy $Ce[Pc(OC_5H_{11})_8]_2$ and octyloxy $Ce[Pc(OC_8H_{17})_8]_2$ analogues as well as other bis(phthalocyaninato) rare earth complexes.

The bis(naphthalocyaninato) analogues $Ln[Nc(^tBu)_4]_2$ (Ln = La, Ce, Pr, Nd, Eu, Gd, Tb, Y, Er) have also been synthesized in a similar manner using 6-*tert*-butylnaphthalonitrile as the starting material.

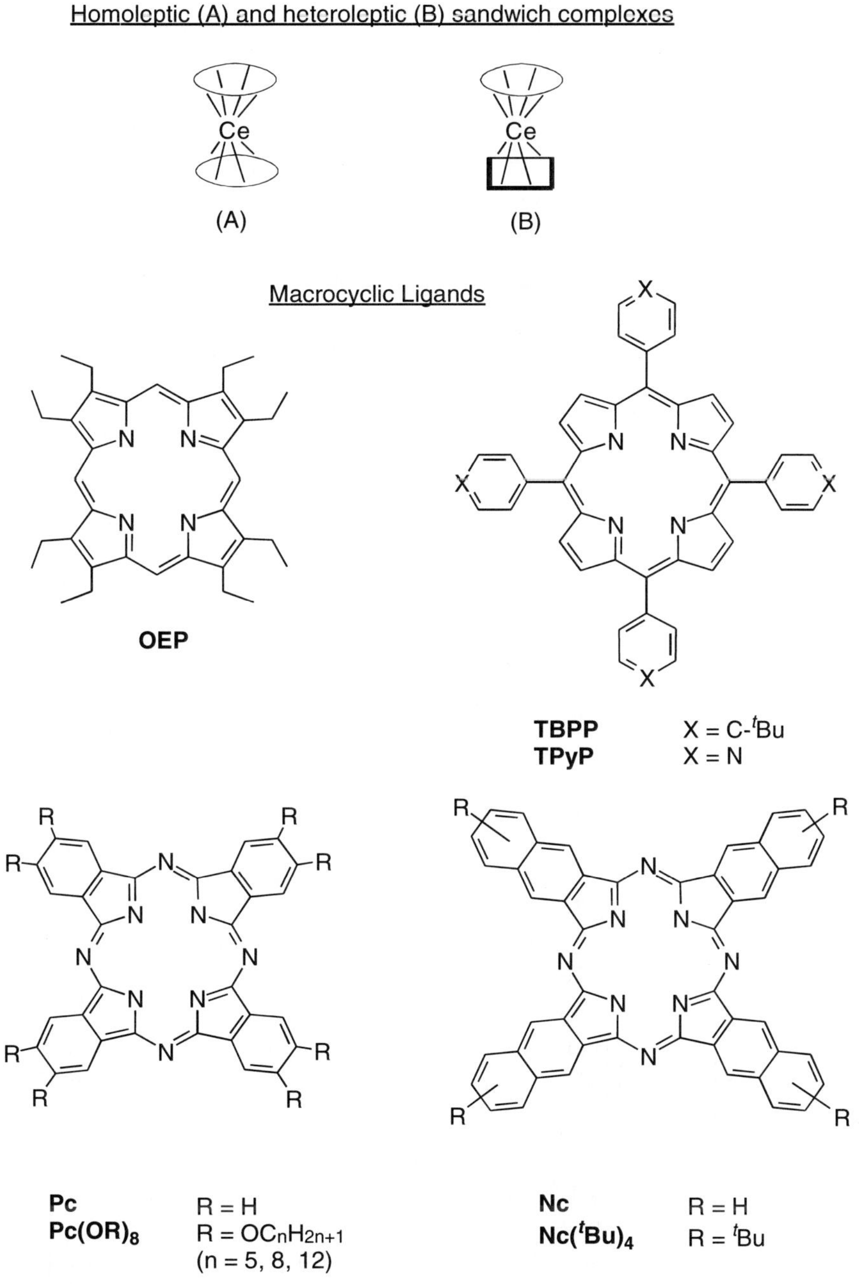

Scheme 10.

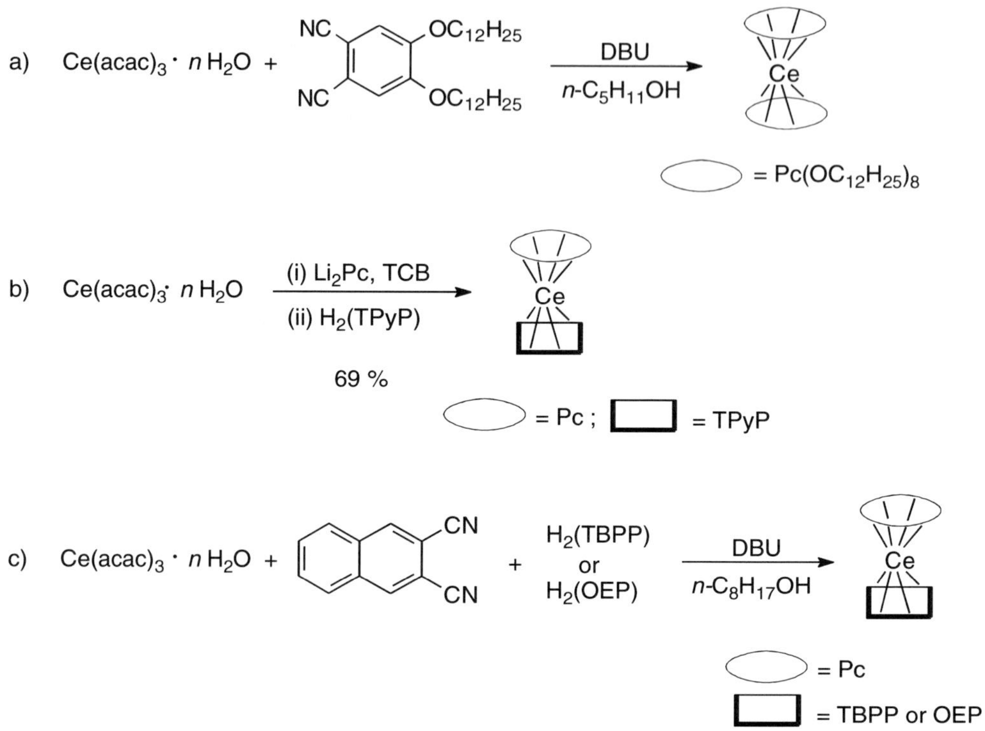

Scheme 11.

The yield of the cerium sandwich complex $Ce[Nc(^tBu)_4]_2$ (71%) was much higher than those of the phthalocyanine counterparts mentioned above (9-49%), following the trend observed for the $Ln[Nc(^tBu)_4]_2$ series, the yield of which decreases gradually with decreasing the size of the metal center [39]. Neutral cerium(IV) sandwich complexes can also be obtained by deliberate oxidation of the anionic cerium(III) precursors. For example, anodic or chemical oxidation with dibenzoylperoxide of solutions of bis(phthalocyaninato)cerate(III) in dichloromethane yielded selectively the sparingly soluble bis(phthalocyaninato)cerium(IV), $Ce(Pc)_2$ as green, needle-like crystals. $Ce(Pc)_2$ is a sandwich complex in which the cerium atom is 8-oordinated by the isoindole nitrogens of the two staggered convex Pc-rings [38].

The mixed ring double-decker Ce(Pc)(TPyP) was prepared by a stepwise procedure as shown in Scheme 11b. Reaction of $Ce(acac)_3{\cdot}nH_2O$ with Li_2Pc in 1,2,4-trichlorobenzene (TCB) followed by the addition of $H_2(TPyP)$ led to Ce(Pc)(TPyP) in 69% yield together with a substantial amount of the triple-deckers $Ce_2(Pc)(TPyP)_2$ and $Ce_2(Pc)_2(TPyP)$. This procedure is one of the most common synthetic methods to prepare mixed double-deckers M(Pc)(Por). By using a convenient one-pot procedure, the mixed sandwich complexes Ce(Nc)(TBPP) and Ce(Nc)(OEP) were also synthesized. As shown in Scheme 11c, treatment of $Ce(acac)_3{\cdot}nH_2O$ with the corresponding metal-free porphyrin and naphthalonitrile in the presence of DBU in *n*-octanol afforded these complexes in *ca.* 20% yield. It has been found that the cerium sandwich complexes always show different electronic absorption and electrochemical properties from the whole series of tervalent rare earth analogues. To resolve and clarify this issue, a series of cerium double-deckers using a range of tetrapyrrole ligands with different electronic properties have been synthesized according to Scheme 11 and studied systematically with respect to their electrochemical and spectroscopic properties. The results showed that the oxidation state of the cerium center varies from +3 to +4 depending on the electronic nature of the two tetrapyrrole ligands. As revealed by X-ray absorption near-edge structure (XANES) study, the valence of the cerium center is 3.59 for $Ce[Pc(OC_5H_{11})_8]_2$ and 3.68 for Ce(Pc)(TPyP) [TPyP = *meso*-tetra(4-pyridyl)porphyrinate] and Ce(Nc)(OEP), while that in $Ce[Nc(^tBu)_4]_2$ is virtually 3 [39].

4. "Pseudo-Organometallic" Cerium(IV) Compounds

4.1. Alkoxides and Siloxides

Tetraalkoxides of cerium form a long known and well established class of lanthanide(IV) compounds which can be condsidered as "pseudo-organometallic".The effect of difference in both size and charge should favor the formation of monomeric Ce(IV) alkoxides rather than their trivalent analogues. Ce^{4+} is smaller than Ce^{3+} (0.87 Å *vs.* 1.01 Å effective ionic radii for 6-coordination according to Shannon [40]) and can accommodate four alkoxide ligands as counter ions. Additionally, "hard" ligands like the alkoxide ligands should stabilize the high oxidation state Ce^{4+}. Lanthanide alkoxides have been widely used as molecular precursors in materials science. Although Ce(IV) alkoxides have been known for many years, their preparation was often laborious and time consuming [41], until the readily available and inexpensive tetravalent cerium reagent ceric ammonium nitrate $(NII_4)_2[Ce(NO_3)_6]$ (= CAN), was utilized for their synthesis. Scheme 12 illustrates how CAN can be successfully employed as starting material for the preparation of homoleptic cerium(IV) alkoxides [42].

Not only can the homoleptic tetraalkoxides be made using this synthetic approach but also a range of Ce^{4+} nitrate, mixed nitrate-alkoxide, and alkoxide complexes have become available by careful control of the reaction stoichiometry. Scheme 13 illustrates this approach. The low costs of the starting materials and control of the functionalities present rendered this method an appealing route into further Ce^{4+} chemistry [42].

Another access to cerium(IV) tetraalkoxides involves treatment of the readily available hexachlorocerate(IV) precursor with 4 equivalents of an appropriate alcohol followed by reaction with ammonia (Scheme 14). [25,41].

There is also one interesting report about Ce^{4+} alkoxides being formed *via* controlled oxidation reaction of cerium(III) alkoxide precursors, *e.g.* by reactions of cerium(III) tris[tri(*tert*-butyl)methoxide] with either an organic peroxide or benzoquinone. For example, treatment of $Ce(OC^tBu_3)_3$ with dibenzoylperoxide according to Scheme 15 afforded a mixed cerium(IV) alkoxide carboxylate, which was, however, stable only in solution and decomposed to cerium(III) species when the reaction was carried out on a preparative scale [43].

Inherent instability was also observed for the mixed alkoxide derivative $Ce(OC^tBu_3)_2(O^tBu)_2$ which could be generated by treatment of $Ce(OC^tBu_3)_3$ with 2 equivalents of di-*tert*-butyl peroxide (Scheme 16).

$$(NH_4)_2Ce(NO_3)_6 + 4\ ROH + 6\ NaOMe \longrightarrow Ce(OR)_4 + 2\ NH_3 + 6\ NaNO_3 + 6\ MeOH$$

Scheme 12.

$$(NH_4)_2[Ce(NO_3)_6] + n\ NaO^tBu \xrightarrow{THF} (^tBuO)_{n-2}Ce(NO_3)_{6-n}(THF)_2 + 2\ NH_3 + n\ NaNO_3 + 2\ HO^tBu \quad n = 2\text{-}6$$

Scheme 13.

$$(pyH)_2[CeCl_6] + 4\ ROH + 6\ NH_3 \longrightarrow Ce(OR)_4 + 2\ py + 6\ NH_4Cl$$

Scheme 14.

$$2\ Ce(OC^tBu_3)_3 + PhC(O)OOC(O)Ph \xrightarrow[25\ °C]{C_6D_6} 2\ Ce(OC^tBu_3)_3[OC(O)Ph]$$

Scheme 15.

$$Ce(OC^tBu_3)_3 + 2\ ^tBuOO^tBu \xrightarrow{C_6H_6} Ce(OC^tBu_3)_2(O^tBu)_2 + [^tBu_3COOC^tBu_3]$$

Scheme 16.

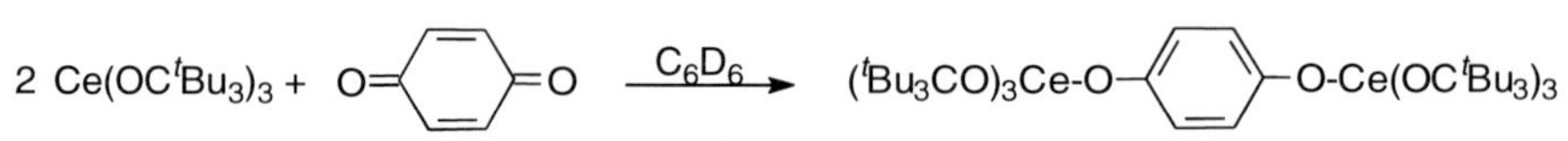

Scheme 17.

The mixed cerium(IV) alkoxide decomposed on attempted isolation under formation of the dimeric cerium(III) alkoxide product $[Ce(OC^tBu_3)_2(O^tBu)]_2$, highlighting the potential instability of the Ce^{4+} state in these cases [43].

The titration of $Ce(OC^tBu_3)_3$ with benzoquinone was the cleanest and most easily characterized oxidation of any in this study. The stoichiometry and NMR spectra were consistent with formation of a binuclear complex bridged by a hydroquinonediolate as shown in Scheme 17. The resultant complex was somewhat less soluble than the starting alkoxide and precipitated from solution as single, well-formed, medium-brown plates suitable for X-ray diffraction. The crystals, once isolated, gave back purple solutions in C_6D_6 that yielded the same NMR data as seen in the titration [43].

Simple homoleptic cerium(IV) tetraalkoxides normally comprise oligomeric structures in the solid state [25,41]. Especially the tetramethoxide and the tetraethoxide are presumably polymeric as they are insoluble even in polar organic solvents [25]. Lower aggregates are formed upon adduct formation with *O*-donor ligands. For example, the tetra(isopropoxide) forms a dinuclear adduct with propan-2-ol. In the dimer $Ce_2(O^iPr)_8(HO^iPr_2)_2$ each cerium is octahedrally coordinated. Thermolysis of $Ce_2(O^iPr)_8(HO^iPr_2)_2$ afforded the tetranuclear oxo-cluster $Ce_4(\mu_4\text{-}O)(\mu_3\text{-}O^iPr_2)_2(\mu\text{-}O^iPr)_8(O^iPr_3)_{14}$. Due to their oligomeric structures, most of the homoleptic cerium(IV) tetraalkoxides are non-volatile and do not sublime without decomposition. A notable exception is the tetra-isopropoxide, which can be sublimed under vacuum at *ca.* 200 °C.

$$(NH_4)_2[Ce(NO_3)_6] + 6\ NaOCMe_2{}^iPr \longrightarrow Ce(OCMe_2{}^iPr)_4 + 6\ NaNO_3 + 2\ NH_3 + 2\ HOCMe_2{}^iPr$$

Scheme 18.

$$Ce(OCMe_2{}^iPr)_4 + 2\ DMAP \longrightarrow Ce(OCMe_2{}^iPr)_4(DMAP)_2$$

DMAP = 4-(dimethylamino)pyridine

$$2\ Ce(OCMe_2{}^iPr)_4 + 3\ Hacac \longrightarrow Ce_2(CCMe_2{}^iPr_2)_5(acac)_3 + 3\ HOMe_2{}^iPr$$

Scheme 19.

The cerium(IV) alkoxide complex $Ce(OCMe_2{}^iPr)_4$, a volatile, non-fluorinated source of cerium, was used as a chemical vapor deposition (CVD) precursor to cerium oxide films. This complex was prepared according to Scheme 18 and isolated as a yellow liquid on a multigram scale in 50-60% yield after vacuum distillation [44].

A conventional thermal chemical vapor deposition process deposited cerium(IV) oxide films from $Ce(OCMe_2{}^iPr)_4$ on silicon, glass, quartz, lanthanum aluminum oxide (001), and nickel (001) substrates at low substrate temperatures (<550 °C). The films were highly oriented when the depositions were carried out on lanthanum aluminum oxide and textured nickel substrates.

An X-ray crystallographic study of $Ce(OCMe_2{}^iPr)_4$, which melts at just below room temperature, shows that it is a loosely bound dimer in the solid state with five-coordinate cerium centers and two bridging alkoxide ligands. The synthesis and X-ray crystal structures of $Ce(OCMe_2{}^iPr)_4(DMAP)_2$ (DMAP = 4-(dimethylamino)pyridine) and $Ce_2(OCMe_2{}^iPr)_5(acac)_3$ have also been reported (Scheme 19). The cerium atoms in $Ce(OCMe_2{}^iPr)_4(DMAP)_2$ and $Ce_2(OCMe_2{}^iPr)_5(acac)_3$ have, respectively, distorted octahedral and pentagonal bipyramidal coordination geometries [44].

The same synthetic route starting with CAN as precursor has been employed for the preparation of the first homoleptic triorganosiloxides of tetravalent cerium. The reaction shown in Scheme 20 proceeded in the way that gas (NH_3) evolved, a white precipitate ($NaNO_3$) formed, and the product was isolated in about 70% yield as a light yellow powder.

$$(NH_4)_2[Ce(NO_3)_6] + 6\ NaOSiPh_3 \xrightarrow{DME} Ce(OSiPh_3)_4(DME) + 6\ NaNO_3 + 2\ NH_3 + 2\ Ph_3SiOH$$

Scheme 20.

$$(NH_4)_2[Ce(NO_3)_6] + 6\,NaOSiEt_3 \xrightarrow{CH_3CN} Ce(OSiEt_3)_4 + 6\,NaNO_3 + 2\,NH_3 + 2\,Et_3SiOH$$

Scheme 21.

The reaction was carried out in DME to facilitate isolation of the product as the DME adduct, $Ce(OSiPh_3)_4(DME)$, and the product was identical with the crystallographically characterized product isolated by reaction of Ph_3SiOH with $Ce(O^iPr)_4$ [45].

The same preparative route also allowed the synthesis of the corresponding cerium(IV) tetrakis(triethylsiloxide) (Scheme 21). This compound was isolated as a yellow-brown powder in 83% yield [45].

Quite in contrast to the alkoxides and simple siloxides, related disiloxanediolate complexes of cerium(IV) form spontaneously upon treatment of cerium(III) precursors with 1,1,3,3-tetraphenyl disiloxanediol (Scheme 22). As a suitable cerium(IV) precursor, the readily available alkoxide nitrate complex $(^tBuO)_3Ce(NO_3)(THF)_2$ was chosen and treated with $(Ph_2SiOK)_2O$ (made *in situ* by deprotonation of 1,1,3,3-tetraphenyl-1,3-disiloxanediol with $KN(SiMe_3)_2$) in a molar ratio of 1:2 according to Scheme 22 (right). From this reaction, the novel heterobimetallic potassium/cerium(IV) disiloxanediolate complex $[\{(Ph_2SiO)_2O\}\{K(THF)_2\}]_2Ce(O^tBu)_2$ was isolated in the form of yellow, block-shaped crystals in 81% yield. Its structure was studied by single-crystal X-ray diffraction. In the molecule, two potassium disiloxanediolate ligands are coordinated to cerium(IV) in a slightly bent coordination geometry. Two *t*-butoxide ligands remain as functionalizable groups coordinated to Ce.

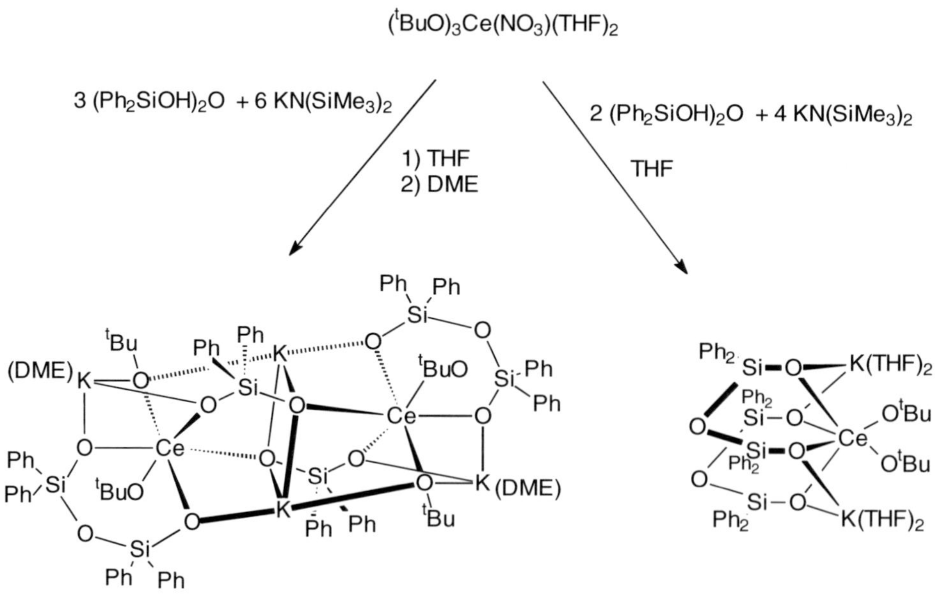

Scheme 22.

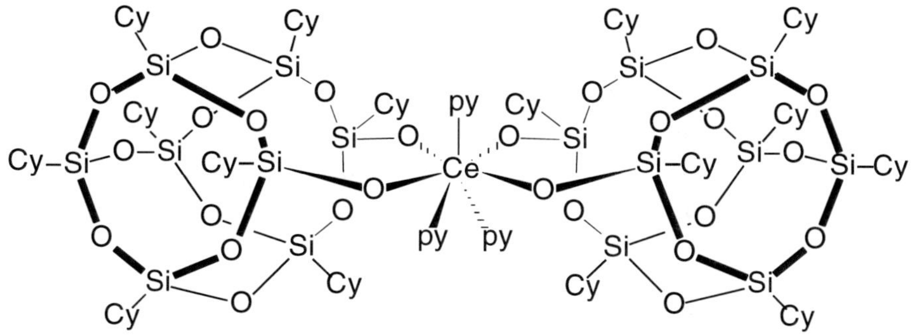

Scheme 23.

The central cerium is in a distorted octahedral coordination environment. A very similar distorted octahedral coordination of tetravalent cerium was found in $Ce(OSiPh_3)_4(DME)$ [45]. One phenyl ring of each disiloxanediolate ligand is engaged in an intramolecular η^1-π-coordination to potassium (K(1)-C(22) 3.395(4) Å). [46]. The same reaction carried out in a molar ratio of 1:3 took a completely different course (Scheme 22, left). In this case, a light yellow compound could be isolated in high yield (70%), which was shown by X-ray crystallography to be the novel heterobinuclear potassium/cerium(IV) disiloxanediolate cluster $[\{(Ph_2SiO)_2O\}_2\{(DME)KO^tBu\}\{(Ph_2SiO_2)K\}Ce]_2$. The X-ray study revealed that in this compound not only the expected dianions of 1,1,3,3-tetraphenyl-1,3-disiloxanediol act as ligands but also the dianions of diphenyl-1,1-silanediol, which originate from cleavage of the disiloxanediolate. Such transformations are not uncommon in the chemistry of silanediolates, disiloxanediolates, and trisiloxanediolates. In the unprecedented centrosymmetric cluster molecule, a planar K_2O_2 ring forms the central structural unit. The central cerium atoms are in a distorted octahedral coordination environment. Two disiloxanediolate dianions act as chelating ligands towards each Ce(IV) and simultaneously bridge two potassium ions with the cerium.

An unusual μ^3,μ^2-bridging coordination mode was found for the dianionic diphenyl-1,1-silane-diolate ligands. One oxygen from each of the two $Ph_2SiO_2^{2-}$ ligands bridges one Ce and two potassium ions, while the other oxygen forms a μ^2-bridge to Ce and K. The overall coordination geometry around each cerium is slightly distorted octahedral.

Four *t*-butoxide groups originating from the precursor $(^tBuO)_3Ce(NO_3)(THF)_2$ are also present in the molecular structure of $[\{(Ph_2SiO)_2O\}_2\{(DME)KO^tBu\}\{(Ph_2SiO_2)K\}Ce]_2$, two of them bridging (K, K) and the other two terminally bonded to cerium. Given the high isolated yield of ca. 70% it is clear that despite its highly unusual structure compound $[\{(Ph_2SiO)_2O\}_2\{(DME)KO^tBu\}\{(Ph_2SiO_2)K\}Ce]_2$ is the main product of the reaction of $(^tBuO)_3Ce(NO_3)(THF)_2$ with three equivalents of $(Ph_2SiOK)_2O$. The coordinative unsaturation of the potassium ions (coordination numbers 4 and 5, resp.) is relieved by η^1- and η^2-π-interactions with neighboring phenyl substituents. The respective K···C distances are in the range of 3.173 - 3.527(4)Å [46].

In a similar manner, the isolation of an unprecedented cerium(IV) silsesquioxane complex (Scheme 23) has been achieved. Treatment of $Ce[N(SiMe_3)_2]_3$ with two equivalents of the incompletely condensed silsesquioxane disilanol derivative $Cy_8Si_8O_{11}(OH)_2$ in diethyl ether in the presence of an excess of pyridine exclusively afforded the diamagnetic complex $(Cy_8Si_8O_{13})_2Ce(py)_3$.

The same compound, albeit in somewhat lower yield, could also be prepared by direct reaction of anhydrous $CeCl_3$ with two equivalents of $Cy_8Si_8O_{11}(OH)_2$ in THF/pyridine mixture. Quite surprisingly in both cases cerium was quantitatively oxidized to the tetravalent oxidation state. An X-ray diffraction study revealed the presence of the first example of a metallasilsesquioxane derived from the octameric disilanol ligand $[(c\text{-}C_6H_{11})_8Si_8O_{13}]^{2-}$. The central Ce atom is coordinated by four oxygen atoms of two siloxane ligands and three nitrogens of three pyridine molecules resulting in a distorted pentagonal bipyramidal arrangement [47].

4.2. Amides

Amido complexes were among the first "pseudo-organometallic" compounds of tetravalent cerium to be reported in the literature. In the course of the synthesis of the first tri-coordinate tris(amido)lanthanides in 1972, it had been noted that the yellow $Ce[N(SiMe_3)_2]_3$ was readily oxidized by minute traces of air as evident by a formation of a brown material on the surface of the yellow crystals [48]. $Ce[N(SiMe_3)_2]_3$ was by far the most air-sensitive compound in the entire series of $Ln[N(SiMe_3)_2]_3$ derivatives. Thus this compound appeared to be a logical choice as starting material for the synthesis of cerium(IV) silylamido complexes. However, various oxidizing agents including elemental chlorine or bromine only led to the formation of decomposition products. The list of other failed alternative oxidants that have been compiled for this cerium amide is quite lengthy: I_2, *N*-bromosuccinimide, *N*-chlorosuccinimide, $AgBF_4$, AgCN, $Hg(C_6F_5)_2$, $PbCl_2$, and ${}^tBuOO{}^tBu$ were all tried and in all cases only unreacted $Ce[N(SiMe_3)_2]_3$ was recovered. Only the use of the uncommon reagent $TeCl_4$ allowed for the first time the isolation of $[(Me_3Si)_2N]_3CeCl$. It furnished well-formed purple/black needles, although the workup procedure had to be carried out rapidly as delays resulted in the deposition of colorless blocks of $[Ce\{N(SiMe_3)_2\}_2(\mu\text{-}Cl)(THF)]_2$ as a by-product in significant amounts. The analogous bromo-derivative $[(Me_3Si)_2N]_3CeBr$ was made by treatment of the cerium(III) silylamide with Ph_3PBr_2 (Scheme 24) [49,50].

Most recently it has been reported that phenyliodine(III) dichloride, $PhICl_2$, can be successfully employed as an innovative and superior reagent for the one-electron oxidation of cerium(III) precursors. The hypervalent organoiodine(III) compound $PhICl_2$ has several significant advantages: (a) it is readily accessible from inexpensive, commercially available starting materials; (b) solid $PhICl_2$ can easily be employed in exact stoichiometric amounts; and (c) liquid, highly soluble, and volatile iodobenzene is the only iodine-containing byproduct. All these assumptions could be verified by a test reaction with the silylamide $Ce[N(SiMe_3)_2]_3$. Treatment of a toluene solution of $Ce[N(SiMe_3)_2]_3$ with the calculated amount of $PhICl_2$ in toluene (Scheme 25) resulted in an immediate color change from orange to purple. Pure $[(Me_3Si)_2N]_3CeCl$ could be isolated in 10% yield in a very simple manner by crystallization from the concentrated reaction mixture at -20 °C. The organoiodine(III) route could be also utilized to access novel five-coordinate cerium(IV) amido species.

Dark red-brown $[(Me_3Si)_2N]_3CeCl(N{\equiv}CC_6H_4OMe\text{-}p)$ was isolated in 45% yield by adding *p*-anisonitrile to a toluene solution of $Ce[N(SiMe_3)_2]_3$ prior to treatment with $PhICl_2$ (*cf.* Section 4.3) [51]. Trityl chloride, Ph_3CCl, has also been found to be a very convenient reagent for the oxidative halogenation of cerium(III) amide precursors such as $Ce[N(SiMe_3)_2]_3$ [52].

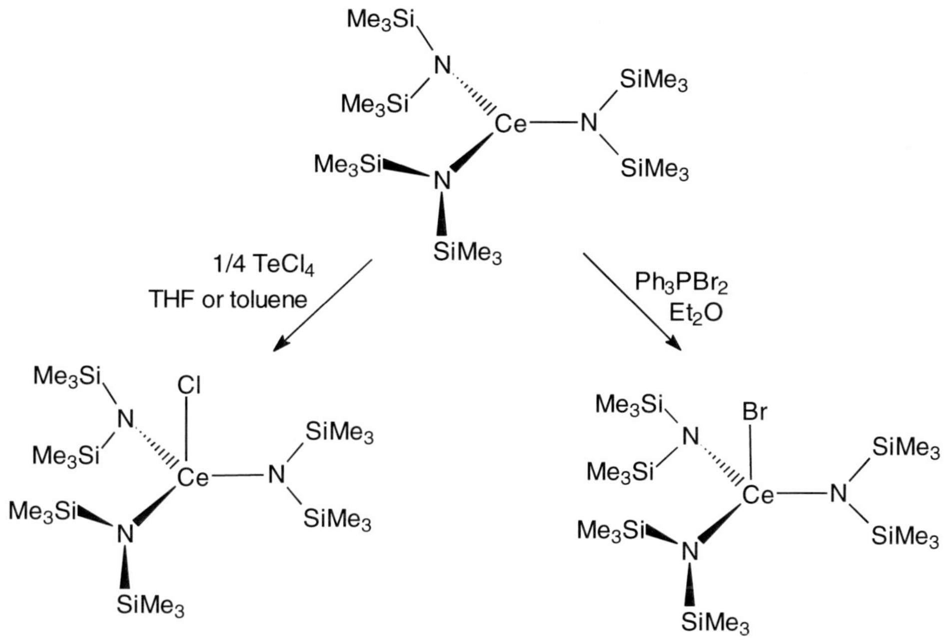

Scheme 24.

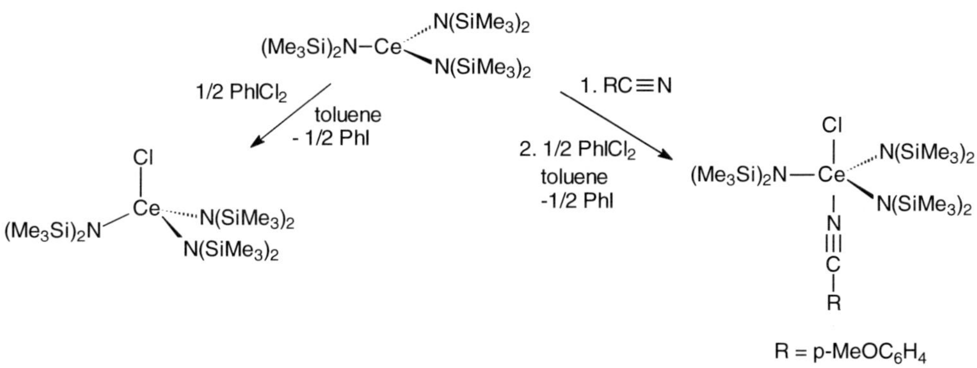

Scheme 25.

Unprecedented crystalline cerium(IV) bis(amides) containing a cyclic $(Ce–O)_n$ core (n = 2, 3, or 4) have been prepared by oxidation of $Ce[N(SiMe_3)_2]_3$ or $Ce(NR"_2)_3$ ($NR"_2$ = $NC(Me)_2(CH_2)_3CMe_2$). The bis(trimethylsilyl)amides were prepared using oxygen gas or 2,6-di(*tert*-butyl)-1,4-benzoquinone/a trace H_2O at low temperature as illustrated in Scheme 26. These complexes were obtained as X-ray quality crystals in modest (38% for the dimer) or low (10% for the trimer) yield. ^1H and $^{13}C\{^1H\}$ NMR spectra in C_6D_6 showed each complex to be diamagnetic. However, the solutions decomposed after a few hours at ambient temperature [53]. In a similar manner, oxidation of the tetramethylpiperidide derivative $Ce(NR"_2)_3$ ($NR"_2$ = $NC(Me)_2(CH_2)_3CMe_2$) with elemental oxygen as shown in Scheme 27 produced a tetrameric bis(amido)cerium(IV) oxide, although only a few single crystals of this compound could be isolated [53].

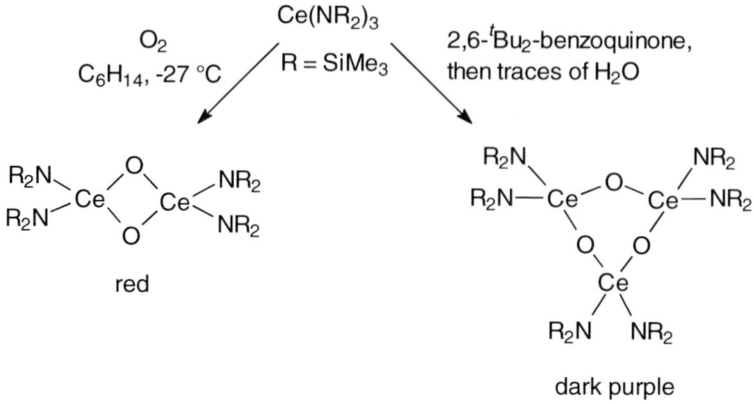

Scheme 26.

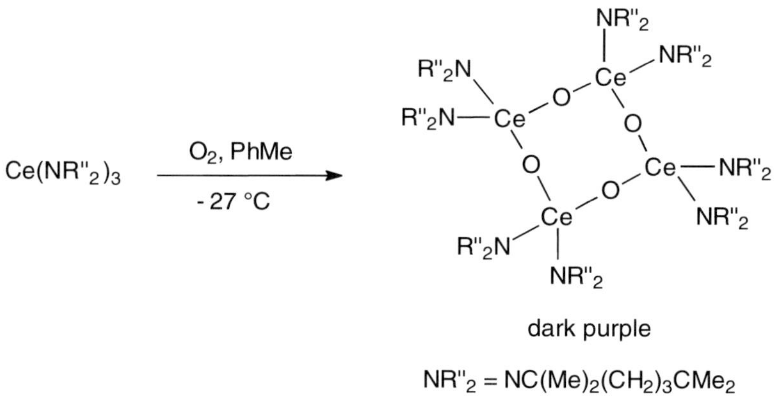

Scheme 27.

The black binuclear complexes $(\mu\text{-}\eta^2\text{:}\eta^2\text{-}O_2)[Ce\{N(SiMe_3)_2\}_2]_2\cdot 2C_nH_{2n+2}$ (n = 5, 6), obtained in a similar manner from $Ce[N(SiMe_3)_2]_3$ and dry oxygen, have either a bridging per- or superoxide ligand, but the O–O bond length data were not definitive. The preparation of related novel sodium and potassium bis(trimethylsilyl)amido (oxo)cerates(IV) is outlined in Scheme 28. Each of these compounds was obtained from oxidation of a mixture of the cerium(III) amide $Ce[N(SiMe_3)_2]_3$ and the appropriate metal amide $MN(SiMe_3)_2$ (M = Na, K) [53]. Facile oxidation of $Ce(NCy_2)_3(THF)$ or $Ce(NCy_2)_2(\mu\text{-}NCy_2)_2Li(THF)$ (Cy = cyclohexyl) with dry air produced the first homoleptic Ce(IV) amide $Ce(NCy_2)_4$ as illustrated in Scheme 29. Both cerium(III) precursors were extremely air-sensitive in solution and if the smallest amount of air was accidentally introduced during the work-up, a blue coloration appeared in the solution immediately, later yielding black microcrystals. When a measured amount of dry air was added to toluene or THF solutions of $Ce(NCy_2)_3(THF)$ or $Ce(NCy_2)_2(\mu\text{-}NCy_2)_2Li(THF)$ the diamagnetic Ce(IV) amide $Ce(NCy_2)_4$ was isolated in a moderate yield. $Ce(NCy_2)_4$ is only sparingly soluble in hexane, benzene, toluene or THF, producing deep blue solutions, which could be stored in a vacuum-sealed tube without decomposition for several months at room temperature (in contrast to the solution instability of $Ce[N(SiMe_3)_2]_3X$ (X = Cl, Br).

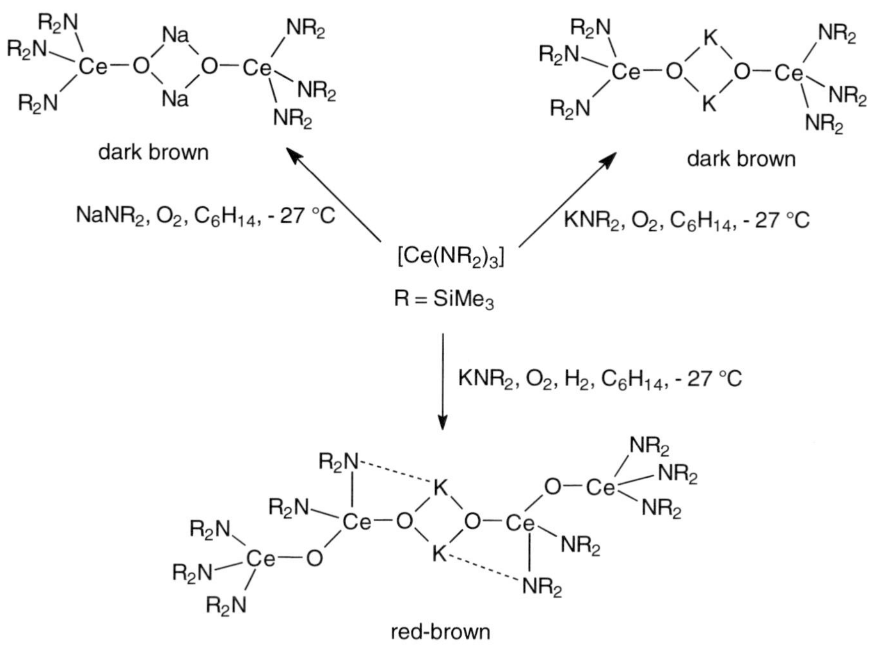

Scheme 28.

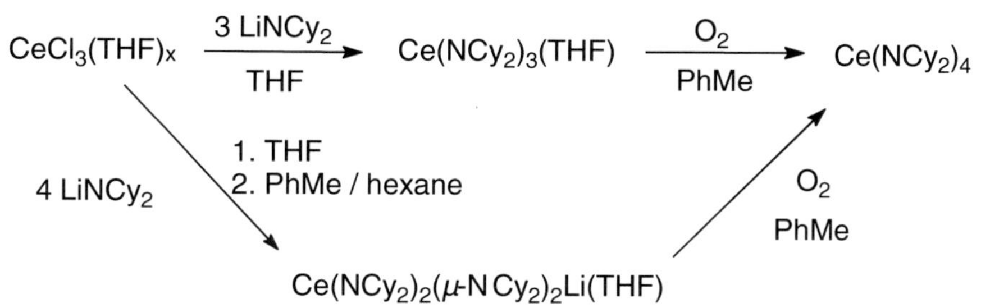

Scheme 29.

The solubility increased upon heating to 70 °C thus allowing recrystallization of $Ce(NCy_2)_4$ without noticeable decomposition even at this temperature [54].

The synthesis and characterization of the related first neutral cerium dialkyl dithiocarbamate complexes, using a novel oxidative displacement of the amido ligands of $Ce[N(SiMe_3)_2]_3$ by tetraalkylthiuram disulfides $[R_2NC(S)S]_2$ (R = Me, Et) in THF solution, has also been reported. The crystalline, toluene-soluble cerium(IV) dithiocarbamate $Ce(\kappa^2\text{-}S_2CNEt_2)_4$ was made by briefly bubbling oxygen gas through a suspension of the yellow $Ce(\kappa^2\text{-}S_2CNEt_2)_3$ in toluene at ambient temperature. The reaction was rapid, as evident from the almost immediate colour change to the black tetravalent complex. Some evidence has also been obtained for the formation of the cerium(IV) complex $Ce[N(SiMe_3)_2]_2(\kappa^2\text{-}S_2CNMe_2)_2$. The cerium(IV) complex $Ce(\kappa^2\text{-}S_2CNEt_2)_4$ has the metal coordinated to eight sulfur atoms of four planar chelating S_2CNC_2 moities and its geometry is intermediate between dodecahedral and square prismatic. The mean Ce–S bond length of 2.803 Å in $Ce(\kappa^2\text{-}S_2CNEt_2)_4$ compares with the 2.950 Å in the Ce(III) complex $Ce[N(SiMe_3)_2]_3(THF)_2$ [55].

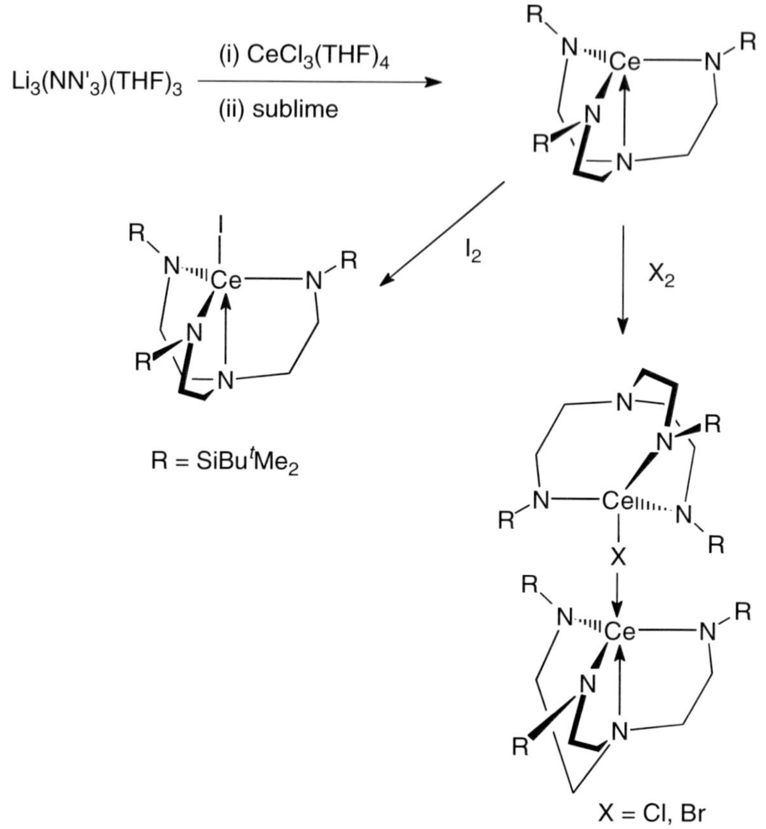

Scheme 30.

Particularly remarkable was the stabilization of cerium(IV) in the presence of an iodide ligand, as neither CeI_4 nor stable salts containing the hexaiodocerate(IV) anion, $[CeI_6]^{2-}$, have ever been isolated. This was achieved by reaction of triamidoamine complex Ce(NN'$_3$) [NN'$_3$ = $N(CH_2CH_2NSiMe_2{}^tBu)_3$] with iodine to give (NN'$_3$)CeI. The base-free trivalent precursor complex Ce(NN'$_3$) was readily synthesized according to Scheme 30 by treatment of $CeCl_3(THF)_4$ with $Li_3(THF)_3$(NN'$_3$) in dry THF followed by sublimation *in vacuo*. This bright yellow-orange complex decomposes rapidly in air but is otherwise quite stable. The reaction of Ce(NN'$_3$) as a solution in pentane with 0.5-1.5 equiv of elemental chlorine led to an immediate color change to purple and deposition of large purple crystals of the fascinating mixed valence Ce(+3/+4) complex (μ-Cl)[Ce(NN'$_3$)$_2$]. Similarly, reaction of Ce(NN'$_3$) with bromine gave purple (μ-Br)[Ce(NN'$_3$)$_2$]. Most unexpectedly, however, the weakest oxidizing agent iodine gave the purple cerium(IV) iodide complex, (NN'$_3$)CeI (Scheme 30) [56].

4.3. Amidinates

Amidinate and guanidinate anions are nitrogen analogues of the carboxylates. They have been widely employed as spectator ligands in main group and transition metal coordination chemistry, with the latter encompassing both early and late transition metals as well as the lanthanides and actinides.

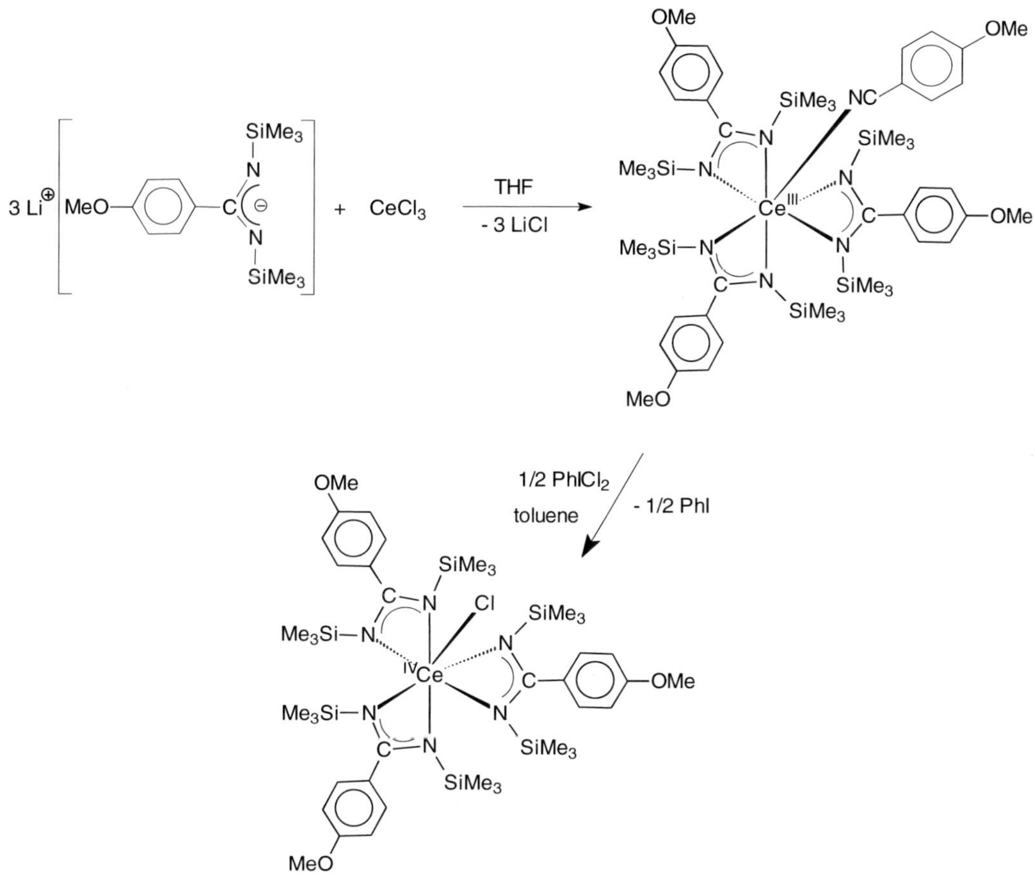

Scheme 31.

These chelating anions are generally readily accessible using commercially available or easily prepared starting materials. Furthermore, their steric and electronic properties can be readily modified in a wide range through variation of the substituents on the carbon and nitrogen atoms. These properties combined make the amidinate anions clearly almost as versatile as the ubiquitous cyclopentadienyl ligands [57,58].

Most recently it was discovered that amidinate ligands are also capable of stabilizing novel cerium(IV) species. Thus they belong to the limited number of organic ligands which allow the stabilization of lanthanide complexes in all possible oxidation states (+2, +3, and +4).

For example, treatment of anhydrous cerium(III) trichloride with three equivalents of the lithium amidinate precursor Li[*p*-$MeOC_6H_4C(NSiMe_3)_2$] first afforded the bright yellow *p*-anisonitrile adduct of the cerium(III) amidinate (Scheme 31). Oxidation to the corresponding cerium(IV) amidinate [*p*-$MeOC_6H_4C(NSiMe_3)_2]_3CeCl$ was readily achieved using the reagent phenyliodine(III) dichloride (*vide supra*). This way the almost black cerium(IV) species could be isolated by simple crystallization directly from the concentrated reaction mixture. A single-crystal X-ray analyses clearly established the presence of the first amidinate complex of tetravalent cerium [51].

5. Organocerium(IV) Complexes

5.1. Cyclopentadienyl Complexes

For cerium the oxidation state +4 is remarkably stable because Ce^{4+} has the electronic configuration of xenon. Thus cerium should be by far the best candidate for making stable organolanthanide(IV) complexes. Historically, however, this was a notoriously difficult task, and it was a fundamental question persisting for decades in organolanthanide chemistry whether tetravalent organocerium complexes would exist at all. It is interesting to note that early reports on the synthesis of organometallic cerium(IV) complexes published in the 1970's were all found to be irreproducible or were shown to yield Ce^{3+} species. In particular, reactions of $(pyH)_2[CeCl_6]$ with cyclopentadienyl or indenyl reagents invariably led to reduction processes and formation of Ce^{3+} products. Even until today well-characterized organocerium(IV) complexes remain scarce. Early reports [59] about the synthesis of $CeCp_4$ and $Ce(C_9H_7)_4$ (C_9H_7 = indenyl) have later been refuted [60]. The same is true for many other compounds of the types Cp_3CeX and $(C_9H_7)_2CeX_2$ (X = H, alkyl, aryl, BH_4, Cl, NH_2, N_3, CN, NCO, NCS, NO_2, NO_3, SR, and O_2CR).

The first well-characterized cyclopentadienyl complexes of tetravalent cerium were the alkoxide derivatives $Cp_3Ce(OR)$ (R = ^{i}Pr, ^{t}Bu) and $Cp_2Ce(O^tBu)_2$. The first compound of this type, $Cp_3Ce(O^iPr)$, was obtained in minute quantities (4% yield) upon treatment of cerium(IV) tetra-isopropoxide with magnesocene according to Scheme 32. The new complex $Cp_3Ce(O^iPr)$ could be isolated by sublimation from the reaction mixture in the form of black microcrystals [61].

About ten years later, a more convenient route has been developed which involves treatment of the same alkoxide precursor with 3 equivalents of Me_3SnCp according to Scheme 33 [62]. This method uses an organotin compound which is a weaker reducing agent than the alkali or alkaline earth metal cyclopentadienides, and is thus less likely to result in reduction to Ce^{3+}. With 69% isolated yield the new synthetic route represented a major improvement over the previously reported synthesis [61]. The product was shown to be a genuine Ce^{4+} complex by NMR, elemental analysis, and cyclic voltammetry/controlled potential coulometry in $Bu_4N^+BF_4^-$/THF.

$$3\,Cp_2Mg + 2\,Ce(O^iPr)_4 \xrightarrow[20\,°C]{benzene} 2\,Cp_3Ce(O^iPr) + 3\,Mg(O^iPr)_2$$

Scheme 32.

$$Ce(O^iPr)_4 + 3\,R_3SnCp \xrightarrow[toluene]{65\,°C} Cp_3Ce(O^iPr) + 3\,R_3Sn(O^iPr)$$

R = CH_3, 69% yield
R = ^{n}Bu, 20% yield

Scheme 33.

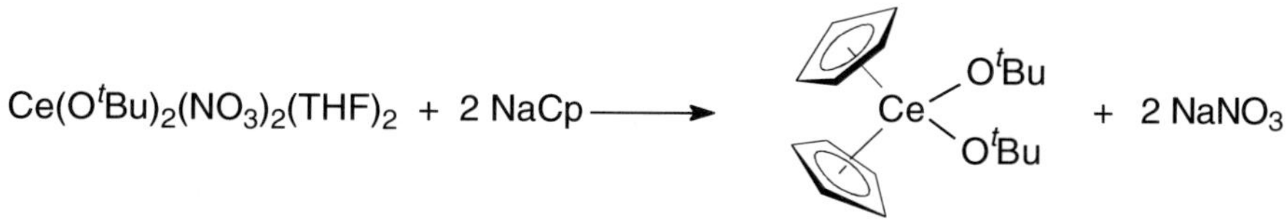

Scheme 34.

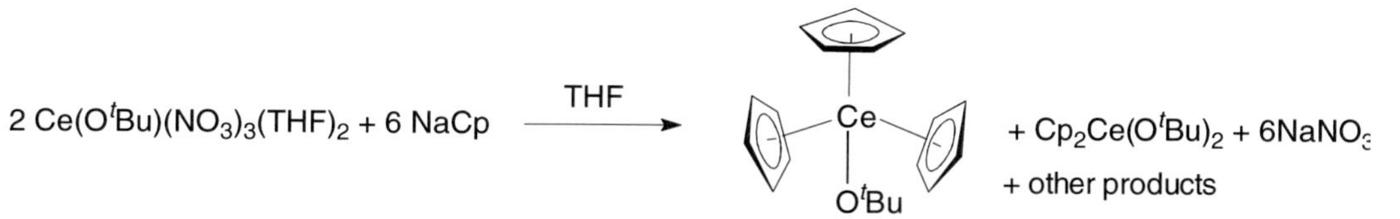

Scheme 35.

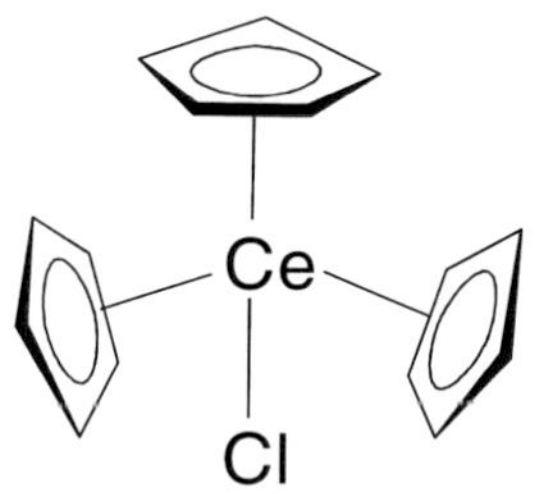

Scheme 36.

The latter data indicated that $Cp_3Ce(O^iPr)$ is a relatively strong oxidizing agent ($E_{1/2}$ = +0.32 *vs* SSCE) and that one-electron reduction is both electrochemically and chemically reversible. Electronic structure calculations at the nonrelativistic X_α-DVM level suggested non-negligible Ce-Cp f orbital covalency and yielded transition state ionization energies in good agreement with He I/He II photoelectron spectra [62]. An alternative synthetic route utilizes the readily available mixed-ligand alkoxide/nitrate complexes [3] of tetravalent cerium as starting materials. For example, the dark brown compound $Cp_2Ce(O^tBu)_2$ can be prepared in high yield (90%) by treatment of $Ce(O^tBu)_2(NO_3)_2(THF)_2$ with 2 equivalents of NaCp as shown in Scheme 34 [63].

It was also found that $Ce(O^tBu)(NO_3)_3(THF)_2$ reacts with 3 equivalents of NaCp to form $Cp_3Ce(O^tBu)$ (Scheme 35) admixed with $Cp_2Ce(O^tBu)_2$ from which it can be separated by crystallization and isolated in 50% yield as a black microcrystalline powder. The molecular structure of $Cp_3Ce(O^tBu)$ was unambiguously verified by X-ray diffraction [63].

Recently, a further example of the versatility of $PhICl_2$ as an oxidant was demonstrated by the synthesis and structural characterization of tris(cyclopentadienyl) cerium(IV) chloride (Scheme 36), the long sought-after cerium(IV) analogue of Cp_3UCl. A material thought to be Cp_3CeCl was first described in 1971 [59] but that report was later refuted [60]. Addition of 0.5 equivalents of $PhICl_2$ to a toluene solution of Cp_3Ce produced an immediate color change from yellow to black. Black crystals were isolated, after removing the volatiles, from a concentrated toluene solution at –35 °C in 15% yield. Diamagnetic Cp_3CeCl displays a singlet at δ 4.74 in the ^{1}H NMR spectrum in C_6D_6. X-ray diffraction analysis revealed these crystals to be Cp_3CeCl.

Thus Cp_3CeCl is not only the first new tris(cyclopentadienyl)cerium(IV) complex reported in more than 20 years but also the first chloro-funtional species of this type for which a promising derivative chemistry can be envisioned [51].

5.2. Carbene Complexes of Tetravalent Cerium

An interesting recent addition to organocerium(IV) chemistry is the successful synthesis and characterization of Ce^{4+} complexes containing a bidentate *N*-heterocyclic carbene alkoxide ligand, $[C\{(N^iPr)CHCHN\}CH_2CMe_2O]^-$. Treatment of $CeI_3(THF)_4$ with three equivalents of KL, $K[OCMe_2CH_2(1\text{-}C\{NCHCHN^iPr\})]$, afforded CeL_3 in good yield as a dark yellow microcrystalline powder (Scheme 37). Treatment of CeL_3 with $TeCl_4$, PBr_2Ph_3 or I_2 (reagents previously used for the oxidation of Ce^{3+} amides) did not yield the desired oxidation products. However, treatment of CeL_3 with benzoquinone afforded tetravalent CeL_4, arising from oxidation and ligand redistribution (Scheme 26). The oxidation could also be effected by XeF_2 and by $[Cp_2Fe][OTf]$, but with lower yields. Subsequent treatment of CeL_4 with 2 equivalents of the borane 9-BBN, 9-borabicyclo[3.3.1]nonane, afforded highly crystalline yellow $L_2Ce(L\text{-}9BBN)_2$ (Scheme 37) [64].

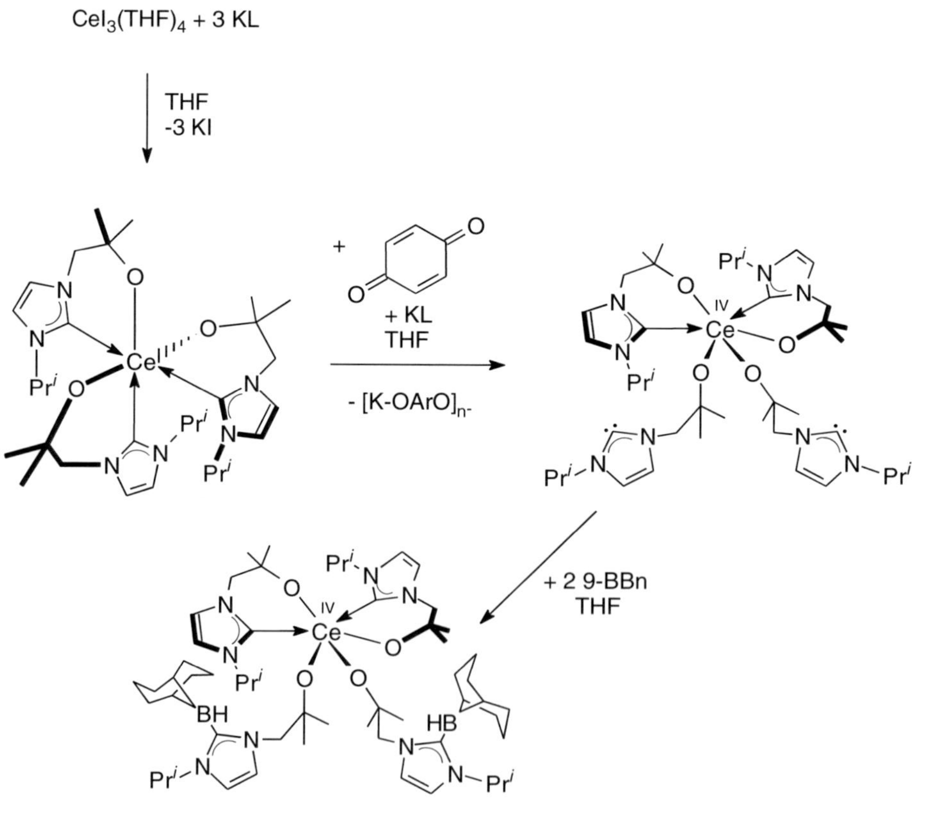

Scheme 37.

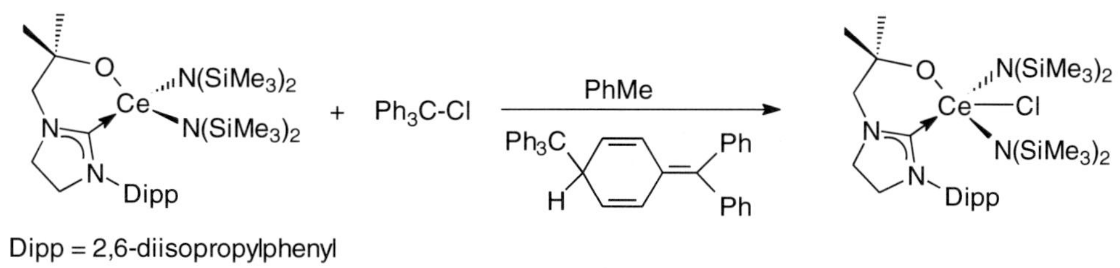

Scheme 38.

More recently it was found that trityl chloride is an excellent reagent to achieve the oxidative halogenation of cerium(III) carbene precursors. Scheme 38 illustrates a typical reaction. It was found that the yellow Ce(III) precursor could be oxidized cleanly by one equivalent of Ph_3CCl in toluene to give the red-orange cerium(IV) carbene complex. Attempts to achieve this oxidation using $TeCl_4$ failed in this case [52].

5.3. Cerocenes

Cerocene, $Ce(C_8H_8)_2$ (Scheme 39), is a particularly interesting metallocene that is formally a lanthanide(IV) analogue of uranocene, $U(C_8H_8)_2$, but combines a strong oxidizing agent, Ce^{4+}, and a strong reducing agent, cyclooctatetraene (COT) dianion, in the same molecule. The first successful synthesis of cerocene resulted surprisingly from the use of a reducing agent, triethylaluminum, in a reaction using cerium(IV) isopropoxide with neutral cyclooctatetraene as a reactant and solvent. Cerocene was isolated as a black-red, pyrophoric solid [61]. A more straightforward synthesis involves oxidation of the anionic cerium(III) sandwich complex $K[Ce(C_8H_8)_2]$ with AgI.

The longer known anionic derivative $K[Ce(C_8H_8)_2]$ can be regenerated by reduction of cerocene with potassium metal. Yet another efficiently modified synthesis for unsubstituted cerocene involves oxidation of $K[Ce(C_8H_8)_2]$ with allyl bromide [65].

In order to cope with the pyrophoric nature of the parent cerocene, synthetic routes to silyl-substituted cerocenes have been developed. These cerocene derivatives can be isolated as deep purple materials which are readily soluble even in non-polar organic solvents. Scheme 40 illustrates a typical reaction sequence leading to 1,1',4,4'-tetrakis(trimethylsilyl)cerocene [66].

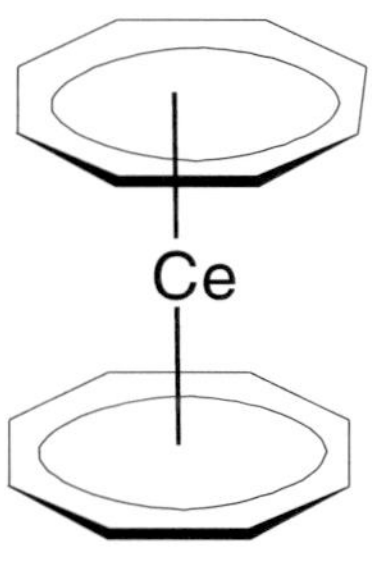

Scheme 39.

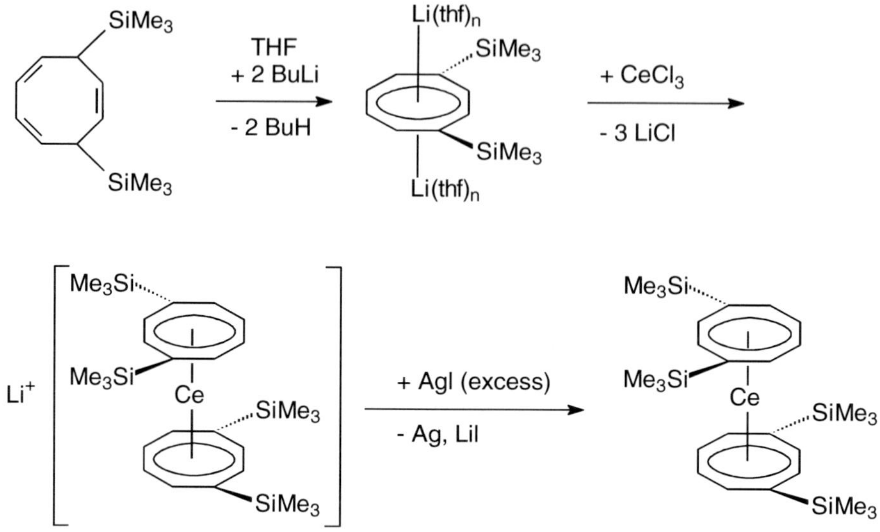

Scheme 40.

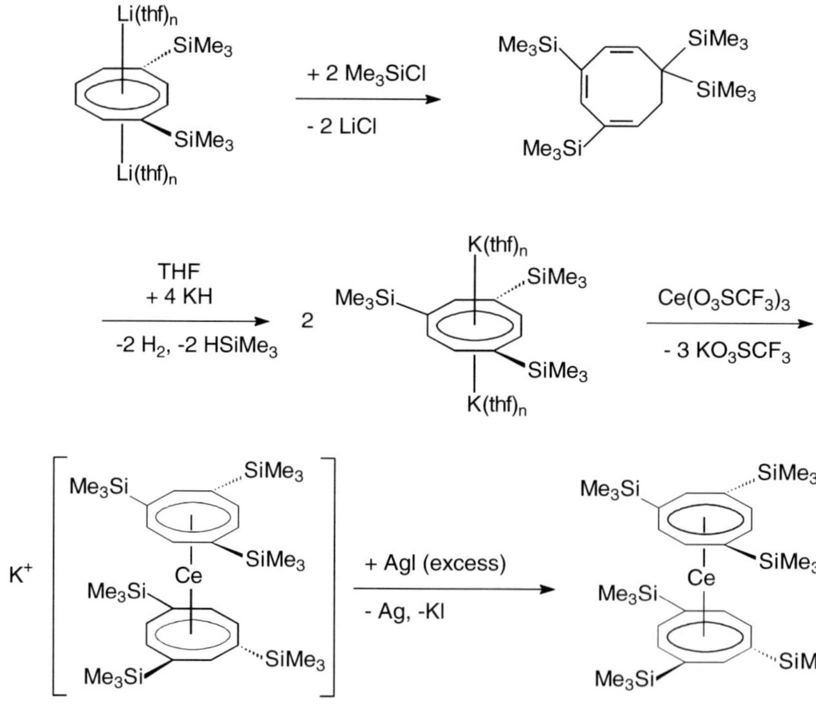

Scheme 41.

In this case, silver iodide was found to be the reagent of choice for achieving the oxidation of the anionic intermediate. 1,1',4,4'-Tetrakis(trimethylsilyl)cerocene was isolated as a dark purple oil which is stable in air for short periods of time.

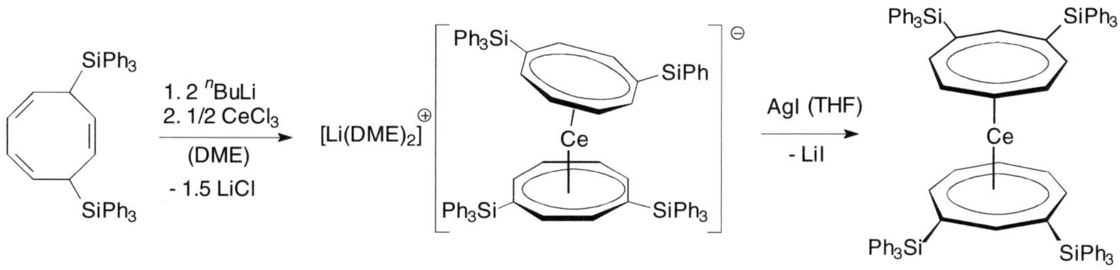

Scheme 42.

A crystalline poly-substituted cerocene derivative was synthesized as illustrated in Scheme 41 by employing the newly developed 1,3,6-tris(trimethylsilyl)cyclooctatetraenyl ligand. The resulting 1,1',3,3',6,6'-hexakis(trimethylsilyl)cerocene forms dark purple, low-melting crystals [66].

Subsequent work in this field has provided further characterization of cerocene and various ring-substituted analogues. Crystal structures of $Ce(C_8H_7Me)_2$ [67] and $Ce[C_8H_5(SiMe_3)_3]_2$ [66] showed that cerocenes have a uranocene-type sandwich structure with ring centroid-Ce-ring centroid angles of 176° in both cases. In all cases the cyclooctatetraenyl ligands are η^8-coordinated to cerium. There is theoretical and spectroscopic evidence (*e.g.* self-consistent field calculations) that the ground state in cerocene is almost entirely $4f^1$ and that this formally tetravalent cerium sandwich complex is better described as a cerium(III) compound such as $Ce^{3+}[(C_8H_8)_2]^{3-}$ [68,69].

Surprising structural features have recently been discoverd for related cerium sandwich complexes containing the superbulky 1,4-bis(triphenylsilyl)cyclooctatetraenide ligand (= COT^{BIG}) [70]. First the anionic cerium(III) sandwich complex $[Li(DME)_2][Ce(COT^{BIG})_2]$ was prepared by reaction of $CeCl_3$ with 2 equivalents of $Li_2(COT^{BIG})$ made *in situ* in DME (Scheme 42). A single-crystal X-ray analysis showed that in contrast to all previously reported $[Ln(COT)_2]^-$ or $[Ln(COT'')_2]^-$ derivatives the anion in $[Li(DME)_2][Ce(COT^{BIG})_2]$ is significantly bent (tilt angle between the eight membered rings: 15.1°). Knowing the bent structure of the anionic sandwich complex, it was tempting to see if the bending would increase upon oxidation. AgI oxidation of $[Li(DME)_2][Ce(COT^{BIG})_2]$ proceeded smoothly with formation of the neutral cerocene $Ce(COT^{BIG})_2$ (Scheme 31, dark purple needles, 63%).

X-ray diffraction analysis of $Ce(COT^{BIG})_2$ unexpectedly revealed the presence of strictly co-parallel rings, but the molecule now contains the 1,3-isomer of the COT^{BIG} ligand. Obviously steric pressure becomes too severe upon oxidation to the neutral cerocene that $SiPh_3$ group migration is induced for relief, resulting in a sterically more relaxed and highly symmetrical molecular structure of $Ce(COT^{BIG})_2$ as a result of this unprecedented rearrangement [70].

Conclusion and Future Outlook

Especially the synthesis and full characterization of organolanthanide(IV) complexes remain a very difficult and often unpredictable task. Thus far, organolanthanide chemistry in the oxidation state +4 remains entirely limited to cerium. This can be traced back to the very highly positive normal potentials of the other tetravalent lanthanide ions Pr^{4+}, Nd^{4+}, Tb^{4+}, and

Dy^{4+} (*e.g.* Pr: +2.86 V) which make them very strong oxidizing agents. Not even well-defined "pseudo-organometallics" (*i.e.* alkoxides, amides and related compounds) of tetravalent lanthanide ions other than Ce^{4+} have ever been isolated. Only alkoxides and amides of tetravalent cerium form fairly well-investigated classes of compounds. In contrast, true organocerium(IV) containing Ce-C bonds remain scarce. The oldest compounds of this type are cerocene and its derivatives, although the electronic structure of these species is far from being fully understood. The chemistry of cerium(IV) complexes containing *N*-heterocyclic carbene ligands is just emerging, and even cyclopentadienyl complexes of the type Cp_3CeX (X = Cl, OR) are limited to a handful of compounds. The recently reported chloro derivative Cp_3CeCl[12] could turn out to be the starting point for a more diverse derivative chemistry. The successful synthesis of "non-classical" organolanthanide(IV) compounds of Pr^{4+}, Nd^{4+}, Tb^{4+} or Dy^{4+} would certainly constitute a major break-through in this area. However, given the current knowledge, the design of suitable ligands for such highly oxidizing species will be very difficult if not impossible.

REFERENCES

[1] Cotton, S. *Lanthanide and Actinide Chemistry*, John Wiley and Sons, Ltd., Chichester, 2006.

[2] Deshpande, S.; Patil, S.; Kuchibatla, S.; Seal, S. *Appl. Phys. Lett.* 2005, *87*, 133113/1-133113/3.

[3] Abanades, S.; Flamant, G. *Solar Energy* 2006, *80*, 1611-1623.

[4] Bera, P.; Hegde, M. S. *Catal. Lett.* 2002, *79*, 75-81.

[5] Valenzuela, R. X.; Bueno, G.; Solbes, A.; Sapina, F.; Martinez, E.; Corberan, V. C. *Topics Catal.* 2001, *15*, 181-188.

[6] Kang, Z. C.; Wang, Z. L. *Adv. Mater.* 2003, *15*, 521-526.

[7] Yu, X. J.; Xie, P. B.; Su, Q. D. *Phys. Chem. Chem. Phys.* 2001, *3*, 5266-5269.

[8] Spanier, J. E.; Robinson, R. D.; Zheng, F.; Chan, S. W.; Herman, I. P. *Phys. Rev. B* 2001, *64*, 245407/1-245407/8.

[9] Adschiri, T.; Hakuta, Y.; Sue, K.; Arai, K. *J. Nanoparticle Res.* 2001, *3*, 227-235.

[10] Wang, Z. L.; Feng, X. *J. Phys. Chem. B* 2003, *107,* 13563-13566.

[11] Zhong, L.-S.; Hu, J.-S.; Cao, A.-M; Liu, Q.; Song, W.-G.; Wan, L.-J. *Chem. Mater.* 2007, *19*, 1648-1655.

[12] Si, R.; Zhang, Y.-W.; You, L.-P.; Yan, C.-H. *Angew. Chem.* 2005, *117*, 3320-3324.

[13] Yu, T.; Joo, J.; Park, Y. I.; Hyeon, T. *Angew. Chem.* 2005, *117*, 7577-7580.

[14] Sutradhar, N.; Sinhamahapatra, A.; Pahari, S.; Jayachandran, M.; Subramanian, B.; Bajaj, H. C.; Panda, A. B. *J. Phys. Chem. C* 2011, *115*, 7628-7637.

[15] Ivanov, V. K.; Polezhaeva, O. S.; Tret'yakov, Yu. D. *Russ. J. Gen. Chem.* 2010, *80*, 604-617.

[16] Silva, G. A. *Nature Nanotechnol.* 2006, *1*, 92-94.

[17] Ansari, A. A.; Kaushik, A. *J. Semiconductors* 2010, *31*, 033001/1-033001/4.

[18] Hay, N. E.; Kochi, J. K. *J. Inorg. Nucl. Chem.* 1968, *30*, 884-886.

[19] Guenin, M.; Mentzen, B.; Bressat, R.; Claudel, B. *J. Chim. Phys. Physico-Chim. Biol.* 1969, *66*, 595-597.

[20] Imamoto, T.; Koide, Y. Hiyama, S. *Chem. Lett.* 1990, *19*, 1445-1446.
[21] Casari, B. M.; Eriksson, A. K.; Langer, V. *Z. Naturforsch. B: Chem. Sci.* 2007, *62*, 771-777.
[22] Zinov'ev, A. A.; Shchirova, N. A. *Zh. Neorg. Khim.* 1960, *5*, 540-546.
[23] Hu, Z.; Kaindl, G.; Müller, B. G. *J. Alloys. Compd.* 1997, *246*, 177-185.
[24] Jacob, K.; Thiele, K.-H. *Z Anorg Allg Chem.* 1986, *536*, 147-152.
[25] Bradley, D. C.; Chatterjee, A. K.; Wardlaw, W. *J. Chem. Soc.* 1956, 2260-2264.
[26] Beineke, T. A.; Delgaudion, J. *Inorg. Chem.* 1968, *7*, 715-721.
[27] Evans, W. J.; Deeming, T. J.; Olofson, J. M.; Ziller, J. W. *Inorg. Chem.* 1989, *28*, 4027-4034.
[28] Xu, Y.; Feng, S.; Pang, W.; Pang, G. *Chem. Commun.* 1996, 1305-1306.
[29] Barnes, J. C.; Blyth, C. S.; Paton, J. D.; Smith, I. G. B. *Lanth. Actin. Res.* 1990, *3*, 181-193.
[30] Pook, N.-P.; Adam, A. *Cryst. Res. Technol.* 2008, *43*, 1197-120.
[31] Ciampolini, M.; Mani, F.; Nardi, N. *J. Chem. Soc., Dalton Trans.* 1977, 1325-1328.
[32] Delarosa, M. J.; Bousman, K. S.; Welch, J. T.; Toscano, P. J. *J. Coord. Chem.* 2002, *55*, 781-793.
[33] Becht, M.; Gerfin, T.; Dahmen, K. H. *Chem. Mater.* 1993, *5*, 137-144.
[34] Pan, M.; Meng, G. Y.; Xin, H. W.; Chen, C. S.; Peng, D. K.; Lin, Y. S. *Thin Solid Films* 1998, *324*, 89-93.
[35] Dröse, P.; Gottfriedsen, J. *Z. Anorg. Allg. Chem.* 2008, *634*, 87-90.
[36] Dröse, P.; Gottfriedsen, J.; Hrib, C. G.; Jones, P. G.; Edelmann, F. T. *Z. Anorg. Allg. Chem.* 2011, *637*, 369-373.
[37] Buchler, J. W.; De Cian, A.; Fischer, J.; Kihn-Botulinski, M.; Paulus, H.; Weiss, R. *J. Am. Chem. Soc.* 1986, *108*, 3652.
[38] Jiang, J.; Ng, D. K. P. *Acc. Chem. Res.* 2009, *42*, 79-88.
[39] Bian, Y.; Jiang, J.; Tao, Y.; Choi, M. T. M.; Li, R.; Ng, A. C. H.; Zhu, P.; Pan, N.; Sun, X.; Arnold, D. P.; Zhou, Z.-Y.; Li, H.-W.; Mak, T. C. W.; Ng, D. K. P. *J. Am. Chem. Soc.* 2003, *125*, 12257-12267.
[40] Shannon, R. D. *Acta Cryst.* 1976, *A32*, 751-767.
[41] Mehrotra, R. C.; Kapoor, P. N.; Batwara, J. M. *Coord. Chem. Rev.* 1980, *31*, 67-91.
[42] Evans, W. J.; Deming, T. J.; Olofson, J. M.; Ziller, J. W. *Inorg. Chem.* 1989, *28*, 4027-4034.
[43] Sen, A.; Stecher, H. A.; Rheingold, A. L. *Inorg. Chem.* 1992, *31*, 473-479.
[44] Suh, S.; Guan, J.; Mîinea, L. A.; Lehn, J.-S. M.; Hoffman, D. M. *Chem. Mater.* 2004, *16,* 1667-1673.
[45] Gradeff, P. S.; Yünlü, K.; Deming, T. J.; Olofson, J. M.; Doedens, R. J.; Evans, W. J.; *Inorg. Chem.* 1990, *29*, 420-424.
[46] Giessmann, S.; Blaurock, S.; Lorenz, V.; Edelmann, F.T. *Inorg. Chem.* 2007, *46*, 8100-8101.
[47] Gun'ko, Yu. K.; Reilly, R.; Edelmann, F. T.; Stalke, D. *Angew. Chem.* 2001, *113*, 1319-1321; *Angew. Chem. Int. Ed.*, 2001, *40*, 1279-1281
[48] Bradley, D. C.; Ghotra J. S.; Hart, F. A. *J. Chem. Soc., Dalton Trans.* 1973, 1021-1023.
[49] Eisenstein, O.; Hitchcock, P. B.; Hulkes, A. G.; Lappert, M. F.; Maron, L. *Chem. Commun.* 2001, 1560-1561.
[50] Hitchcock, P. B.; Hulkes, A. G.; Lappert, M. F. *Inorg. Chem.* 2004, *43*, 1031-1038.

[51] Dröse, P.; Crozier, A. R.; Lashkari, S.; Gottfriedsen, J.; Blaurock, S.; Hrib, C. G.; Maichle-Mössmer, C.; Schädle, C.; Anwander, R.; Edelmann, F. T. *J. Am. Chem. Soc.* 2010, *132*, 14046-14047.

[52] Arnold, P. L.; Turner, Z. R.; Kaltsoyannis, N.; Pelekanaki, P.; Bellabarba, R. M.; Tooze, R. P. *Chem. Eur. J.*, 2010, *16*, 9623–9629.

[53] Coles, M. P.; Hitchcock, P. B.; Khvostov, A. V.; Lappert, M. F.; Li, Z.; Protchenko, A. V. *Dalton Trans.* 2010, *39*, 6780-6788.

[54] Hitchcock, P. B.; Lappert, M. F.; Protchenko, A. V. *Chem. Commun.* 2006, 3546-3548.

[55] Hitchcock, P. B.; Hulkes, A. G.; Lappert, M. F.; Li, Z. *Dalton Trans.* 2004, 129-136.

[56] Morton, C.; Alcock, N. W.; Lees, M. R.; Munslow, I. J.; Sanders, C. J.; Scott, P. *J. Am. Chem. Soc.* 1999, *121*, 11255-11256.

[57] F. T. Edelmann, *Adv. Organomet. Chem.* 2008, *57*, 183-352.

[58] Edelmann, F. T. *Chem. Soc. Rev.* 2009, *38*, 2253-2268.

[59] Kalsotra, B. L.; Multani, R. K.; Jain, B. D. *Isr. J. Chem.* 1971, *9*, 569-572.

[60] Deacon, G. B.; Tuong, T. D.; Vince, D. G. *Polyhedron* 1983, *2*, 969-970.

[61] Greco, A.; S. Cesca, G. Bertolini, *J. Organomet. Chem.* 1976, *113*, 321-330.

[62] Gulino, A.; Casarin, M.; Ponticello, V. P.; Gaudiello, J. G.; Mauermann, H.; Fragalà, I.; Marks, T. J. *Organometallics* 1988, *7*, 2360-2364.

[63] Evans, W. J.; Deming, T. J.; Zlller, J. W. *Organometallics* 1989, *8*, 1581-1583.

[64] Casely, I. J.; Liddle, S. T.; Blake, A. J.; Wilson, C.; Arnold, P. L. *Chem. Commun.* 2007, 5037–5039.

[65] Streitwieser, A.; Kinsley, S. A.; Jenson, C. H.; Rigsbee, J. T. *Organometallics* 2004, *23*, 5169-5175.

[66] Kilimann, U.; Herbst-Irmer, R.; Stalke, D.; Edelmann, F. T. *Angew. Chem.*, 1994, *106*, 1685-1687; *Angew. Chem. Int. Ed. Engl.* 1994, *33*, 1618-1621.

[67] Boussie, T. R.; Eisenberg, D. C.; Rigsbee, J.; Streitwieser, A.; Zalkin, A. *Organometallics* 1991, *10*, 1922-1928.

[68] Dolg, M.; Fulde, P.; Stoll, H.; Preuss, H.; Chang, A.; Pitzer, R. M. *Chem. Phys.* 1995, *195*, 71-82.

[69] Edelstein, N. M.; Allen, P. G.; Bucher, J. J.; Shuh, D. K.; Sofield, C. D.; Kaltsoyannis, N.; Maunder, G. H.; Russo, M. R.; Sella, A. *J. Am. Chem. Soc.* 1996, *118,* 13115-13116.

[70] Lorenz, V.; Schmiege, B. M.; Hrib, C. G.; Ziller, J. W.; Edelmann, A.; S. Blaurock, Evans, W. J.; Edelmann, F. T. *J. Am. Chem. Soc.* 2011, *133*, 1257-1259.

In: Cerium
Editors: Aleksey Izyumov and Gulnaz Plaksin

ISBN: 978-1-62257-670-8

Chapter 4

COMPARATIVE TOXICITY OF CERIUM AND OF OTHER RARE EARTH ELEMENTS (REEs) IN PLANT AND INVERTEBRATE TEST SYSTEMS

Giovanni Pagano[1,*], Franca Tommasi[2] and Marco Guida[1]
[1]Federico II Naples University, Environmental Hygiene, Naples, Italy
[2]"Aldo Moro" Bari University, Department of Plant Biology, Bari, Italy

ABSTRACT

The widespread and growing use of REEs in a number of technological applications requires a comparative evaluation of REE-associated effects on environmental health, in view of optimizing the choice between the technological requirements for individual REEs and their expected effects on environmental and human health. A review of the published literature in the last decade points to an imbalance between the dataset accumulated for Ce and La, and a definite scarcity of information on the toxicity of many REEs. Moreover, only few comparative studies have focused on the relative toxicities of several REEs.

Along with a well-established literature in toxicity testing by means of plant and invertebrate bioassay systems, a limited number of independent studies have investigated the Ce- and REE-associated effects in plant and invertebrate bioassays. The reported effects were evaluated in a wide range of endpoints, including growth and viability, cytogenetic effects, developmental defects, and in either triggering or counteracting oxidative stress. Indications for a hormetic, non-monotonic concentration-related trend were provided by some recent studies, providing mechanistic hints for the coexistence of toxic and beneficial effects (e.g. on plant growth), and of antioxidant/prooxidant effects. The current information should prompt further investigations providing a reliable classification and ranking of REE-associated relative toxicities.

Along with a well-established literature in toxicity testing by means of plant and invertebrate bioassay systems, a limited number of independent studies have investigated the Ce- and REE-associated effects in plant and invertebrate bioassays. The reported

* Giovanni Pagano, Federico II University, Department of Biological Sciences, Section of Hygiene, Via Mezzocannone 16, I-80134 Naples, Italy; E-mail: gbpagano@tin.it; Tel.: 39-335-7907261.

effects were evaluated in a wide range of endpoints, including growth and viability, cytogenetic effects, developmental defects, and in either triggering or counteracting oxidative stress. Indications for a hormetic, non-monotonic concentration-related trend were provided by some recent studies, providing mechanistic hints for the coexistence of toxic and beneficial effects (e.g. on plant growth), and of antioxidant/prooxidant effects. The current information should prompt further investigations providing a reliable classification and ranking of REE-associated relative toxicities.

INTRODUCTION

The utilization of REEs in a broad array of industrial processes has been established in the recent decades, making REEs indispensible ingredients in a number of technological applications, e.g. in automobile, electronic, glass and ceramics, and nuclear industries (Dickson 2006; Cordier 2012). Thus, both human and environmental health concern is raised about REE utilization, mining, and manufacturing; moreover, further health concern may arise from REE mixtures used as fertilizers in China (Xu et al. 2002; Chen et al. 2004; Gong et al. 2011; Xu and Chen 2011), and from a number of medical uses of some REEs as, e.g., in orthopedic or in radiodiagnostic applications (Schmidt et al. 2007; Feyerabend et al. 2010; Zhou et al. 2011). In spite of the pervasive and growing use of REEs in an extensive number of technologies, the state-of-the-art of REE-associated toxicity is quite scarce and far from the knowledge wealth accumulated for other environmental/occupational agents as, e.g., heavy metals.

REE-Associated Toxicity or Safety: State-of-the-Art

When comparing the published toxicity reports on individual REEs, outstanding differences appear in the numbers of publications on some more investigated REEs, such as Ce and La vs. other REEs that are almost disregarded in terms of toxicity reports. As shown in Table I, in the last decade (retrieved by MedLine, 2000 to April 2012) only Ce and La were investigated for toxicity endpoints in >50 reports (56 and 51, respectively). Fewer publications focused on Gd (35 reports) and Nd (10). As for the other REEs, the scanty available publications range from seven (Tb and Y) to one or zero (Pr, Pm, Sa, Er, Tm and Lu).

An evaluation of the published literature on REE-associated effects (toxicity, safety or beneficial outcomes) shows striking information gaps both regarding laboratory studies and, even more so, information from human health studies. With the unique exception of Gd and La (reviewed by Tonolini and Bianco 2011; Navaneethan et al. 2011) that are utilized in medical applications, one is faced with a questionable void of information about human health effects of the other REEs.

Parallel to the above mentioned imbalance of published toxicity reports, the data arising from REE toxicity testing in mammals, fish and invertebrates, plants and plant cells, cell cultures, and microrganisms (Table I), all concur to display some – however limited – information on Ce, La and Gd, whereas only scanty reports are devoted to the other REEs, down to one or zero reports for six REEs (Pr, Pm, Sa, Er, Tm and Lu). The current lack of published information might raise the question about the impact, if any, of these elements in environmental and human health, faced with the recognized utilization of these REEs in a number of technologies.

Health Effects of Cerium and Other REEs in Alternative Toxicity Testing

A line of studies has focused on the toxicity, if any, of Ce and La by utilizing alternative bioassay models, i.e. fish and invertebrates, plants, cell cultures, and microrganisms. The published information on Ce-associated toxicity is summarized in Table II, including studies that utilized fish and invertebrates, namely echinoids, crustaceans, insects and nematodes.

Studies from fish bioassays *(Carassius gibelio, Cyprius carpio,* and *Danio rerio)* reported on acute toxicity of CeO_2 nanoparticles (nano-CeO_2) in a micromolar range (Van Hoecke et al. 2009; Gaiser et al. 2012). A study by Ling and Hong (2010) reported on a protective effect of $CeCl_3$ i.p. injection in *C. gibelio* against Pb-induced oxidative stress in the fish liver, by counteracting the accumulation of lipid peroxide, reactive oxygen species (ROS), and by diminishing Pb-induced inhibition of superoxide dismutase (SOD), catalase (CAT), ascorbate peroxidase (AscPx), glutathione peroxidase (GPx), and of glutathione and ascorbic acid levels.

Other studies reported on various adverse effects in echinoids, crustaceans, insects and nematodes. Two studies reported on Ce-associated toxicity in the sea urchin *Paracentrotus lividus*. Oral et al. (2010) found concentration-related developmental defects in *P. lividus* embryos reared in $Ce(SO_4)_2$, up to 100% embryotoxicity at 10^{-5} M Ce(IV). Moreover, cytogenetic anomalies were observed in Ce(IV)-exposed embryos, according to biphasic (hormetic) trends. Following sperm exposure to Ce(IV), only minor loss of fertilizing capacity was obeserved, yet the offspring displayed a significant concentration-related increase in developmental defects, thus suggesting Ce(IV)-induced transmissible damage (Oral et al. 2010). Concurrent results on the effects of nano-CeO_2 in the offspring of exposed sperm were obtained in an independent study (Prof. Carla Falugi, personal communication). A study of nano-CeO_2-associated effects in *P. lividus* embryos reported a dysfunction of innate immunity in a set of coelomocyte stress markers, including two heat shock proteins and two cholinesterases (Falugi et al. 2011).

Huang et al. (2010) investigated the effects of $Ce(SO_4)_2$ in the fruit fly *Drosophila melanogaster,* in a micro- to millimolar range, and found decreased lifespan and reproductive output that were associated with oxidative stress markers as increased lipid peroxidation (MDA content), and decreased expression of SOD and CAT activities.

A study by Lee et al. (2009) reported on the geno- and ecotoxicity of CeO_2, SiO_2 and TiO_2 nanoparticles on *Daphnia magna* and *Chironomus riparius*, and found that nano-CeO_2 induced DNA damage (evaluated by Comet assay) and caused mortality in a micromolar range. Accordingly, other studies of nano-CeO_2 in the micromolar range found growth inhibition in *D. magna* (Van Hoecke et al. 2009) as well as an increase of mortality (Gaiser et al. 2012).

Table I. Toxicity reports on individual REE, published 2000 to April 2012

Element	Atomic number	No. reports (2000 – Apr 2012)	Toxicity data					
			Humans	Mammals	Fish and invertebrates	Plants and plant cells	Cell lines and cell-free	Microrganisms
Scandium	21	2	1	0	0	1	0	0
Yttrium	39	7	0	2	1	0	3	1
Lanthanum	57	51	3	14	5	13	10	6
Cerium	58	56	0	15	9	7	17	8
Praseodymium	59	1	0	0	0	0	1	0
Neodymium	60	10	0	5	0	0	4	1
Promethium	61	0	0	0	0	0	0	0
Samarium	62	1	0	0	1	0	0	0
Europium	63	2	0	1	0	0	1	0
Gadolinium	64	35	10	9	3	0	12	1
Terbium	65	7	0	0	0	6	1	0
Dysprosium	66	5	0	0	0	0	3	2
Holmium	67	3	0	1	1	1	0	0
Erbium	68	1	0	0	1	0	0	0
Thulium	69	1	0	0	1	0	0	0
Ytterbium	70	3	0	1	2	0	0	0
Lutetium	71	1	0	0	0	0	0	1

Source: MedLine.

Studies of REE radionuclides (e.g. used in radiotherapy or diagnostics) were omitted in this table. Also omitted were review papers.

Table II. Cerium-associated toxic (or protective) effects reported in fish and invertebrate test systems

Test systems	Agents	Toxicity endpoints	References
Silver crucian carp *(Carassius gibelio)*	Ce(III) (1.5 mg/kg, i.p.)	↓ ROS accumulation; relieved $PbCl_2$-induced inhibition of antioxidant enzyme activities and decreased antioxidant levels in fish liver	Ling and Hong 2010
Sea urchin *(Paracentrotus lividus)*	$Ce(SO_4)_2$ [10^{-7} to 10^{-5} M]	↑ embryonic mortality; mitotoxic effect and induction of mitotic aberrations; hormetic trends	Oral et al. 2010
Sea urchin *(Paracentrotus lividus)*	nano-CeO_2 (50 and 105 nm) [$7x10^{-7}$ to $7x10^{-5}$ M]	inhibition of coelomocyte innate immunity; down-regulation of heat shock proteins, and cholinesterases	Falugi et al. 2011
Drosophila melanogaster	$Ce(SO_4)_2$ [$3x10^{-6}$ to $3x10^{-3}$ M]	↓ lifespan; ↓ reproductive output; ↑ MDA content; ↓ SOD and CAT activities	Huang et al. 2010
Daphnia magna and *Chironomus riparius* larvae	nano-CeO_2 (15 and 30 nm) [$7x10^{-6}$ M]	↑ mortality; ↑ DNA damage (Comet assay)	Lee et al. 2009
Daphnia magna	nano-CeO_2 [up to $7x10^{-5}$ M]	↑ mortality in micromolar range	Gaiser et al. 2012
Daphnia magna, Thamnocephalus platyurus, and embryos of *Danio rerio*	nano-CeO_2 (14 to 29 nm) [1.4 to $7x10^{-3}$ M]	growth inhibition in micromolar range	Van Hoecke et al. 2009
Caenorhabditis elegans	nano-CeO_2 [1-100 nM]	↓ lifespan; ROS accumulation; oxidative damage	Zhang et al. 2011
Caenorhabditis elegans	nano-CeO_2 (15 and 45 nm)	↑ expression of the cyp35a2 gene; ↓ fertility and survival	Roh et al. 2010

Table IIIa. Cerium-associated protective effects reported in plants

Test systems	Agents	Toxicity endpoints	References
Arabidopsis thaliana	Ce or La nitrates 0.5-50 μM	↑ Floral initiation and reproductive growth	He and Lo 2000
Arabidopsis thaliana	Ce(III) 20 μM; La(III); Nd(III) 500 μM	↑ Photochemical reaction and O_2 evolution and improved photosynthesis	Xiaoqing et al. 2009
Arabidopsis thaliana Tobacco	Ce(III) 0.1-1 μM	Improved transgene integration	Boyko et al. 2011
Armoracia rusticana	^{141}Ce(III)	Macronutrient metabolism disturbances	Guo et al. 2007
Crocus sativus	Ce(III) 40 μM; La(III) 60 μM	↑ Crocin production; ↑ Callus growth	Chen et al. 2004
Cucumber and corn	nano-CeO_2 0-30 mM	↑ Root growth	López Moreno et al. 2010a,b
Oryza sativa	Ce nitrates 0-20 mM	↑Seed germination	Fashui 2002
Oryza sativa	Ce(III)	Mitochondrial swelling	Dai et al. 2011
Peach plantlet in vitro	Ce(III)	↑ Nitrogen metabolism	Weiping et al. 2003
Soybean	Ce(III) 15 μM	Alleviated UV stress	Liang et al. 2011
Spinacia oleracea	Ce(III) 5 μg/mL	↑ Photosynthesis and chlorophyll content; Ca depletion on the conversion efficiency of light under Ca deficiency condition;	Fashui et al. 2002
Spinacia oleracea	Ce(III) 15 mM	Ca substitute in chloroplasts	Huang et al. 2008
Spinacia oleracea	Ce(III) 15 μM	↑ Growth; calcium deficiency	Chao et al. 2008
Spinacia oleracea	Ce(III) 15 μM	↑ Photosynthesis in Ca-deficient chloroplasts	Hao et al. 2008
Spinacia oleracea	Ce(III)15 μM	Mg deficiency improvement;	Ze et al. 2009
Spinacia oleracea	Ce(III) 15 μM	↑ Enzymes of CO_2 assimilation	Yuguan et al. 2009
Zea Mays	Ce(III) 15 μM;	Mg deficiency improvement	Gong et al. 2011
Zizyphus jujuba	Ce(IV)	Preserving fruits from degradation; ↓ residues of organophosphorus pesticides	Wu et al. 2010

Table IIIb. Cerium-associated toxic effects reported in plants

Test systems	Agents	Toxicity endpoints	References
Wheat	La(III) and Ce(III) ($3.5\text{-}17.5\times10^{-6}$ M)	↓ Growth and mineral nutrition	Hu et al. 2002
Triticum durum	REE mixtures La(III)	↓ seed germination; ↓ mitotic index in root tips	d'Aquino et al. 2009
Daucus carota	Y, Ce, Pm, and Gd	Ca competition on plasma membrane transporters	Ozaki et al. 2005
Zea mays; Medicago sativa	nano-CeO_2 ($0\text{-}3\times10^{-3}$ M)	↓ seed germination; ↓ root growth	López Moreno et al. 2010a
Zea mays	La, Ce>0.2×10^{-3} M	↓ growth; ↓ shoot elongation	Diatloff et al. 2008
Mungbean *(Vigna radiata)*			
Rice and *Arabidopsis*	Single-wall carbon nanotubes	Oxidative stress; apoptosis	Shen et al. 2010
Rice	Ce(IV) (2×10^{-5} M)	↑ROS production; ↓antioxidant activity	Xu and Chen 2011
Soybean *(Glycine max)*	nano-CeO_2 ($1.5\text{-}3\times10^{-3}$ M)	Genotoxic effects (RAPD assay)	López Moreno et al. 2010b
Taxus cuspidata	Ce(IV) (10^{-3} M)	↑ROS production; apoptosis	Yang et al. 2009
Taxus tricuspidata cells	Ce (IV) (10^{-3} M)	Apoptosis	Ge et al. 2006
Lemna minor	REE mixture ($5\text{-}10\times10^{-3}$ M)	Toxicity; ↑ antioxidants; ↓ chilling stress response	Ippolito et al. 2010
Lycopersicon esculentum	REE mixture	chlorosis; ↑ antioxidant levels; ↓↑ concentration-related lipid peroxidation	Ippolito et al. 2011
	La(III) $5\text{-}10\times10^{-3}$ M	↑H_2O_2 production; ↓ drought stress response	

Table IV. Cerium-associated action mechanisms in cell-free systems and in vitro toxicity

Test systems	Agents	Toxicity endpoints	References
L-dopa autoxidation	nano-CeO_2 [$1.4x10^{-3}$ M]	↑ ROS production	Luna-Velasco et al. 2011
Ferricytochrome C assay	nano-CeO_2 [$4.5x10^{-3}$ M]	SOD mimetic antioxidant	Heckert et al. 2008
Saccharomyces cerevisiae	nano-CeO_2 [up to $7x10^{-3}$ M]	no toxicity to yeast without dispersant; moderate toxicity (47%) with polyacrylate dispersant only at millimolar concentration	García-Saucedo et al. 2011
Anabaena CPB4337 and *Pseudokirchneriella subcapitata*	nano-CeO_2 [$7x10^{-6}$ to $2x10^{-4}$ M]	cytotoxicity	Rodea-Palomares et al. 2011
Lung epithelial cells (BEAS-2B)	nano-CeO_2 (15 to 45 nm) [$4x10^{-5}$ to $3x10^{-4}$ M]	activation of caspase-3 and apoptotic cytotoxicity; ↑ ROS production; ↓ GSH; induction of heme oxygenase-1, catalase, GSH S-transferase, and thioredoxin reductase	Park et al. 2008
Human lung cancer cells	nano-CeO_2 (20 nm) [$2.5x10^{-5}$ to $1.7x10^{-4}$ M]	↑ ROS production; ↓ GSH and α-tocopherol ↑ MDA; ↑ lactate dehydrogenase	Lin et al. 2006
J774A.1 mouse macrophages	nano-CeO_2 (5 to 10 nm) [10^{-5} M]	↓ oxidative stress and ↓ iNOS protein expression	Hirst et al. 2009
Insulinoma betaTC-tet cells	nano-CeO_2 (20 nm)	free radical scavengers	Tsai et al. 2007
Nerve cells HT22	nano-CeO_2 (6 to 12 nm) [$1.4x10^{-4}$ M]	neuroprotection independent of particle size	Schubert at al. 2006

Table V. Studies reporting on comparative toxic (or protective) effects following exposures to ≥2 REE

Test systems	Agents	Toxicity endpoints and ranking	References
HeLa cells and *Caenorhabditis elegans*	nano-Y(III), Yb(III) and -Tm(III)	no toxicity in protein expression, life span, egg production, egg viability, and growth rate up to 7×10^{-4} M	Zhou et al. 2011
ICR mice (i.p. injection)	La (III), Ce (III) and Nd(III)	Ce(III) and Nd(III) exhibited higher oxidative stress and toxicity on brain than La(III), and Ce(III) caused more severe brain injuries and oxidative stress than Nd(III)	Zhao et al. 2011
ICR mice (gavage)	La (III), Ce (III) and Nd(III)	oxidative damage in hepatic nuclei and mitochondria; ↓ SOD, CAT, GPx and GSH; ↑ MDA,	Huang et al. 2011
ICR mice (i.p. injection)	La(III), Ce(III) and Nd(III)	liver histopathological changes and liver malfunctions; changes in mRNA and in expression of inflammatory cytokines (NF-κB, MIF, TNF-α, IL-1β, IL-6, IL-4, and IL-10)	Fei et al. 2011
ICR mice (intranasal instillation)	La(III), Ce(III) and Nd(III)	⌄ SOD, CAT, AscPx, TAC, GSH and Asc, which were greatest in Ce(III) treatment, medium in Nd(III), and least in La(III)	Li et al. 2010
Sea urchin embryos *(Paracentrotus lividus)*	Ce(IV) and La(III)	^ Developmental defects [Ce(IV) and La(III)] and embryotoxicity [Ce(IV) only]; ↑ mitotoxicity and induction of mitotic aberrations [Ce(IV) only]	Oral et al. 2010

Ce- and REE-Associated Effects in Plants and Plant Cells

The subject of Ce- and REE-associated effects in plants appears complex and controversial. REE-enriched fertilizers have been used in China since the 1980s for soil and leaf dressing of crops and food integrators to improve animal growth (He et al. 2001). REE-enriched fertilizers are reported to be REE mixtures, mainly as nitrates, obtained by extracting REEs from their ores using nitric acid (Xu et al. 2002). A body of literature points to positive effects of REE mixtures, mainly containing Ce and La, on different plant species, as shown in Table IIIa.

Positive effects on crop production following treatments with REEs are largely reported in the Chinese literature (Hu et al. 2004), and a number of physiological responses have also been reported in different plant species (Fashui et al. 2000; Chen et al. 2001; Hu et al. 2002a,b; 2004). REEs were found to enhance seed germination, root growth, chlorophyll content, and plant resistance (Hu et al. 2004). It has also been reported that La(III) promotes plant tolerance to environmental stresses by increasing antioxidant systems (Zhang et al. 2003; Arora et al. 2002). However, the results of field trials and laboratory studies are controversial and the results depend on plant species, physiological stage and type of treatments (Diatloff et al. 1995a,b,c; He and Loh, 2000; von Tucher and Schmidhalter, 2005). Less information is available about cerium effects. Table IIIa summarizes the published data about cerium-associated protective effects in some species. An increase of root growth in cucumber and corn was reported (López Moreno et al. 2010a,b); a stimulation of reproductive growth and floral initiation occurred in *Arabidopsis thaliana* (He and Lo 2000; 2002). A report by Chen et al. (2004) found that both Ce(III) and La(III) exert stimulatory effects on *Crocus sativus* by either enhancing crocin production [Ce(III)] or growth [La(III)]. Other reports found that cerium treatments alleviate magnesium deficiency (Gong et al. 2011; Zhao et al. 2012; Ze et al. 2009), calcium deficiency (Chao et al. 2008; Hao et al 2008), and UV stress (Liang et al. 2011). Consistently, Fashui (2002) reported that Ce(III) treatment of aged *Oryza sativa* seeds contrasts seed aging by promoting seed germination and antioxidant enzyme activities as SOD and GPx. In addition, some data indicated that cerium enhanced photosynthesis either in plants or in isolated chloroplasts and carbon assimilation (Fashui et al. 2002; Yuguan et al. 2009; Xiaoqing et al. 2009), mitochondrial metabolism (Dai et al 2011), and nitrogen metabolism (Weiping et al. 2003). As reported in Table IIIa, cerium treatments seem to promote also transgene integration in some species (Boyko et al. 2011) and preserve fruit degradation due to pesticide accumulation (Wu et al. 2010).

On the other hand, some data indicate Ce-associated negative effects and toxicity.The published information on Ce-associated plant toxicity is summarized in Table IIIb. The application of nanoceria causes decrease of seed germination in maize and root decrease in alfalfa and tomato (López Moreno et al. 2010a). In mungbean Ce at concentrations $> 0.2 \times 10^{-3}$ M induced decrease of growth (Diatloff et al. 2008). In addition, single-wall carbon nanotubes induced oxidative stress and programmed cell death in rice and *Arabidopsis* (Shen et al. 2010). Genotoxic effects were reported in soybean (López Moreno et al. 2010b); increase in ROS production in rice, as well as inhibition of some antioxidants in rice (Xu and Chen 2011), and impairment in macronutrient metabolism in horseradish (Guo et al. 2007) were reported. In *Taxus tricuspidata* (Yang et al. 2009) and *T. tricuspidata* cells (Ge et al. 2006), increased ROS production and apoptosis were induced by Ce(IV), along with decrease of seed germination, and alteration of root apex mitotic index in *Triticum durum* (d'Aquino et al. 2009). Toxicity and worsening of stress responses were induced in *Lemna minor* (Ippolito et al. 2009) and tomato (Ippolito et al. 2011) following treatment with REE mixture. Ozaki et al. (2005) reported on a competition of Y, Ce, Pm and Gd with Ca in plasma membrane transport. Few data are available about the REE effects in non seeds plants as fern and mosses, although in a fern, *Dicranopteris dichotoma,* a REE binding-peptide was reported (Wang et al. 2003).

In Vitro Studies of Ce-associated Effects

As shown in Table IV few, quite conflicting studies reported on the action mechanisms and effects of cerium in yeast, unicellular algae, green algae, cell cultures, and cell-free systems. These latter showed that L-dopa autoxidation and, respectively, ferricytochrome C assay either resulted in prooxidant or antioxidant effects (Luna-Velasco et al. 2011; Heckert et al. 2008). Yeast cultures, *Saccharomyces cerevisiae,* exposed to nano-CeO_2, failed to exhibit any toxicity without dispersant, while displayed moderate toxicity (47%) with polyacrylate dispersant (Dispex A40) only at millimolar concentration (7×10^{-3} M) (García-Saucedo et al. 2011).

Nano-CeO_2 was tested in four aquatic biota, finding significant chronic toxicity to *Pseudokirchneriella subcapitata* that was observed with 10% effective concentrations (EC10s) between 2.6 and 5.4 mg/L (van Hoecke et al. 2009). Rodea-Palomares et al. (2011) tested nano-CeO_2 in the blue-green alga *Anabaena* and in the unicellular green alga *P. subcapitata,* finding Ce-induced cytotoxicity in the micromolar range. Another set of reports investigated Ce-associated effects in mammalian cell cultures, either finding lack of toxicity and antioxidant effects (Hirst et al. 2009; Tsai et al. 2007; Schubert at al. 2006), or prooxidant effects and cytotoxicity (Park et al. 2008; Lin et al. 2006).

Comparative Studies of Multiple REE-associated Effects

As noted above (Table I), the published literature on REE toxicity is mostly confined to Ce, La, and Gd, while only scanty published information is available for the other REEs. That statement may be reinforced if one seeks any published papers comparing the effects of different REEs, by utilizing all bioassay information from both animal and in vitro studies. As shown in Table V, only few papers were retrieved reporting on the effects of ≥2 REEs, and even less reports provided comparative toxicity information only on 2 or 3 REEs.

Some of these reports failed to find REE-induced toxicity outcomes (Zhou et al. 2011), or found protective antioxidant effects (Schubert et al. 2006).

A number of reports pointed to Ce as displaying highest toxicity compared to the other tested REEs as, e.g., La (Zhao et al. 2011; Oral et al. 2010; Li et al. 2010; Feyerabend et al. 2010; Nakamura et al. 1997; Palmer et al. 1987). Other studies suggested that both Ce and La displayed highest toxicity compared to other REEs (Huang et al. 2011; Ye et al. 2008), or failed to provide a comparative toxicity ranking among the tested REEs (Fei et al. 2011). Thus the currently published information might allow the statement that Ce should be regarded as the most toxic REE (Table V). Nevertheless, a warning should be raised about the substantial void of information on the toxicities of other as yet inexplored REEs that are however relevant in terms of industrial applications and, hence, with possible concern in terms of environmental and human health.

Three papers reporting more extensive investigations on REE-associated toxicities should be regarded as positive examples in view of the required demand of comparative information in ranking REE toxicities (Feyerabend et al. 2010; Blaise et al. 2008; Nakamura et al. 1997).

Feyerabend et al. (2010) investigated the relative toxicities of eight REEs (Ce, La, Y, Nd, Dy, Pr, Gd, and Eu) in three cell lines (human osteosarcoma cell line MG63, human umbilical cord perivascular (HUCPV) cells and mouse macrophages (RAW264.7), by assaying the induction of apoptosis and expression of TNF-α and IL-1α. The study found that La and Ce resulted in the highest cytotoxicity among the tested elements, followed by Pr and Nd. One could assume that the elegant paper by Feyerabend and co-workers should be followed in the wake of the definite demand of information on the comparative health effects of REEs.

Another relevant report in this context, the study by Blaise et al. (2008) evaluated the toxicity of nine metallic nano-powders (NPs, including four REEs, i.e. Y, Sa, Er, and Ho) and of two organic nano- powders. Bioassays were conducted across several taxonomic groups, i.e. bacteria, micro-algae, and primary and secondary consumers (micro-invertebrates and fish). When mixed with a certified reference material (CRM) sediment, two solid phase assays and an elutriate assay, five NPs (including Sa, Er and Ho oxides) increased both CRM sediment toxicity and its elutriate toxicity.

An early, and as yet valid report by Nakamura et al. (1997) investigated the in vivo toxicity (by i.v. injection in rats) of seven REEs, i.e. Ce, Y, Pr, Eu, Dy, Yb, and Lu, within a realistic dosage range (approx. 60 and 120 μmol/kg). The results showed liver, spleen, and lung accumulation of the tested REEs. Y, Eu, Dy, and Yb in liver were highest at 8 hr to 2 days, then decreased gradually; hepatic Pr levels remained high; Ce and Pr induced severe hepatotoxicity.

REEs: Another Case for Hormesis and for the Antioxidant/Prooxidant Dilemma

As mentioned above (Tables II-V), a number of independent reports on Ce and other REEs have focused on the roles for redox mechanisms in supporting results that either showed oxidative stress-related adverse effects (Fei et al. 2011; Huang et al. 2010; Huang et al. 2011; Li et al. 2010; Lin et al. 2006; Luna-Velasco et al. 2011; Park et al. 2008; Zhang et al. 2011; Zhao et al. 2011) or, at the opposite, the data showed antioxidant mechanisms deemed to be related to the observed protective and detoxifying outcomes (Heckert et al. 2008; Hirst et al. 2009; Ling and Hong 2010; Schubert at al. 2006; Tsai et al. 2007).

The phenomenon through which the same effector (either a chemical or a physical agent) displays non-monotonic, or biphasic concentration- or dose-related trends, with observed shifts from a positive stimulation to inhibition, has been long recognized and is termed hormesis (Calabrese 2010; Maynard 2011). Thus, the apparent dilemmas such as protective vs. toxic, or antioxidant vs. prooxidant mechanisms for REE-associated action might be another case of the broadly recognized hormesis phenomenon. An antioxidant/prooxidant concentration-related shift has been reported for a number of agents as, e.g., ascorbic acid

(Chen et al. 2008; Shenbagam and Nalini 2011). In two independent studies of REE toxicity, we had observed non-monotonic concentration-related trends in Ce-induced effects, such as the induction of mitotic aberrations and in developmental defects in sea urchin embryos following sperm exposure (Oral et al. 2010); two other studies investigated the effects of a mix of different REEs and of lanthanum nitrate on catalase and antioxidant systems involved in the ascorbate–glutathione cycle were investigated in the common duckweed, *Lemna minor* and in tomato, *Lycopersicon esculentum* (Ippolito et al. 2009; 2011). A remarkable increase in glutathione content as well as in enzymatic antioxidants was observed before the appearance of the stress symptoms in treated plants; this finding suggested a biphasic effect, from stimulation of defence systems to growth inhibition (Ippolito et al. 2009; 2011). An early report by Wang et al. (2000) showed dual effects of REEs in *Tetrahymena shanghaiensis*, and a recent paper by Wang et al. (2012) reported on the non-monotonic response induced by La in *Vicia faba* seedlings submitted to Cd-induced oxidative damage. The results showed that the supplementation with low La concentrations (<120 μM) decreased metal uptake (Cd, Ca, Cu, Zn, Mn and Fe) and increased antioxidant activities (SOD, CAT, guaiacol peroxidase and ascorbate peroxidase), and heat shock protein 70 (HSP 70) production in the roots. However, the supplementation with higher La levels (>120 μM) showed adverse effects, i.e. increased Cd contents in the roots, as well as decreased antioxidant activities and HSP 70, and increased carbonylated proteins. Thus, La supplementation contributed to biphasic effects, i.e. stimulated antioxidation at lower concentrations and pro-oxidation at higher concentrations against Cd-induced oxidative stress in the roots (Wang et al. 2012).

The above reports thus suggest that Ce, La and possibly other REEs may both induce protective, antioxidant effects, and toxic, prooxidant effects according to non-monotonic, hormetic mechanisms.

Conclusion

Cerium and the other REEs represent a formidable challenge in environmental toxicology, as far as they have recently become indispensible means in an overwhelming set of industrial applications, while the current knowledge on REE-associated health effects are, at the best for Ce and La, quite limited and require extensive investigations. In the case for other REEs, whose toxicity is almost unknown, while broadly utilized in the industry, the appraisal of their health effects should be regarded as a mandatory task in the nearest future.

References

Arora, A.; Sairam, R. K. and Srivastava, G. C. (2002). Oxidative stress and antioxidative system in plants. *Curr. Sci.*, 82, 1227–1238.

Blaise, C.; Gagné, F.; Férard, J. F. and Eullaffroy, P. (2008). Ecotoxicity of selected nano-materials to aquatic organisms. *Environ. Toxicol.*, 23, 591-598.

Boyko, A.; Matsuoka, A. and Kovalchuk, I. (2011). Potassium chloride and rare earth elements improve plant growth and increase the frequency of the *Agrobacterium tumefaciens*-mediated plant transformation. *Plant Cell Rep.*, 30, 505-518.

Calabrese, E. J. (2010). Hormesis is central to toxicology, pharmacology and risk assessment. *Hum. Exp. Toxicol.*, 29, 249-261.

Chao, L.; Bofu, P.; Weiqian, C.; Yun, L.; Hao, H. et al. (2008). Influences of calcium deficiency and cerium on growth of spinach plants. *Biol. Trace Elem. Res.*, 121, 266-275.

Chen, W. J.; Tao, Y.; Gu, Y. H. and Zhao, G. W. (2001). Effect of lanthanide chloride on photosynthesis and dry matter accumulation in tobacco seedlings. *Biol. Trace Elem. Res.*, 79, 169–176.

Chen, S. A.; Zhao B.; Wang, X.; Yuan, X. and Wang, Y. (2004). Promotion of the growth of *Crocus sativus* cells and the production of crocin by rare earth elements. *Biotechnol. Lett.*, 26, 27-30.

Chen, Q.; Espey, M. G.; Sun, A. Y.; Pooput, C. and Kirk, K. L. (2008). Pharmacologic doses of ascorbate act as a prooxidant and decrease growth of aggressive tumor xenografts in mice. *Proc. Natl. Acad. Sci. U. S. A.*, 105, 11105-11109.

Cordier, D. J. (2012). Rare Earths. http://minerals.usgs.gov/minerals/pubs/commodity/rare_earths/mcs-2012-raree.

d'Aquino, L.; de Pinto, M. C.; Nardi, L.; Morgana, M. and Tommasi, F. (2009). Effect of some light rare earth elements on seed germination, seedling growth and antioxidant metabolism in *Triticum durum*. *Chemosphere*, 75, 900–905.

Dai, J.; Liu, J. J.; Zhou, G. Q.; Zhang, Y. Z. and Liu, Y. (2011). Effect of Ce(III) on heat production of mitochondria isolated from hybrid rice. *Biol. Trace Elem. Res.*, 143, 1142-1148.

Diatloff, E.; Smith, F. W. and Asher, C. J. (1995a). Rare earth elements and plant growth: I. Effects of lanthanum and cerium on root elongation of corn and mungbean. *J. Plant Nutr.*, 18, 1963–1976.

Diatloff, E.; Smith, F. W. and Asher, C. J. (1995b). Rare earth elements and plant growth: II. Responses of corn and mungbean to low concentrations of lanthanum in dilute, continuously flowing nutrient solutions. *J. Plant Nutr.*, 18, 1977–1989.

Diatloff, E.; Smith, F. W. and Asher, C. J. (1995c). Rare earth elements and plant growth: III. Responses of corn and mungbean to low concentrations of cerium in dilute, continuously flowing nutrient solutions. *J. Plant Nutr.*, 18, 1991–2003.

Diatloff, E.; Smith, F. W. and Asher, C. J. (2008). Effects of lanthanum and cerium on the growth and mineral nutrition of corn and mungbean. *Ann. Bot.*, 101, 971-982.

Dickson, E. (2006). Rare earths. www.resourceworld.com, pp. 1-20.

Falugi, C.; Aluigi, M. G.; Chiantore, M. C.; Privitera, D.; Ramoino, P. et al. (2012). Toxicity of metal oxide nanoparticles in immune cells of the sea urchin. *Mar. Environ. Res.*, 76, 114-121.

Fashui, H.; Zhenggui, W. and Guiwen, Z. (2000). Effect of lanthanum on aged seed germination of rice. *Biol. Trace Elem. Res.*, 75, 205–213.

Fashui, H. (2002). Study on the mechanism of cerium nitrate effects on germination of aged rice seed. *Biol. Trace Elem. Res.*, 87, 191-200.

Fashui, H.; Ling, W.; Xiangxuan, M.; Zheng, W. and Guiwen, Z. (2002). The effect of cerium (III) on the chlorophyll formation in spinach. *Biol. Trace Elem. Res.*, 3, 263-276.

Fei, M.; Li, N.; Ze, Y.; Liu, J.; Wang, S. et al. (2011). The mechanism of liver injury in mice caused by lanthanoids. *Biol. Trace Elem. Res.*, 140, 317-329.

Feyerabend, F.; Fischer, J.; Holtz, J.; Witte, F.; Willumeit, R. et al. (2010). Evaluation of short-term effects of rare earth and other elements used in magnesium alloys on primary cells and cell lines. *Acta Biomater.*, 6, 1834-1842.

Gaiser, B. K.; Fernandes, T. F.; Jepson, M. A.; Lead, J. R.; Tyler, C. R. et al. (2012). Interspecies comparisons on the uptake and toxicity of silver and cerium dioxide nanoparticles. *Environ. Toxicol. Chem.*, 31, 144-154.

García-Saucedo, C.; Field, J. A.; Otero-Gonzalez, L. and Sierra-Álvarez, R. (2011). Low toxicity of HfO_2, SiO_2, Al_2O_3 and CeO_2 nanoparticles to the yeast, *Saccharomyces cerevisiae*. *J. Hazard Mater.*, 192, 1572-1579.

Ge, Z. Q.; Yang, S. and Yuan, Y. J. (2006). Ce(4+) induced down-regulation of ERK-like MAPK and activation of nucleases during the apoptosis of cultured *Taxus cuspidata* cells. *J. Inorg. Biochem.*, 100, 167-177.

Gong, X.; Qu, C.; Liu, C.; Hong, M.; Wang, L. et al. (2011). Effects of manganese deficiency and added cerium on nitrogen metabolism of maize. *Biol. Trace Elem. Res.*, 144, 1240-1250.

Guo, X.; Zhou, Q.; Lu, T.; Fang, M. and Huang, X. (2007). Distribution and translocation of 141Ce(III) in horseradish. *Ann. Bot.*, 100, 1459-1465.

He, Y. and Loh, C. (2000). Cerium and lanthanum promote floral initiation and reproductive growth of *Arabidopsis thaliana*. *Plant Sci.*, 159, 117-124.

He, M. L.; Ranz, D. and Rambeck, W. A. (2001). Study on the performance enhancing effects of rare earth elements in growing and fattening pigs. *J. Anim. Physiol. Anim. Nutr.*, 85, 263–270.

He, Y. W. and Loh, C. S. (2002). Induction of early bolting in *Arabidopsis thaliana* by triacontanol, cerium and lanthanum is correlated with increased endogenous concentration of isopentenyl adenosine (iPAdos). *J. Exp. Bot. Mar.*, 53, 505-512.

Heckert, E. G.; Karakoti, A. S.; Seal S. and Self W.T. (2008). The role of cerium redox state in the SOD mimetic activity of nanoceria. *Biomaterials*, 29, 2705–2709.

Hirst, S. M.; Karakoti, A. S.; Tyler, R. D.; Sriranganathan, N.; Seal, S. et al. (2009). Anti-inflammatory properties of cerium oxide nanoparticles. *Small*, 5, 2848-2856.

Hu, X.; Ding, Z.; Chen, Y.; Wang, X. and Dai, L. (2002a). Bioaccumulation of lanthanum and cerium and their effects on the growth of wheat *(Triticum aestivum* L.*)* seedlings. *Chemosphere*, 48, 621-629.

Hu, X.; Ding, Z.; Wang, X.; Chen, Y. and Dai, L. (2002b). Effects of lanthanum and cerium on vegetable growth of wheat *(Triticum aestivum* L.*)* seedlings. *Bull. Environ. Contam. Toxicol.*, 69, 727–733.

Hu, Z.; Richter, H.; Sparovek, G. and Schnug, E. (2004). Physiological and biochemical effects of rare earth elements on plants and their agricultural significance: a review. *J. Plant Nutr.*, 27, 183–220.

Huang, H.; Liu, X.; Qu, C.; Liu, C.; Chen, L. et al. (2008). Influences of calcium deficiency and cerium on the conversion efficiency of light energy of spinach. *Biometals*, 21, 553-561.

Huang, S. F.; Li, Z. Y.; Wang, X. Q.; Wang, Q. X. and Hu, F. F. (2010). Cerium caused life span shortening and oxidative stress resistance in *Drosophila melanogaster*. *Ecotoxicol. Environ. Saf.*, 73, 89-93.

Huang, P.; Li, J.; Zhang, S.; Chen, C.; Han, Y. et al. (2011). Effects of lanthanum, cerium, and neodymium on the nuclei and mitochondria of hepatocytes: accumulation and oxidative damage. *Environ. Toxicol. Pharmacol.*, 31, 25-32.

Ippolito, M. P.; Fasciano, C.; d'Aquino, L.; Morgana, M. and Tommasi, F. (2009). Responses of antioxidant systems after exposition to rare earths and their role in chilling stress in common duckweed *(Lemna minor* L.*)*: A defensive weapon or a boomerang? *Arch. Environ. Contam. Toxicol.*, 58, 42-52.

Ippolito, M. P.; Fasciano, C.; D'Aquino, L. and Tommasi, F. (2011). Responses of antioxidant systems to lanthanum nitrate treatments in tomato plants during drought stress. *Plant Biosystems*, 145, 248-252.

Liang, C.; Zhang, G. and Zhou, Q. (2011). Effect of cerium on photosynthetic pigments and photochemical reaction activity in soybean seedling under ultraviolet-B radiation stress. *Biol. Trace Elem. Res.*, 142, 796-806.

Lee, S. W.; Kim, S. M. and Choi, J. (2009). Genotoxicity and ecotoxicity assays using the freshwater crustacean *Daphnia magna* and the larva of the aquatic midge *Chironomus riparius* to screen the ecological risks of nanoparticle exposure. *Environ. Toxicol. Pharmacol.*, 28, 86-91.

Li, N.; Wang, S.; Liu, J.; Ma, L.; Duan Y. et al. (2010). The oxidative damage in lung of mice caused by lanthanoids. *Biol. Trace Elem. Res.*, 134, 68-78.

Lin, W.; Huang, Y. W.; Zhou, X. D. and Ma, Y. (2006). Toxicity of cerium oxide nanoparticles in human lung cancer cells. *Int. J. Toxicol.*, 25, 451-457.

Ling, Q. and Hong, F. (2010). Antioxidative role of cerium against the toxicity of lead in the liver of silver crucian carp. *Fish Physiol. Biochem.*, 36, 367-376.

López-Moreno, M. L.; de la Rosa, G.; Hernández-Viezcas, J. A.; Castillo-Michel, H.; Botez, C. E. et al. (2010a). Evidence of the differential biotransformation and genotoxicity of ZnO and CeO_2 nanoparticles on soybean *(Glycine max)* plants. *Environ. Sci. Technol.*, 44, 7315-7320.

López-Moreno, M. L.; de la Rosa, G.; Hernández-Viezcas, J. A.; Peralta-Videa, J. R. and Gardea-Torresdey, J. L. (2010b). X-ray absorption spectroscopy (XAS) corroboration of the uptake and storage of CeO_2 nanoparticles and assessment of their differential toxicity in four edible plant species. *J. Agric. Food Chem.*, 58, 3689-3693.

Luna-Velasco, A.; Field, J. A.; Cobo-Curiel, A. and Sierra-Alvarez, R. (2011). Inorganic nanoparticles enhance the production of reactive oxygen species (ROS) during the autoxidation of L-3,4-dihydroxyphenylalanine (L-dopa). *Chemosphere*, 85, 19-25.

Maynard, K. I. (2011). Hormesis pervasiveness and its potential implications for pharmaceutical research and development. *Dose Response*, 9, 377-386.

Nakamura, Y.; Tsumura, Y.; Tonogai, Y.; Shibata, T. and Ito, Y. (1997). Differences in behavior among the chlorides of seven rare earth elements administered intravenously to rats. *Fundam. Appl. Toxicol.*, 37, 106-116.

Navaneethan, S.; Palmer, S. C.; Vecchio, M.; Craig, J. C.; Elder, G. J. et al. (2011). Phosphate binders for preventing and treating bone disease in chronic kidney disease patients. *Cochrane Database Syst. Rev.*, 16, CD006023.

Oral R.; Bustamante, P.; Warnau, M.; D'Ambra, A.; Guida, M. et al. (2010). Cytogenetic and developmental toxicity of cerium and lanthanum to sea urchin embryos. *Chemosphere*, 81, 194-198.

Ozaki,T.; Ambe, S.; Abe, T. and Francis, A. J. (2005). Competitive inhibition and selectivity enhancement by Ca in the uptake of inorganic elements (Be, Na, Mg, K, Ca, Sc, Mn, Co, Zn, Se, Rb, Sr, Y, Zr, Ce, Pm, Gd, Hf) by carrot *(Daucus carota* cv. U.S. *harumakigosun). Biol. Trace Elem. Res.*, 103, 69-82.

Palmer, R. J.; Butenhoff, J. L. and Stevens, J. B. (1987). Cytotoxicity of the rare earth metals cerium, lanthanum, and neodymium in vitro: comparisons with cadmium in a pulmonary macrophage primary culture system. *Environ. Res.*, 43, 142-156.

Park, E. J.; Choi, J.; Park, Y. K. and Park, K. (2008). Oxidative stress induced by cerium oxide nanoparticles in cultured BEAS-2B cells. *Toxicology*, 245, 90-100.

Rodea-Palomares, I.; Boltes, K.; Fernández-Piñas, F.; Leganés, F.; García-Calvo, E. et al. (2011). Physicochemical characterization and ecotoxicological assessment of CeO_2 nanoparticles using two aquatic microorganisms. *Toxicol. Sci.*, 119, 135-145.

Roh, J. Y.; Park, Y. K.; Park, K. and Choi, J. (2010). Ecotoxicological investigation of CeO_2 and TiO_2 nanoparticles on the soil nematode *Caenorhabditis elegans* using gene expression, growth, fertility, and survival as endpoints. *Environ. Toxicol. Pharmacol.*, 29, 167-172.

Schmidt, S.; Chevallier, P.; Bessoud, B.; Meuwly, J. Y.; Felley, C. et al. (2007). Diagnostic performance of MRI for detection of intestinal fistulas in patients with complicated inflammatory bowel conditions. *Eur. Radiol.*, 17, 2957-2963.

Schubert, D.; Dargusch, R.; Raitano, J. and Chan, S. W. (2006). Cerium and yttrium oxide nanoparticles are neuroprotective. *Biochem. Biophys. Res. Commun.*, 342, 86-91.

Shen, C. X.; Zhang, Q. F.; Li, J.; Bi, F. C. and Yao, N. (2010). Induction of programmed cell death in *Arabidopsis* and rice by single-wall carbon nanotubes. *Am. J. Bot.*, 97, 1602-1609.

Shenbagam, M. and Nalini, N. (2011). Dose response effect of rutin a dietary antioxidant on alcohol-induced prooxidant and antioxidant imbalance - a histopathologic study. *Fundam. Clin. Pharmacol.*, 25, 493-502.

Tonolini, M. and Bianco R. (2011). Acute adverse reactions to radiographic iodinated and gadolinium-based contrast media: incidence, risk factors and premedication: from published evidence to a practical approach. *Clin. Ter.*, 162, 591-594.

Tsai, Y. Y.; Oca-Cossio, J.; Agering, K.; Simpson, N. E.; Atkinson, M. A. et al. (2007). Novel synthesis of cerium oxide nanoparticles for free radical scavenging. *Nanomed.*, 2, 325-332.

van Hoecke, K.; Quik, J. K.; Mankiewicz-Boczek, J.; De Schamphelaere, K. A. C.; Elsaesser, A. et al. (2009). Fate and effects of CeO_2 nanoparticles in aquatic ecotoxicity tests. *Environ. Sci. Technol.*, 43, 4537–4546.

von Tucher, S. and Schmidhalter, U. (2005). Lanthanum uptake from soil and nutrient solution and its effects on plant growth. *J. Plant Nutr. Soil Sci.*, 168, 574–580.

Wang, Y.; Zhang, M. and Wang, X. (2000). Population growth responses of *Tetrahymena shanghaiensis* in exposure to rare earth elements. *Biol. Trace Elem. Res.*, 75, 265-275.

Wang, H.; Shan, X. Q.; Zhang, S. and Wen, B. (2003). Preliminary characterization of a light-rare-earth-element-binding peptide of a natural perennial fern *Dicranopteris dichotoma. Anal. Bioanal. Chem.*, 376, 49-52.

Wang, C.; Luo, X.; Tian, Y.; Xie, Y.; Wang, S. et al. (2012). Biphasic effects of lanthanum on *Vicia faba* L. seedlings under cadmium stress, implicating finite antioxidation and potential ecological risk. *Chemosphere*, 86, 530-537.

Weiping, S.; Fashui, H.; Zhigang, W.; Yuzhen, Z.; Fugen, G. et al. (2003). Effects of cerium on nitrogen metabolism of peach plantlet in vitro. *Biol. Trace Elem. Res.*, 95, 259-268.

Wu, H.; Wang, D.; Shi, J.; Xue, S. and Gao, M. (2010). Effect of the complex of zincII and ceriumIV with chitosan on the preservation quality and degradation of organophosphorus pesticides in Chinese jujube *(Zizyphus jujuba* Mill. cv. *Dongzao). J. Agric. Food Chem.*, 58, 5757-5762.

Yang, S.; Lu, S. H. and Yuan, Y. J. (2009). Cerium elicitor-induced phosphatidic acid triggers apoptotic signaling development in *Taxus cuspidata* cell suspension cultures. *Chem. Phys. Lipids*, 159, 13-20.

Ye, Y.; Wang, L.; Huang, X.; Lu, T.; Ding, X. et al. (2008). Subcellular location of horseradish peroxidase in horseradish leaves treated with La(III), Ce(III) and Tb(III). *Ecotoxicol. Environ. Saf.*, 71, 677-684.

Yuguan, Z.; Min, Z.; Luyang, L.; Zhe, J.; Chao, L. et al. (2009). Effects of cerium on key enzymes of carbon assimilation of spinach under magnesium deficiency. *Biol. Trace Elem. Res.*, 131, 154-164.

Xiaoqing, L.; Hao, H.; Chao, L.; Min, Z. and Fashui, H. (2009). Physico-chemical property of rare earths-effects on the energy regulation of photosystem II in *Arabidopsis thaliana. Biol. Trace Elem. Res.*, 130, 141-151.

Xu, X.; Zhu, W.; Wang, Z. and Witkamp, G. J. (2002). Distributions of rare earths and heavy metals in field-grown maize after application of rare earth-containing fertilizer. *Sci. Total Environ.*, 293, 97-105.

Xu, Q. M. and Chen, H. (2011). Antioxidant responses of rice seedling to Ce under hydroponic cultures. *Ecotoxicol. Environ. Saf.*, 74, 1693-1699.

Ze, Y.; Yin, S.; Ji, Z.; Luo, L.; Liu, C. et al. (2009). Influences of magnesium deficiency and cerium on antioxidant system of spinach chloroplasts. *Biometals*, 22, 941-949.

Zhang, L.; Zeng, F.; Xiao, R. (2003). Effect of lanthanum ions (La^{3+}) on the reactive oxygen species scavenging enzymes in wheat leaves. *Biol. Trace Element Res.*, 91, 243–252.

Zhang, H.; He, X.; Zhang, Z.; Zhang, P.; Li, Y. et al. (2011). Nano-CeO_2 exhibits adverse effects at environmental relevant concentrations. *Environ. Sci. Technol.*, 45, 3725-3730.

Zhao, H.; Cheng, Z.; Hu, R.; Chen, J.; Hong, M. et al. (2011). Oxidative injury in the brain of mice caused by lanthanid. *Biol. Trace Elem. Res.*, 142, 174-189.

Zhao, H.; Zhou, Q.; Zhou, M.; Li, C.; Gong, X. et al. (2012). Magnesium deficiency results in damage of nitrogen and carbon cross-talk of maize and improvement by cerium addition. *Biol. Trace Elem. Res.*, DOI: 10.1007/s12011-012-9340-x.

Zhou, J. C.; Yang, Z. L.; Dong, W.; Tang, R. J.; Sun, L. D. et al. (2011). Bioimaging and toxicity assessments of near-infrared upconversion luminescent NaYF4:Yb,Tm nanocrystals. *Biomaterials*, 32, 9059-9067.

In: Cerium
Editors: Aleksey Izyumov and Gulnaz Plaksin
ISBN: 978-1-62257-670-8

Chapter 5

CERIUM-BASED CATALYSTS FOR THE SELECTIVE CATALYTIC REDUCTION OF NO_X WITH NH_3

Fudong Liu**[*]**, Wenpo Shan and Hong He
Research Center for Eco-Environmental Sciences,
Chinese Academy of Sciences, Beijing, P. R. China

ABSTRACT

Selective catalytic reduction of NO_x with NH_3 (NH_3-SCR) is one of the most promising NO_x control technologies for lean burn engines, particularly for diesel engines, to meet the increasingly stringent standards for NO_x emission worldwide. Due to some inevitable disadvantages of the well-established vanadium-based catalyst (V_2O_5-WO_3 (MoO_3)/TiO_2) for NH_3-SCR of NO_x from stationary sources, including the narrow operation temperature window, the large N_2O formation at high temperatures, the easy sublimation and biological toxicity of vanadium pentoxide, environmentally-benign and vanadium-free catalysts for NH_3-SCR reaction are requisite to be developed urgently. Recently, cerium-based oxides with high oxygen storage capacity and excellent redox property attracted much attention for their use as environmentally-benign NH_3-SCR catalysts. In this chapter, we will comprehensively introduce our recent research results about the cerium-based catalysts for NH_3-SCR process, including the supported type CeO_x/TiO_2 catalyst prepared by conventional impregnation method and mixed oxide $CeTiO_x$ catalyst prepared by a novel homogeneous precipitation method. The effect of WO_x addition into the $CeTiO_x$ catalyst will also be studied in detail, which can improve the NH_3-SCR activity in the whole temperature range and enhance the N_2 selectivity in the high temperature range simultaneously. The total substitution of titanium species by tungsten species in $CeWO_x$ catalyst can further improve the SCR activity and N_2 selectivity to a certain extent, with nearly 100% NO_x conversion in a wide temperature range from 250 to 425 °C even under an extremely high gas hour space velocity (GHSV) of 500,000 h^{-1}. The excellent catalytic performance of this catalyst is attributed to a synergistic effect between Ce and W species. The cerium-based catalysts introduced in this chapter can be used as potential candidates for practical application in the $deNO_x$

[*] E-mail address: fdliu@rcees.ac.cn; lfd1982@gmail.com

process for diesel engines, which is an important developing direction for the technological applications of cerium materials.

Keywords: Selective catalytic reduction; Diesel engines; Nitrogen oxides; Cerium-based catalysts; Environmentally-benign catalysts; Practical application

1. Introduction

Nitrogen oxides (NO_x), which are mainly composed of NO and NO_2, is a major type of air pollutant influencing the atmospheric physical-chemical process and endangering human health [1-6], which should be controlled strictly worldwide for environmental protection purpose. The NO_x emissions resulting from human activity are mainly attributed to agricultural production and the usage of fossil fuels [5], and the NO_x emission amount due to the latter factor is greatly increased with the development of world economy and increasing of world population. The NO_x emissions from the combustion of fossil fuels can also be categorized as stationary sources including coal-fired power plants, industrial kilns and industrial boilers *etc.*, and mobile sources including the gasoline engine cars and diesel engine trucks *etc.* In economically developed region, the NO_x emission from mobile sources even exceeds that from stationary sources owing to the enormous increasing of vehicle amount. With the rising challenges from energy crisis and global warming, diesel engines become more and more competitive due to its high fuel efficiency. However, the removal of NO_x from diesel engine exhaust under oxygen-rich conditions remains one of the major challenges in the field of environmental catalysis [6-8].

Selective catalytic reduction (SCR) of NO_x with NH_3 or urea is one of the most promising after-treatment technologies for NO_x abatement from diesel engines under oxygen-rich conditions [9]. The catalyst plays an important role in the core of NH_3-SCR technology, which requires showing high NH_3-SCR activity, N_2 selectivity, hydrothermal stability and acceptable durability to various co-existing poisoning components including H_2O, SO_2 and hydrocarbons *etc.* The well-established NH_3-SCR catalyst system for the deNO_x process from stationary sources is WO_3 or MoO_3 promoted V_2O_5/TiO_2, which presents high SCR activity, SO_2 durability [7,10,11] and has been introduced into the diesel vehicle market. A typical diesel emission control system using SCR technology usually includes a diesel particular filter (DPF) to meet higher emission control standards worldwide, which may expose the SCR catalysts to high temperatures (> 650 °C) during its regeneration process [12]. The traditional vanadium-based catalysts cannot keep active under such high temperatures because the melting point of V_2O_5 is only 690 °C. The toxicity of active vanadium species, the narrow operation temperature window, the high conversion of SO_2 to SO_3, together with the large N_2O formation at high temperatures, have restrained its wider application, and vanadium-based SCR catalysts for NO_x removal from mobile sources will be gradually removed from the vehicle market over the next few years [9]. Therefore, great efforts have been made to develop environmentally-benign NH_3-SCR catalysts for controlling the NO_x emissions from diesel engines under complex reaction conditions, including the Fe, Cu-exchanged zeolite catalysts (*e.g.* Fe-ZSM-5, Fe-BEA, Cu-ZSM-5, Cu-SSZ-13 *etc.*) [13-16] and the Fe, Cu, Mn, Ce-based oxide catalysts (*e.g.* γ-Fe_2O_3, $FeTiO_x$, $Fe_2(SO_4)_3/TiO_2$, $Fe_2O_3/WO_3/ZrO_2$, CuO/cordierite-Al_2O_3, CuO_x/WO_x-ZrO_2, MnO_x/TiO_2, $(Fe_{3-x}Mn_x)_{1-\delta}O_4$,

NiO/sulfate modified CeO_2-ZrO_2, CeO_2-WO_3/TiO_2 *etc.*) [17-29]. All the above-mentioned environmentally-benign catalysts show relatively high deNO_x efficiency in different temerature ranges, among which the Ce-based catalysts show good NH_3-SCR activity and N_2 selectivity at medium or high temperatures above 200 °C possibly due to the high oxygen storage capacity and excellent redox property of cerium species.

Based on our recent research results, in this chapter, the supported type CeO_x/TiO_2 catalyst prepared by conventional impregnation method and mixed oxide $CeTiO_x$ catalyst prepared by a novel homogeneous precipitation method will be comprehensively introduced [30-32]. The novel homogeneous precipitation method exhibits obvious advantage in the preparation process of $CeTiO_x$ catalyst, therefore the further improvement of this catalyst system is mainly based on this novel method, *i.e.* the addition of WO_x into the $CeTiO_x$ catalyst (denoted as $CeWTiO_x$) to improve the NH_3-SCR activity in the whole temperature range and enhance the N_2 selectivity in the high temperature range simultaneously [33]. Furthermore, to obtain better SCR activity, more excellent thermal stability along with more outstanding durability to gas hour space velocity (GHSV) for the volume reduction of SCR catalytic converter, the total substitution of titanium species in $CeWTiO_x$ catalyst by tungsten species in $CeWO_x$ catalyst is investigated in detail [34]. Under the same test conditions, the $CeWO_x$ catalyst shows much better SCR performance than V_2O_5-WO_3/TiO_2 and Fe-ZSM-5 catalysts, which have been industrially and commercially used for NH_3-SCR of NO_x. A synergistic effect between Ce and W species in $CeWO_x$ catalyst is clearly observed, making this catalyst a potential candidate for practical application in the deNO_x process for diesel engines. The usage of cerium-based catalysts in NH_3-SCR of NO_x for diesel engines is a research hotspot in the field of environmental catalysis, which is also an important developing orientation for the technological applications of cerium materials.

2. Experimental Methods

2.1. Catalyst Preparation and Activity Test

The supported type CeO_x/TiO_2 catalysts were prepared by conventional impregnation method, using an anatase type TiO_2 as support and an aqueous solution of cerium nitrate as catalyst precursor. After impregnation, the excess water was removed in a rotary evaporator at 80 °C. The samples were first dried at 110 °C for 12 h followed by calcination at 500 °C in air for 3 h. The weight ratio of Ce was measured with respect to the TiO_2 support, *i.e.* 2, 5, 10, 20, 70 wt.% Ce/TiO_2. Pure TiO_2 was prepared through a densification process. The mixture of TiO_2 and deionized water was dried and also calcined at 500 °C in air for 3 h. Pure CeO_2 was prepared by direct decomposition of $Ce(NO_3)_3 \cdot 6H_2O$ which was calcined at 500 °C in air for 3 h.

Previous studies have already shown that the catalyst preparation methods can significantly influence the catalytic performance of Ce-Ti containing catalysts, and the well dispersion of active component CeO_2 on TiO_2 is very important for the improvement of NH_3-SCR activity [30,35,36]. Homogeneous precipitation, characterized by a highly uniform increase in the pH of metal salts solution achieved by thermal decomposition of some organic alkalia, is a suitable catalyst preparation method that can produce highly dispersed mixed metal oxides. Therefore, in this study, we prepared a series of mixed oxide $CeTiO_x$ catalysts by homogeneous precipitation method using urea as precipitator and compared with the supported type CeO_x/TiO_2 catalysts. In the preparation process, the aqueous solutions of $Ce(NO_3)_3{\cdot}6H_2O$ and $Ti(SO_4)_2$ were mixed with required molar ratios (Ce/Ti = 0.1, 0.2, 0.5 and 1.0). Then, excessive urea was added into the mixed solution. The mixed solution was thereafter heated to 90 °C and held there for 12 h under vigorous stir. After filtration and washing with deionized water, the resulting precipitant was dried at 100 °C overnight and subsequently calcined at 500 °C for 5 h in air condition. The obtained catalysts were denoted as Ce_aTiO_x, where "*a*" denotes the Ce/Ti molar ratio. Pure CeO_2 and TiO_2 were also prepared using the same method.

Based on the optimal $Ce_{0.2}TiO_x$ catalyst, the WO_x species was introduced into the catalyst system. The WO_x doped catalysts were denoted as $Ce_{0.2}W_aTiO_x$ ("*a*" represents the W/Ti molar ratio; a = 0.1, 0.2, 0.3, 0.5). The aqueous solutions of $Ce(NO_3)_3{\cdot}6H_2O$, $Ti(SO_4)_2$ and $(NH_4)_{10}W_{12}O_{41}$ were mixed with required molar ratios ($H_2C_2O_4{\cdot}2H_2O$ was used to facilitate the solution of $(NH_4)_{10}W_{12}O_{41}$). Excessive urea was then added into the mixed solution. The mixed solution was heated to 90 °C and held there for 12 h under vigorous stir. After filtration and washing with deionized water, the resulting precipitant was dried at 100 °C overnight and subsequently calcined at 500 °C for 5 h in air condition.

To investigate the effect of total substitution of titanium species in $CeTiO_x$ catalyst by tungsten species, serial $CeWO_x$ mixed oxide catalysts were prepared by homogeneous precipitation method using $Ce(NO_3)_3{\cdot}6H_2O$ and $(NH_4)_{10}W_{12}O_{41}$ as precursors. The detailed preparation procedures were exactly the same as described above, and the resulting catalysts were denoted as $Ce_aW_bO_x$, where "*a*/*b*" denotes the Ce/W molar ratio, such as $Ce_2W_1O_x$, $CeWO_x$ and $Ce_1W_2O_x$ calcined at 500 °C for 5 h and $CeWO_x$-800 calcined at 800 °C for 1 h for thermal stability test. Pristine CeO_x and WO_x were also prepared as reference samples using the same method.

Before all NH_3-SCR activity tests, the powder catalysts were pressed, crushed, and sieved to 40-60 mesh. The SCR activity tests of the sieved powder catalysts were all carried out in a fixed-bed quartz flow reactor at atmospheric pressure. The reaction conditions were controlled as follows: 500 ppm NO, 500 ppm NH_3, 5 vol.% O_2, 5 vol.% CO_2 (when used), 5 vol.% H_2O (when used), 100 ppm SO_2 (when used), balance N_2, and 500 ml/min total flow rate. Different GHSV were obtained by changing the volume of catalyst. The effluent gas, including NO, NH_3, NO_2 and N_2O was continuously analyzed by an online NEXUS 670-FTIR spectrometer equipped with a gas cell with 0.2 dm^3 volume. The FTIR spectra were collected at least after 1 h when the SCR reaction reached a steady state, and the NO_x conversion and N_2 selectivity were calculated accordingly [19].

2.2. Characterization Methods

The surface areas of the catalysts were obtained by N_2 adsorption/desorption analysis at -196 °C using a Quantachrome Quadrasorb SI-MP. Prior to the N_2 physisorption, the catalysts were degassed at 300 °C for 4 h. Surface areas were determined by BET equation in 0.05-0.35 partial pressure range.

The surface morphology and elemental composition of the catalysts were studied using a scanning electron microscope (SEM, Hitachi, S-3000N) combined with an energy dispersive X-ray (EDX) attachment. The accelerating voltage was 5.0 kV.

Powder X-ray diffraction (XRD) measurements of the catalysts were carried out on a computerized PANalytical X'Pert Pro diffractometer with Cu Kα (λ = 0.15406 nm) radiation. The data of 2θ from 20 to 80 ° were collected at 8 °/min with the step size of 0.07 °.

Visible Raman spectra of the catalysts were collected at room temperature on a Spex 1877 D triplemate spectrograph with spectral resolution of 2 cm^{-1}. A 532 nm DPSS diode-pump solid semiconductor laser was used as the excitation source and the power output was about 40 mW. Before measurements, the samples were well ground and mounted into a spinning holder to avoid thermal damage during scanning. The Raman signals were collected with conventional 90 ° geometry and the time for recording each spectrum was about 1000 ms. All Raman spectra used in this chapter were original and unsmoothed.

The XPS of catalysts were recorded on a scanning X-ray microprobe (PHI Quantera, ULVAC-PHI, Inc.) using Al *Kα* radiation (1486.7 eV). Binding energies were calibrated using C 1s peak (BE = 284.8 eV) as standard.

The *in situ* DRIFTS experiments were performed on an FTIR spectrometer (Nicolet Nexus 670) equipped with a smart collector and an MCT/A detector cooled by liquid nitrogen. The reaction temperature was controlled precisely by an Omega programmable temperature controller. Prior to each experiment, the sample was pretreated at 400 °C for 0.5 h in a flow of 20 vol.% O_2/N_2 and then cooled down to 200 °C. The background spectrum was collected in flowing N_2 and automatically subtracted from the sample spectrum. The reaction conditions were controlled as follows: 300 ml/min total flow rate, 500 ppm NH_3 or 500 ppm NO + 5 vol.% O_2, and N_2 balance. All spectra were recorded by accumulating 100 scans with a resolution of 4 cm^{-1}.

3. Improvement of $CeTiO_x$ Catalyst by Homogeneous Precipitation Method for NH_3-SCR of NO_x

3.1. Supported Type Ce/TiO_2 Catalyst

Due to the unique redox and acid-base properties of cerium oxide material, it has been commonly used as an efficient additive in the three way catalyst (TWC) for the exhaust purification from gasoline engines [37,38]. For the deNO_x process from diesel engines using NH_3-SCR technique, the cerium oxide material has also been used as catalyst promoter or active phase because of its possession of labile oxygen vacancies and easy redox shift between Ce^{3+} and Ce^{4+} species under oxidizing and reducing conditions [39]. For example, Long and Yang [40,41] concluded that the addition of cerium oxide into Fe-ZSM-5 and Fe-Ti-pillared layered clay (Fe-Ti-PILC) catalysts could facilitate the oxidation of NO to NO_2 and thus result in an improvement of NH_3-SCR activity. The novel MnO_x-CeO_2 catalyst developed by Qi and Yang [42] was highly active for NH_3-SCR reaction at low temperatures. Ito *et al.* [43] also found that Ce-exchanged mordenite was active and highly selective for NO_x reduction in a broad temperature range (250-560 °C).

Previous research results showed that cerium oxide material can be a suitable candidate for NO_x reduction with NH_3 based on its unique properties, and TiO_2 is usually used as the support of SCR catalysts due to its excellent SO_2 durability, because TiO_2 essentially reacts with neither SO_3 nor SO_2 above 200 °C [44]. Therefore, in order to combine the advantages of cerium and titanium species into one catalyst for NH_3-SCR reaction, for the first time, we developed a novel Ce/TiO_2 material using conventional impregnation method for this reaction and found that a high catalytic performance was obtained in a relatively wide temperature range with *ca.* 100% N_2 selectivity.

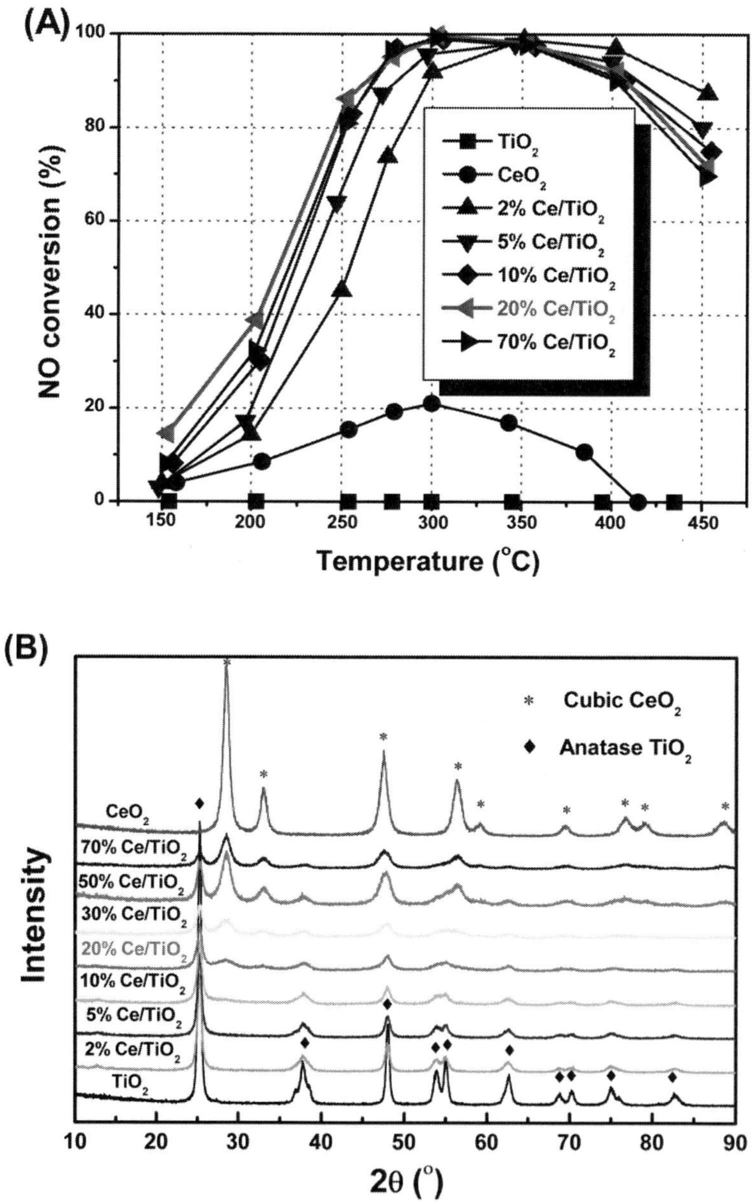

Figure 1. (A) NH_3-SCR activity of Ce/TiO_2 catalysts with different Ce loadings; (B) XRD patterns of Ce/TiO_2 serial catalysts [30]. (Reproduced by permission of Elsevier)

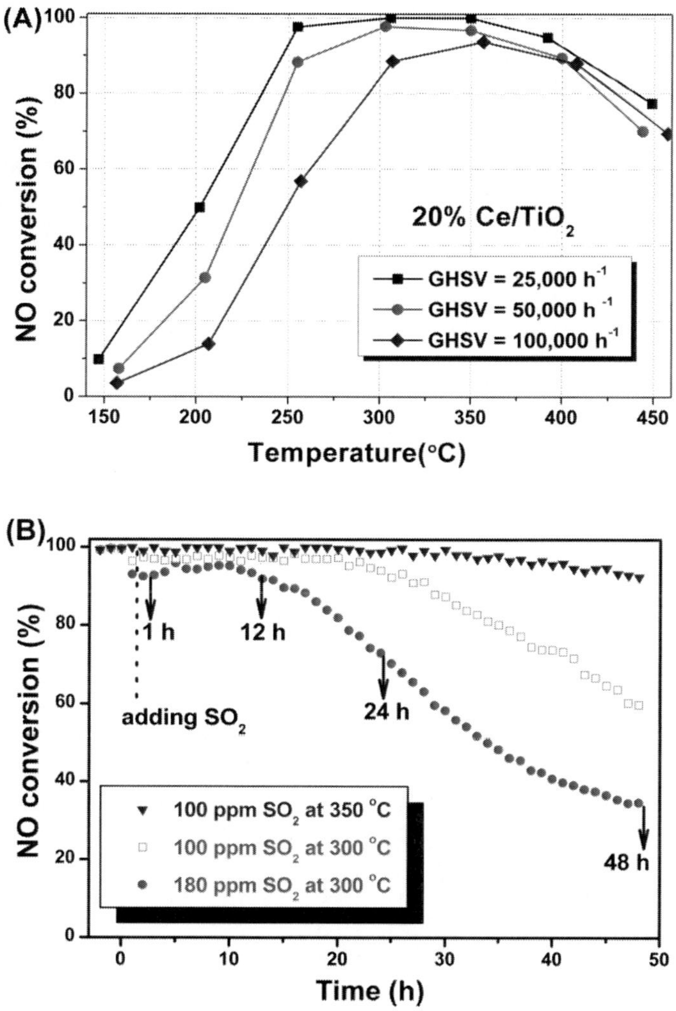

Figure 2. (A) Influence of GHSV [30] and (B) effect of SO_2 on NH_3-SCR activity of 20% Ce/TiO_2 catalyst [45]. (Reproduced by permissions of Elsevier and ACS Publications)

As shown in Figure 1A, the pristine TiO_2 support presented nearly no SCR activity in the whole temperature range that we investigated, and pristine CeO_2 only showed very low activity and rather narrow operation temperature window. Surprisingly, the addition of as low as 2% Ce onto TiO_2 support caused a significant enhancement of NH_3-SCR activity, and with the increase of Ce loading amount from 2% to 20% the SCR activity at low temperatures could be further improved. The further increase of Ce loading from 20% to 70% lowered the NO conversion at both low and high temperatures.

Over the optimal 20% Ce/TiO_2 catalyst, a NO conversion over 90% could be maintained from 275-400 °C with no N_2O by-product detected. The XRD results in Figure 1B showed that with the Ce loading below 20%, no obvious cubic CeO_2 was observed, suggesting that surface Ce species existed as a highly dispersed or amorphous phase. Above 20% Ce loading, clear crystallization of cubic CeO_2 was detected, along with the decrease of BET surface area of Ce/TiO_2 catalysts (*e.g.* 128.6 mg/m^2 for 20% Ce/TiO_2 and 82.6 mg/m^2 for 50% Ce/TiO_2), resulting in the decrease of SCR activity to a certain extent.

For the practical use of NH_3-SCR catalysts, GHSV is an important factor to be considered in order to determine the catalyst volume for the designed catalytic converter. Thus, the influence of GHSV on the SCR activity of 20% Ce/TiO_2 catalyst was investigated, as shown in Figure 2A. With the increase of GHSV from 25,000 to 100,000 h^{-1}, the NH_3-SCR activity especially at low temperatures showed significant decrease, implying that this catalyst was very sensitive to the high GHSV, which should be improved for reducing the catalyst volume for the SCR converter. Another problem of this 20% Ce/TiO_2 catalyst was the deactivation in the presence of SO_2, as the results shown in Figure 2B. It is clearly seen that the lower the reaction temperature, the greater the negative effect of SO_2 was. It seems that the TiO_2 support did not exhibit obvious SO_2 resistance in NH_3-SCR reaction, possibly because this is a supported-type catalyst and the active Ce species is mainly present on the surface covering a majority of surface area of TiO_2. In short summary, the improvement of Ce/TiO_2 NH_3-SCR catalyst is urgently needed to meet the complex reaction conditions for diesel engine exhaust.

3.2. Improvement of Ce-Ti Oxide Catalyst by Homogeneous Precipitation Method

As mentioned above, we significantly improved the activity of Ce-Ti oxide catalyst by homogeneous precipitation method using urea as precipitator. NO_x conversions and N_2 selectivity over Ce_aTiO_x (a = 0.1, 0.2, 0.5, 1.0) catalysts are presented in Figure 3. Pure TiO_2 showed no SCR activity below 400 °C, while $Ce_{0.1}TiO_x$ showed high SCR activity and excellent N_2 selectivity in the temperature range from 225 to 400 °C. With the increase of Ce loading, both $Ce_{0.2}TiO_x$ and $Ce_{0.5}TiO_x$ presented enhanced low temperature activity comparing with $Ce_{0.1}TiO_x$, while the high temperature NO_x conversions decreased to a certain extent.

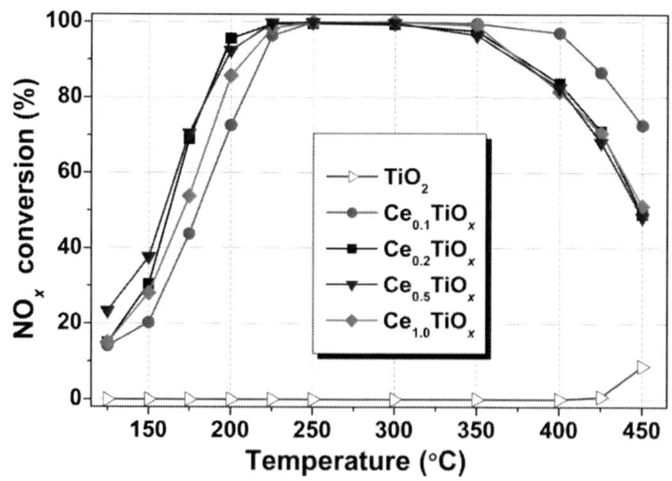

Figure 3. NH_3-SCR activity of Ce_aTiO_x (a = 0.1, 0.2, 0.5 and 1.0) serial catalysts [31]. (Reproduced by permission of John Wiley & Sons, Inc.)

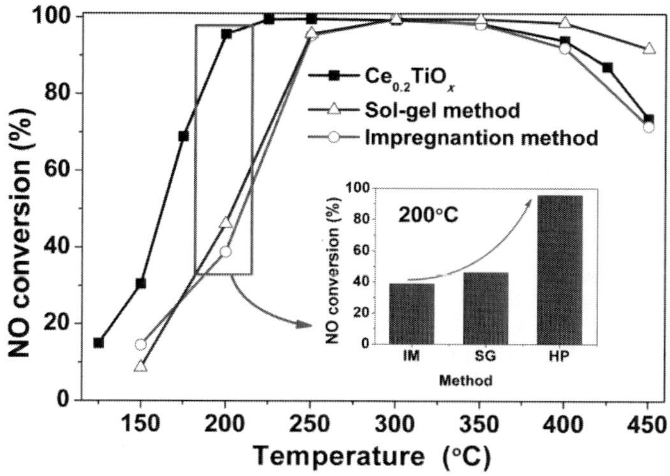

Figure 4. NH_3-SCR activity of Ce-Ti based oxide catalysts prepared by different methods under GHSV of 50,000 h^{-1} [30,31,36]. (Reproduced by permission of John Wiley & Sons, Inc.)

However, further increasing the Ce loading to Ce/Ti = 1.0 resulted in the decrease of NO_x conversion in the low temperature range. Comparing with the previously reported Ce-Ti mixed oxide catalysts prepared by impregnation (20% Ce/TiO_2) [30] and sol-gel (CeTi-SG) [36] methods, the $Ce_{0.2}TiO_x$ catalyst prepared by homogeneous precipitation method presented significantly enhanced catalytic activity, especially in the low temperature range (Figure 4).

3.3. Optimized $Ce_{0.2}TiO_x$ Catalyst

Fixing the Ce/Ti molar ratio at 0.2, we further optimized the concentration of solution, the addition amount of urea, and the reaction time of homogeneous precipitation method. After optimization, the optimized $Ce_{0.2}TiO_x$ catalyst showed over 90% NO_x conversion in a wide temperature range from 250 to 400 °C under a high GHSV of 250,000 h^{-1} (Figure 5). V_2O_5-WO_3/TiO_2 and Fe-ZSM-5 are two types of NH_3-SCR catalysts which have been industrially and commercially used for NO_x abatement from diesel exhaust. To better evaluate the NH_3-SCR performance of the optimized $Ce_{0.2}TiO_x$ catalyst, comparative SCR activity tests over V_2O_5-WO_3/TiO_2 and Fe-ZSM-5 catalysts were also carried out (Figure 5). The NO_x conversion over $Ce_{0.2}TiO_x$ catalyst was almost the same as that over V_2O_5-WO_3/TiO_2, while the N_2 selectivity over $Ce_{0.2}TiO_x$ catalyst was much higher than that over V_2O_5-WO_3/TiO_2 in the high temperature range (not shown here). Compared with Fe-ZSM-5 catalyst, the operation temperature window of $Ce_{0.2}TiO_x$ catalyst was more than 50 °C lower under the same GHSV, with similar N_2 selectivity above 95% in the whole temperature range. To evaluate the thermal stability of the catalyst, the influence of high temperature calcination on the optimized $Ce_{0.2}TiO_x$ and V_2O_5-WO_3/TiO_2 was investigated (Figure 5).

After high temperature calcination at 800 °C, the V_2O_5-WO_3/TiO_2-800 lost almost all the NH_3-SCR activity in the whole temperature range. However, the low temperature activity of $Ce_{0.2}TiO_x$-800 catalyst was still obviously higher than that of Fe-ZSM-5, indicating its high resistance to the thermal shock.

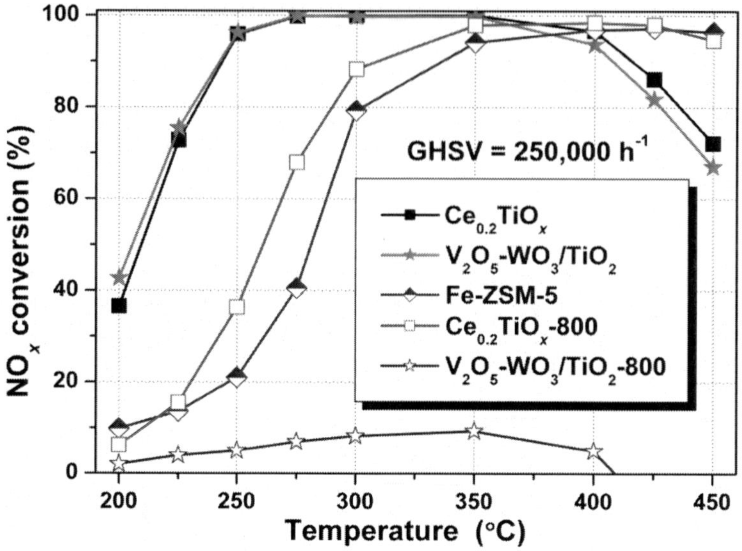

Figure 5. Comparison of NH_3-SCR activity of $Ce_{0.2}TiO_x$ catalyst (calcined at 500 and 800 °C) with those of V_2O_5-WO_3/TiO_2 and Fe-ZSM-5 catalysts [32]. (Reproduced by permission of Elsevier)

In short summary, the optimized $Ce_{0.2}TiO_x$ catalyst is a potential candidate for the catalytic removal of NO_x from mobile sources, especially diesel engines with wide exhaust temperature range and high GHSV.

3.4. Formation Process of the Optimized $Ce_{0.2}TiO_x$ Catalyst

Figure 6 shows that the pH value of the mixed solution increased with preparation time during the homogeneous precipitation process. The initial pH value of the mixed $Ce(NO_3)_3{\cdot}6H_2O$, $Ti(SO_4)_2$, and urea solution was nearly zero due to the strong acidity of $Ti(SO_4)_2$. With urea hydrolysis, the pH value of the solution increased quickly in the first hour, after which a white precipitate was gradually produced and the increase in pH value slowed down. After three hours of heating, the pH value of the solution showed a sharp increase and the precipitate turned to be yellow. The pH value then steadily increased to *ca.* 7.0 and the precipitate gradually turned to be orange. After eight hours of heating, there was no significant change in the pH value of the mixed solution or the color of precipitate.

The Ce/Ti molar ratios of the samples prepared under different precipitation times were determined by SEM-EDX. All the samples consisted of spherical particles with diameters varying from 1 to 5 μm. The Ce/Ti molar ratios in the samples prepared with the precipitation time of (A) 1.5 h, (B) 3 h, (C) 3.5 h, (D) 6.5 h, (E) 10 h and (F) 12 h were 0, 0.005, 0.18, 0.21, 0.23 and 0.23, respectively, indicating a gradual increase in the Ce/Ti molar ratio with precipitation time. Based on the above-mentioned results, the formation process of the optimized $Ce_{0.2}TiO_x$ catalyst was proposed (Figure 7). The Ti species was firstly precipitated when the pH value of the mixed solution was between 1.0 and 2.0, and then the Ce species was precipitated uniformly onto the precipitated Ti species following the rapid increase in pH value to be above 2.0.

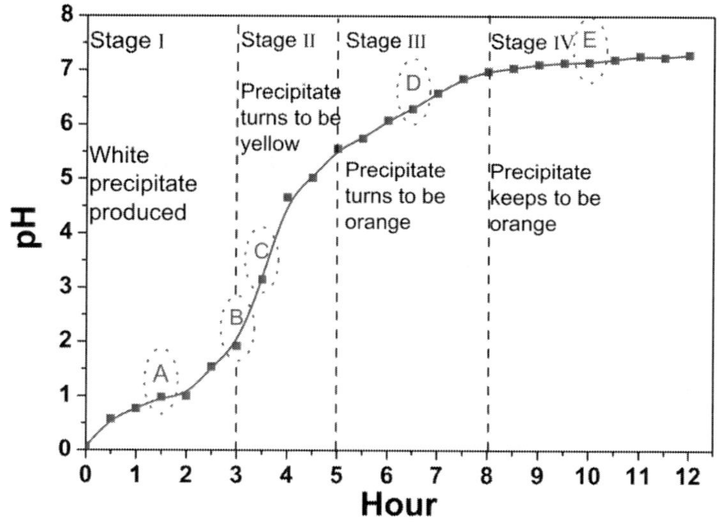

Figure 6. The pH variation of the mixed solution during the preparation of $Ce_{0.2}TiO_x$ catalyst by homogeneous precipitation method [32]. (Reproduced by permission of Elsevier)

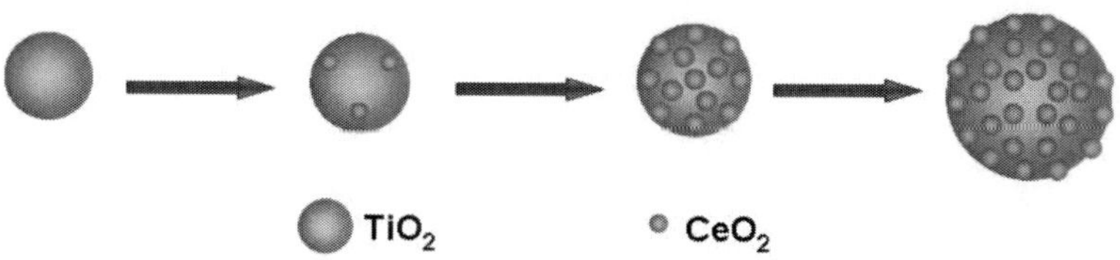

Figure 7. Formation process of $Ce_{0.2}TiO_x$ catalyst using homogeneous precipitation method.

Accompanied with the increase in Ce/Ti molar ratio during the precipitation process, the size of the precipitated spherical particles also increased. The process of homogenous precipitation is very important for the production of highly dispersed CeO_2 on TiO_2, which is probably a main reason for its excellent activity and high resistance to GHSV under SCR conditions.

4. Development of $CeWTiO_x$ Catalyst for NH_3-SCR of NO_x

4.1. NH_3-SCR Activity of $CeWTiO_x$ Catalyst

W species was doped into the $Ce_{0.2}TiO_x$ catalyst as a promoter (Figure 8). Small doping amount of W, with W/Ti molar ratio of 0.1, could increase both of the low temperature activity and high temperature activity of $Ce_{0.2}TiO_x$ catalyst. With the increase of W doping molar ratio from 0.1 to 0.2, the high temperature NO_x conversion was maintained at 100% even up to 450 °C, while the low temperature NO_x conversion did not show any decrease. However, further increasing the relative W doping amount to 0.3 and 0.5 resulted in an obvious decline of low temperature SCR activity, although the high temperature NO_x conversions still maintained being 100%.

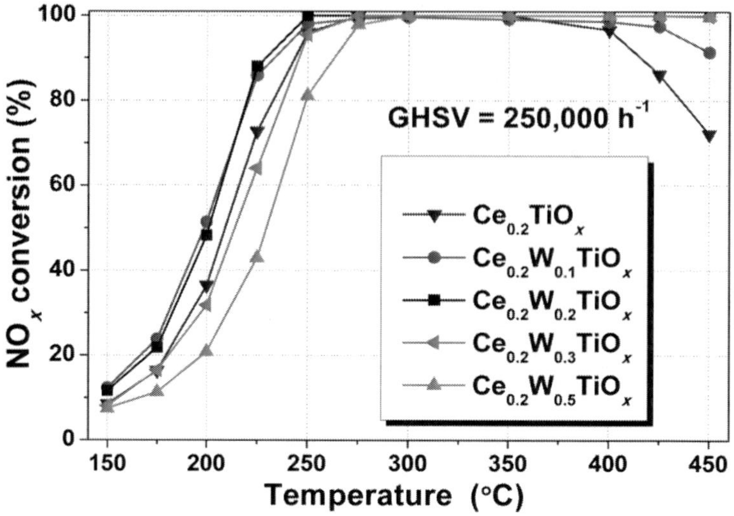

Figure 8. NH_3-SCR activity of $Ce_{0.2}W_aTiO_x$ (a = 0, 0.1, 0.2, 0.3, 0.5) serial catalysts [33]. (Reproduced by permission of Elsevier)

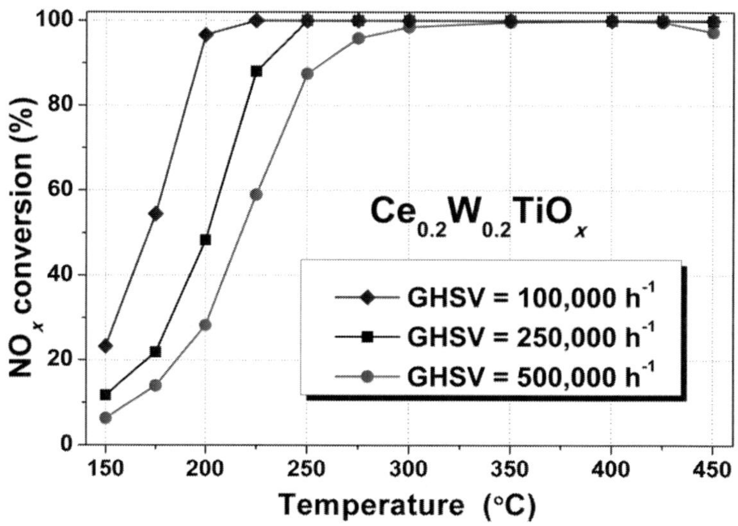

Figure 9. NH_3-SCR activity of $Ce_{0.2}W_{0.2}TiO_x$ catalyst under different GHSVs [33]. (Reproduced by permission of Elsevier)

Therefore, the $Ce_{0.2}W_{0.2}TiO_x$ catalyst presented an optimal NH_3-SCR performance with Ce/W molar ratio being 1:1. The NO_x conversions over $Ce_{0.2}W_{0.2}TiO_x$ catalyst under different GHSVs are shown in Figure 9. Obviously, the increase of GHSV only resulted in some decrease of low temperature SCR activity, while no clear influence on the high temperature activity was observed. The catalyst even showed over 90% NO_x conversion in a wide temperature range from 275 to 450 °C under a rather high GHSV of 500,000 h^{-1}.

4.2. NO Oxidation Activity and NH_3 Oxidation Activity of $CeWTiO_x$ Catalyst

In order to investigate the effects of W species in $Ce_{0.2}W_{0.2}TiO_x$ catalyst, the separate NO oxidation (NO + O_2) and separate NH_3 oxidation (NH_3 + O_2) experiments were also carried out (Figure 10). The NO_2 production in the separate NO oxidation reaction over $Ce_{0.2}W_{0.2}TiO_x$ is obviously higher than that over $Ce_{0.2}TiO_x$ in the low temperature range (Figure 10A). Many studies have shown that, if the SCR catalyst can oxidize NO to NO_2 *in situ*, its low temperature SCR activity will be significantly enhanced due to the effect of "fast SCR" [7]. Therefore, the introduction of W species could enhance the low temperature activity of $Ce_{0.2}W_{0.2}TiO_x$ catalyst by promoting NO oxidation to NO_2 thus facilitating the "fast SCR" reaction.

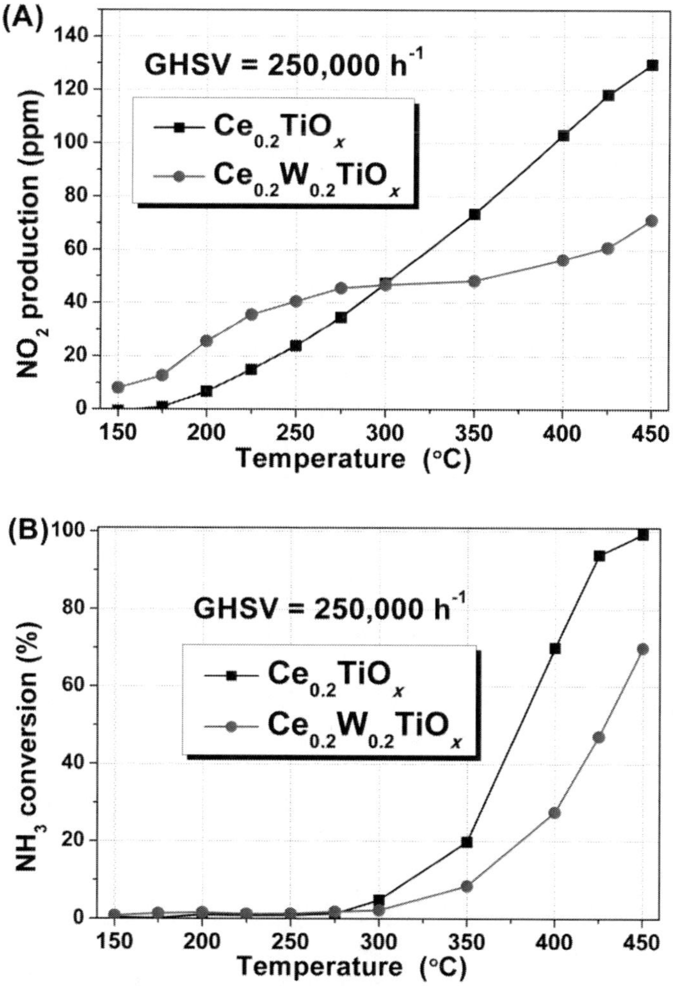

Figure 10. (A) NO_2 productions during separate NO oxidation reaction and (b) NH_3 conversions during separate NH_3 oxidation reaction over $Ce_{0.2}TiO_x$ and $Ce_{0.2}W_{0.2}TiO_x$ catalysts [33]. (Reproduced by permission of Elsevier)

The high temperature NH_3 oxidation ability of $Ce_{0.2}W_{0.2}TiO_x$ is obviously lower than that of $Ce_{0.2}TiO_x$ (Figure 10B), which means that the addition of W species could inhibit the unselective oxidation of NH_3 and thus promote the high temperature activity of $Ce_{0.2}W_{0.2}TiO_x$. Besides, some NO, NO_2 and N_2O were detected at high temperatures in the NH_3 oxidation reaction over $Ce_{0.2}TiO_x$, while none of these by-products were observed over $Ce_{0.2}W_{0.2}TiO_x$, suggesting that the introduction of W species into $Ce_{0.2}W_{0.2}TiO_x$ is also beneficial to the enhancement of N_2 selectivity in NH_3-SCR reaction.

A catalytic cycle for the SCR reaction involving both acid and redox reactions was proposed by Topsøe *et al.* [46,47], and the promotion effects of W species in V_2O_5-WO_3/TiO_2 catalysts were attributed to the increase of both redox properties and acid sites [48-52]. Previous studies by Lietti *et al.* have indicated that the redox functions of catalyst (especially the re-oxidation process of catalytic sites) govern the catalytic activity in the low temperature region, whereas the SCR reaction in the high temperature region is likely to be controlled by the surface acid properties [48-50].

Therefore, in this study, we can expect that the promoted NO oxidation activity at low temperatures (< 300 °C) and the inhibited unselective oxidation of NH_3 (as well as the decreased NO oxidation) at high temperatures (>300 °C) is strongly associated with the impact of W addition on the redox and acid properties of $Ce_{0.2}W_{0.2}TiO_x$ catalyst, respectively.

4.3. Characterizations of CeWTiO$_x$ Catalyst

The BET surface area of $Ce_{0.2}W_{0.2}TiO_x$ (107.9 m^2/g) is lower than that of $Ce_{0.2}TiO_x$ (125.7 m^2/g), yet the SCR activity of $Ce_{0.2}W_{0.2}TiO_x$ is actually much higher than that of $Ce_{0.2}TiO_x$, indicating that some synergistic effect must exist between Ce and W species. According to the XPS results, there are more surface Ce atoms dispersed on $Ce_{0.2}W_{0.2}TiO_x$ than those on $Ce_{0.2}TiO_x$, which indicate that the introduction of W species could induce more abundant CeO_2 crystallite and thus increase the amount of active sites on the catalyst surface. Furthermore, the existence of W species could also contribute to the NH_3-SCR reaction by their promotion effects.

No Ce or W species was observed in $Ce_{0.2}W_{0.2}TiO_x$ either in XRD or in Raman analysis (Figure 11), while only anatase TiO_2 was detected. Although the Ce-W containing species might already exceed the monolayer dispersion on TiO_2, the multi-layer active species probably only contained amorphous or crystallite phase with very small particle size (such as Ce-W solid solution without long-range order).

In such case, it is still reasonable to see no diffraction peaks of Ce or W species. In the XRD analysis, a slightly higher crystallization degree of anatase TiO_2 in the bulk phase was observed in $Ce_{0.2}W_{0.2}TiO_x$ than that in $Ce_{0.2}TiO_x$. However, the Raman spectra apparently showed that the introduction of W species strongly inhibited the crystallization of anatase TiO_2 on the catalyst surface, probably leading to a much higher dispersion degree of active Ce species.

The *in situ* DRIFTS of NH_3 adsorption at 200 °C were measured to examine the change of acidity on the catalyst after W addition (Figure 12). After NH_3 adsorption, obviously stronger NH_3 (1595 and 1208 cm^{-1}) and NH_4^+ (1671 and 1428 cm^{-1}) bands were observed on $Ce_{0.2}W_{0.2}TiO_x$, despite its low surface area, than those on $Ce_{0.2}TiO_x$, indicating that the introduction of W species tremendously increased both of the Lewis and Brønsted acid sites on the catalyst. It is worth noting that the adsorbed H_2O on the surface of $Ce_{0.2}W_{0.2}TiO_x$ showed an obvious consumption at 1625 cm^{-1}, and at the same time much more hydroxyl groups were consumed on $Ce_{0.2}W_{0.2}TiO_x$ than those on $Ce_{0.2}TiO_x$ owing to the interaction with NH_3 adsorbed species to form NH_4^+. This result clearly indicates that H_2O can be readily

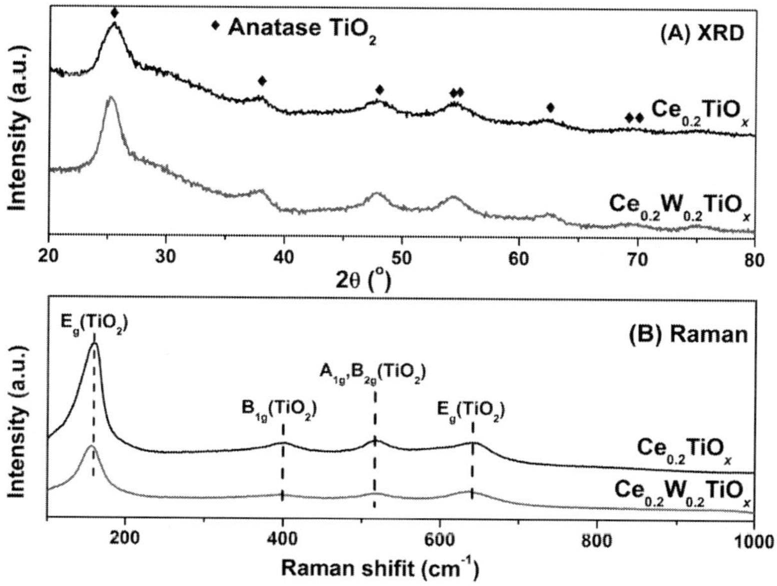

Figure 11. (A) XRD and (B) Raman spectra (λ_{ex} = 532 nm) of $Ce_{0.2}TiO_x$ and $Ce_{0.2}W_{0.2}TiO_x$ catalysts [33]. (Reproduced by permission of Elsevier)

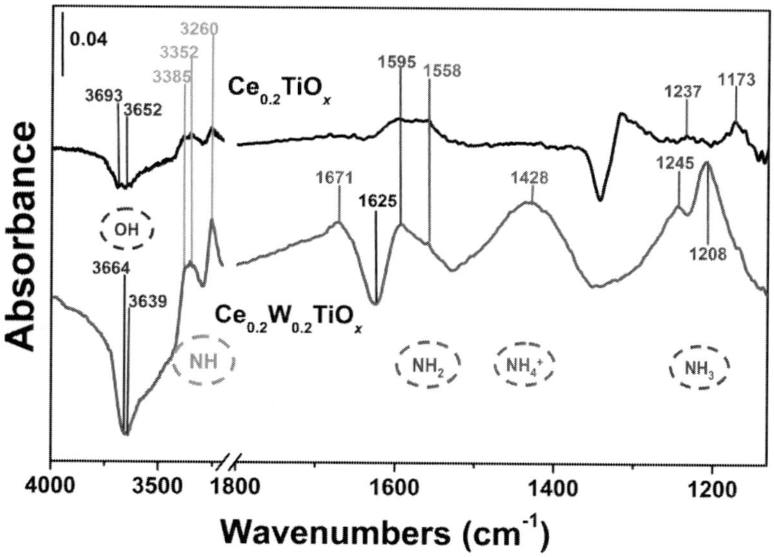

Figure 12. *In situ* DRIFTS of NH_3 adsorption at 200 °C over $Ce_{0.2}TiO_x$ and $Ce_{0.2}W_{0.2}TiO_x$ catalysts [33]. (Reproduced by permission of Elsevier)

activated on the W doped catalyst to form Brønsted acid sites, participating in the NH_3-SCR reaction. It has been indicated that surface acidity plays an important role in the adsorption and activation of NH_3, and thus affects the occurrence of SCR reaction, especially in the high temperature range [50,53]. Therefore, the increased Brønsted and Lewis acid sites by W addition should be an important reason for the enhanced catalytic activity, especially the high temperature activity.

5. EXCELLENT $CeWO_x$ CATALYST FOR NH_3-SCR OF NO_x

5.1. NH_3-SCR Activity of $CeWO_x$ Catalyst

Based on the understanding of W effects in $CeWTiO_x$ catalyst, we further investigated the effect of total substitution of Ti species in the catalyst by W species. Figure 13 shows the NH_3-SCR activity of $Ce_aW_bO_x$ catalysts under a fixed GHSV of 250,000 h^{-1}. Without W doping, the pristine CeO_x showed rather poor SCR activity in the whole temperature range with the maximum NO_x conversion of 26%. With the Ce/W molar ratio being 2:1, the operation temperature window of $Ce_2W_1O_x$ catalyst was greatly broadened, with 100% NO_x conversion from 250 to 400 °C. This result indicates that synergistic effect for NH_3-SCR reaction might exist between Ce and W species. For $CeWO_x$ catalyst, both the low temperature SCR activity (< 250 °C) and high temperature SCR activity (> 400 °C) were further improved to a certain extent. Decreasing the Ce/W molar ratio to 1:2 in $Ce_1W_2O_x$ catalyst led to a severe decline of NO_x conversion in the whole temperature range, which was probably due to the decrease of active Ce sites in catalyst. The pristine WO_x sample also showed poor SCR activity, indicating again that the coexistence of Ce and W species is very important for the promotion of SCR activity.

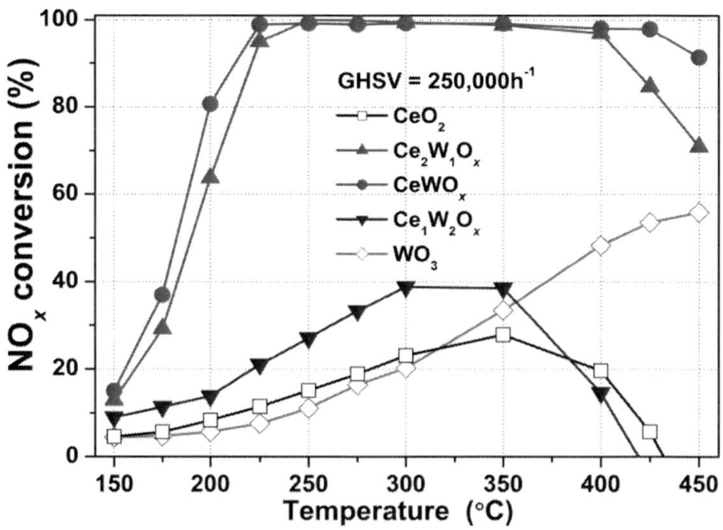

Figure 13. NH_3-SCR activity of $Ce_aW_bO_x$ (a:b = 2:1, 1:1 and 1:2) serial catalysts [34]. (Reproduced by permission of The Royal Society of Chemistry)

Figure 14 shows the NH_3-SCR activity of $CeWO_x$ catalyst under different GHSV. Obviously, the increase of GHSV only resulted in some decrease of low temperature SCR activity, while no obvious influence on the high temperature SCR activity was observed. The NO_x conversion curve slightly shifted to high temperature and nearly 100% NO_x conversion could still be achieved in a wide temperature range from 250 to 425 °C under a rather high GHSV of 500,000 h^{-1}, suggesting that this novel $CeWO_x$ catalyst is highly resistant to the large space velocity, which is crucial for its application on diesel vehicle with limited space on board. In addition, a test of SO_2 effect was also carried out in the presence of 100 ppm SO_2 at 250 °C under a GHSV of 250,000 h^{-1}. The NO_x conversion did not show any decrease in 24 h, indicating the high poisoning resistance of the catalyst.

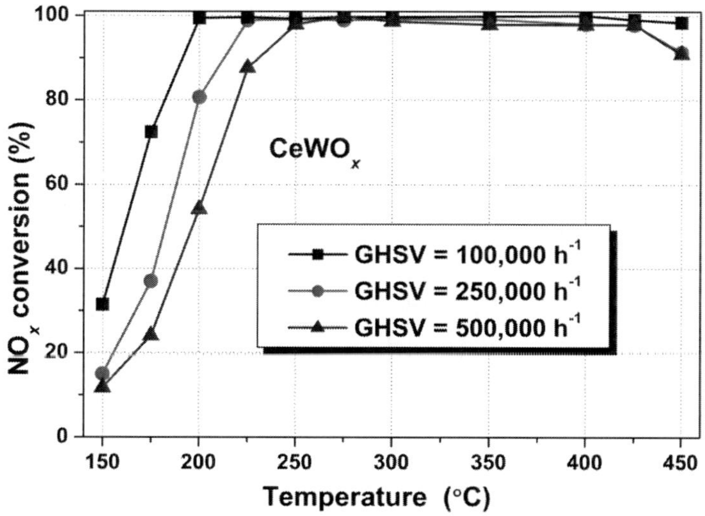

Figure 14. NH_3-SCR activity of $CeWO_x$ catalyst under different GHSV [34]. (Reproduced by permission of The Royal Society of Chemistry)

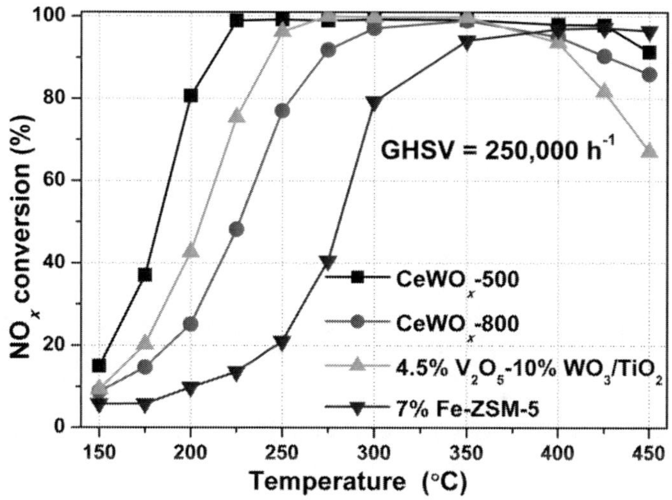

Figure 15. Comparison of NH_3-SCR activity of $CeWO_x$ catalyst (calcined at 500 and 800 °C) with those of V_2O_5-WO_3/TiO_2 and Fe-ZSM-5 catalysts [34]. (Reproduced by permission of The Royal Society of Chemistry)

To better evaluate the NH_3-SCR performance of $CeWO_x$ catalyst, we also carried out the comparative SCR activity tests over V_2O_5-WO_3/TiO_2 and Fe-ZSM-5 catalysts (Figure 15). The $CeWO_x$ catalyst showed much higher SCR activity in the whole temperature range than V_2O_5-WO_3/TiO_2, together with much better N_2 selectivity. The broad operation temperature window and less formation of nitrogenous by-products are important characteristics of $CeWO_x$ catalyst for its practical utilization. Comparing with Fe-ZSM-5 catalyst which has been industrially and commercially used for NH_3-SCR of NO_x from diesel engines, the operation temperature window of $CeWO_x$ catalyst is nearly 100 °C lower under the same GHSV. Even after high temperature calcination at 800 °C, the low temperature activity of $CeWO_x$ was still much higher than that of Fe-ZSM-5, indicating its high resistance to the thermal shock, which is also very important for practical use. This novel $CeWO_x$ catalyst is another potential candidate for the catalytic NO_x removal from diesel engines with wide exhaust temperature range and limited installation space for SCR converter.

5.2. Characterizations of $CeWO_x$ Catalyst

The XRD analysis (Figure 16) suggest that the coexistence of Ce and W species in $CeWO_x$ catalyst greatly lowered the particle size of cerianite CeO_2, and W species itself was mainly in amorphous or highly dispersed state. Highly dispersed nanocrystalline CeO_2 is thought to be the active phase in Ce-Ti catalyst. According to the XRD results in this study, we can deduce that the main active phase in $Ce_aW_bO_x$ is also small CeO_2 crystallite, and the formation of highly dispersed CeO_2 and WO_3 crystallites by the coexistence of Ce and W could extensively increase the amount of active sites on catalyst surface. After normalization by BET surface area, the sequence of the SCR activity is as follows: $Ce_2W_1O_x > CeWO_x > Ce_1W_2O_x$, confirming that the catalyst with the most abundant CeO_2 crystallite exhibited the highest intrinsic activity.

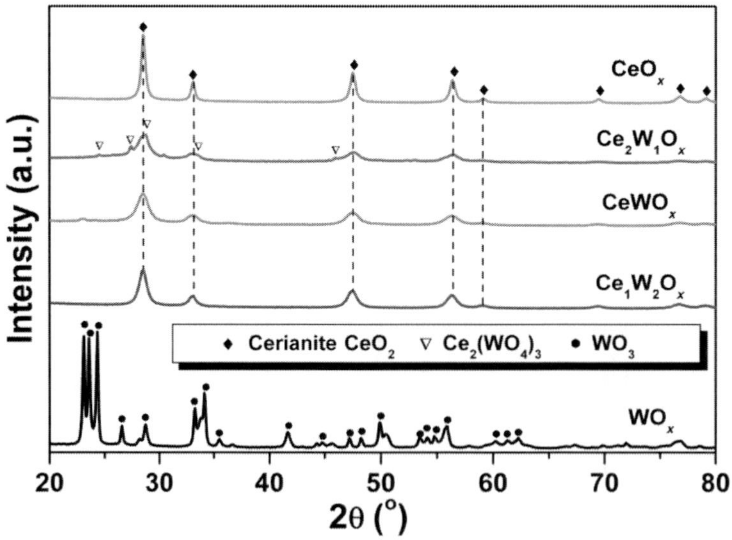

Figure 16. XRD patterns of $Ce_aW_bO_x$ (a:b = 2:1, 1:1 and 1:2) serial catalysts [34]. (Reproduced by permission of The Royal Society of Chemistry)

As the *in situ* DRIFTS shown in Figure 17, the inclusion of W species in $CeWO_x$ greatly enhanced the NH_3 adsorption (*e.g.* 1668 and 1419 cm^{-1} of NH_4^+, 1591 and 1190 cm^{-1} of NH_3), yet showed no obvious inhibition on NO_x adsorption (*e.g.* 1601 and 1232 cm^{-1} of bridging nitrate, 1578 cm^{-1} of bidentate nitrate, 1549 cm^{-1} of monodentate nitrate). Only the difference in surface areas between CeO_x and $CeWO_x$ cannot explain the discrepancy of the reactant adsorption ability, thus some synergistic effect must exist between Ce and W. Moreover, the adsorbed NH_3 species (*i.e.* ionic NH_4^+ and coordinated NH_3) and adsorbed NO_x species (*i.e.* bridging nitrate and monodentate nitrate) on $CeWO_x$ surface were both reactive to each other, contributing to the total $DeNO_x$ efficiency. From this point of view, the coexistence of Ce and W species with synergistic effect in $CeWO_x$ is crucial for achieving high NH_3-SCR performance.

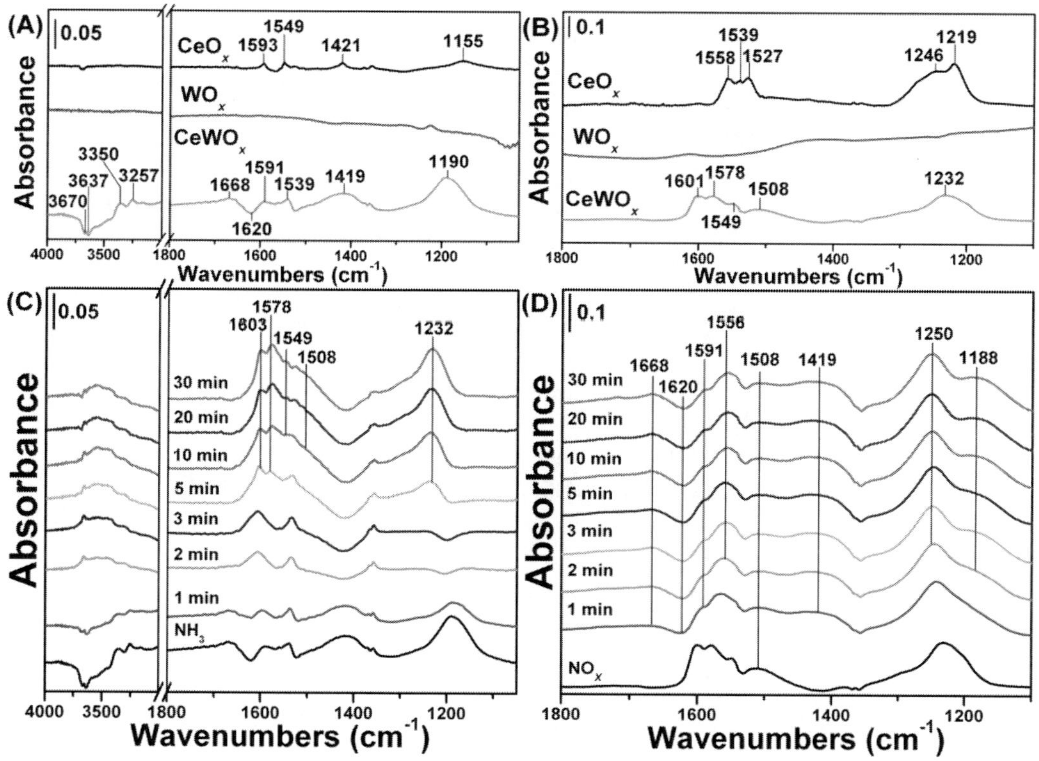

Figure 17. *In situ* DRIFTS of (A) NH_3 adsorption, (B) NO + O_2 adsorption, (C) NO + O_2 reacted with pre-adsorbed NH_3, and (D) NH_3 reacted with pre-adsorbed NO_x at 200 °C on $CeWO_x$ catalyst [34]. (Reproduced by permission of The Royal Society of Chemistry)

CONCLUSION

In this chapter, in order to combine the advantages of cerium and titanium species into one catalyst for NH_3-SCR reaction, for the first time, we developed a novel supported type Ce/TiO_2 material using conventional impregnation method for this reaction and found that a high catalytic performance was obtained in a relatively wide temperature range with 100% N_2

selectivity. However, this catalyst is very sensitive to the high GHSV, which should be improved for reducing the catalyst volume for the SCR converter on diesel engines.

A Ce-Ti mixed oxide catalyst ($Ce_{0.2}TiO_x$) was prepared by a facile homogeneous precipitation method. Comparing with the previously reported Ce-Ti catalysts, this catalyst showed remarkably improved low temperature SCR activity. Further optimization of the preparation method resulted in significantly enhanced high temperature activity and further broadened the operation temperature window. In addition, the SCR activity under high space velocity condition was also obviously improved. The process of homogeneous precipitation is crucial for the production of highly dispersed active CeO_2 on TiO_2, which is very important for the high SCR activity of $Ce_{0.2}TiO_x$ catalyst.

A superior $CeWTiO_x$ mixed oxide catalyst was prepared by doping W into $Ce_{0.2}TiO_x$ catalyst. The $Ce_{0.2}W_{0.2}TiO_x$ catalyst showed both enhanced low temperature activity and high temperature activity simultaneously combined with enhanced N_2 selectivity, comparing with those over the undoped $Ce_{0.2}TiO_x$ catalyst. Investigation on the promotion effects of W speices on the catalyst showed that, although the BET surface area of $Ce_{0.2}W_{0.2}TiO_x$ catalyst was decreased to some extent, the introduction of W species could increase the amount of surface active CeO_2 crystallite and the amount of surface Brønsted and Lewis acid sites, which is beneficial to improve the low temperature activity by facilitating “fast SCR” reaction and enhance both of the high temperature activity and N_2 selectivity by inhibiting the unselective oxidation of NH_3 at high temperatures.

Based on the understanding of W effects in $CeWTiO_x$ catalyst, a novel $CeWO_x$ catalyst, with a Ce/W molar ratio of 1:1, was developed for the SCR of NO_x with NH_3. The $CeWO_x$ catalyst even presented nearly 100% NO_x conversion in a wide temperature range from 250 to 425 °C under an extremely high GHSV of 500,000 h^{-1}. It also exhibited excellent N_2 selectivity, good stability, and high resistance to poisoning. Under the same test conditions, the $CeWO_x$ catalyst showed much better SCR performance than the V_2O_5-WO_3/TiO_2 and Fe-ZSM-5 catalysts, which have been industrially and commercially used for NH_3-SCR of NO_x from diesel engines. Several characterizations clearly showed that, the excellent catalytic performance of the $CeWO_x$ catalyst is attributed to a synergetic effect between Ce and W species.

ACKNOWLEDGMENTS

The authors sincerely thank the former group member Dr. Wenqing Xu for the study on supported type CeO_x/TiO_2 catalyst. The work in this chapter was partially funded by the National Natural Science Foundation of China (50921064, 51108446), the National Basic Research Program of China (2010CB732304), the National High Technology Research and Development Program of China (2012AA062506), and the Specialized Research Foundation for the Gainer of Outstanding Doctoral Thesis and Presidential Scholarship of Chinese Academy of Sciences.

REFERENCES

[1] Crutzen, P.J., Ramanathan, V. *Science,* 2000, *290*, 299.

[2] Finlayson-Pitts, B.J., Pitts, J.N. *Science,* 1997, *276*, 1045.

[3] Galloway, J.N., Townsend, A.R., Erisman, J.W., Bekunda, M., Cai, Z., Freney, J.R., Martinelli, L.A., Seitzinger, S.P., Sutton, M.A. *Science,* 2008, *320*, 889.

[4] Ravishankara, A.R. *Science*, 1997, *276*, 1058.

[5] Bosch, H., Janssen, F. *Catal. Today*, 1988, *2*, 369.

[6] Centi, G., Perathoner, S. Studies in Surface Science and Catalysis. Elsevier. 2007, 1.

[7] Granger, P., Parvulescu, V.I. *Chem. Rev*., 2011, *111*, 3155.

[8] Liu, Z., Woo, S.I. *Catal. Rev*., 2006, *48*, 43.

[9] Johnson, T.V. *Int. J. Engine Res*., 2009, *10*, 275-285.

[10] Busca, G., Lietti, L., Ramis, G., Berti, F. *Appl. Catal. B*, 1998, *18*, 1.

[11] Roy, S., Hegde, M.S., Madras, G. *Appl. Energ., 2009*, *86*, 2283.

[12] Kwak, J.H., Tran, D., Szanyi, J., Peden, C.H.F., Lee, J.H. *Catal. Lett.,* 2012, *142*, 295.

[13] Long, R.Q., Yang, R.T. *J. Am. Chem. Soc*., 1999, *121*, 5595.

[14] Høj, M., Beie, M.J., Grunwaldt, J.D., Dahl, S. *Appl. Catal. B,* 2009, *93*, 166.

[15] Sultana, A., Nanba, T., Haneda, M., Hamada, H. *Catal. Commun*., 2009, *10*, 1859.

[16] Kwak, J.H., Tonkyn, R.G., Kim, D.H., Szanyi, J., Peden, C.H.F. *J. Catal.,* 2010, *275*, 187.

[17] Mou, X., Zhang, B., Li, Y., Yao, L., Wei, X., Su,D., Shen W. *Angew. Chem. Int. Ed*., 2012, *52*, 2479.

[18] Liu, F., He, H., Zhang, C. *Chem. Commun*., 2008, 2043.

[19] Liu, F., He, H., Zhang, C., Feng, Z., Zheng, L., Xie, Y., Hu, T. *Appl. Catal. B: Environ*., 2010, *96*, 408.

[20] Liu, F., He, H. *J. Phys. Chem. C,* 2010, *114*, 16929.

[21] Ma, L., Li, J., Ke, R., Fu, L. *J. Phys. Chem. C*, 2011, *115*, 16929.

[22] Apostolescu, N., Geiger, B., Hizbullah, K., Jan, M.T., Kureti, S., Reichert, D., Schott, F., Weisweiler, W. *Appl. Catal. B: Environ*., 2006, *62*, 104.

[23] Liu, Q., Liu, Z., Su, *J. Catal. Today*, 2010, *158*, 370.

[24] Si, Z., Weng, D., Wu, X., Jiang, Y., Wang, B. *Catal. Sci. Technol*., 2011, *1*: 453.

[25] Thirupathi, B., Smirniotis, P.G. *Appl. Catal. B: Environ*., 2011, *110*, 195.

[26] Yang, S., Wang, C., Li, J., Yan, N., Ma, L., Chang, H. *Appl. Catal. B: Environ*., 2011, *110*, 71.

[27] Si, Z., Weng, D., Wu, X., Yang, J., Wang, B. *Catal. Commun*., 2010, *11*, 1045.

[28] Li, Y., Cheng, H., Li, D., Qin, Y., Xie, Y., Wang, S. *Chem. Commun.,* 2008, 1470.

[29] Chen, L., Li, J., Ge, M., Zhu, R. *Catal. Today*, 2010, *153*, 77.

[30] Xu, W., Yu, Y., Zhang, C., He, H. *Catal. Commun*., 2008, *9*, 1453.

[31] Shan, W. Liu, F., He, H., Shi, X., Zhang, C. *ChemCatChem*, 2011, *3*, 1286.

[32] Shan, W. Liu, F., He, H., Shi, X., Zhang, C. Catal. *Today,* 2012, *184*, 160.

[33] Shan, W. Liu, F., He, H., Shi, X., Zhang, C. *Appl. Catal. B: Environ*., 2012, *115-116*, 100.

[34] Shan, W. Liu, F., He, H., Shi, X., Zhang, C. *Chem. Commun*., 2011, *47*, 8046.

[35] Gao, X., Jiang, Y., Zhong, Y., Luo, Z., Cen, K. *J. Hazard. Mater*., 2010, *174*, 734.

[36] Gao, X., Jiang, Y., Fu, Y., Zhong, Y., Luo, Z., Cen, K. *Catal. Commun.,* 2010, *11*, 465.

[37] Trovarelli, A. *Catal. Rev. Sci. Eng.*, 1996, *38*, 439.
[38] Damyanova, S., Perez, C.A., Schmal, M., Bueno, J.M.C. *Appl. Catal. A,* 2002, *234*, 271.
[39] Reddy, B.M., Khan, A., Yamada, Y., Kobayashi, T., Loridant, S., Volta, J.C. *J. Phys. Chem. B,* 2003, *107*, 5162.
[40] Long, R.Q., Yang, R.T. *J. Am. Chem. Soc.*, 1999, *121*, 5595.
[41] Long, R.Q., Yang, R.T. *Appl. Catal. B,* 2000, *27*, 87.
[42] Qi, G., Yang, R.T. *J. Catal.*, 2003, 217, 434.
[43] Ito, E., Hultermans, R.J., Lugt, P.M., Burgers, M.H.W., Rigutto, M.S., Bekkum, H. Van, Van den Bleek, C.M. *Appl. Catal. B*, 1994, *4*, 95.
[44] Nakajima, F., Hamada, I. *Catal. Today,*1996, *29*, 109.
[45] Xu, W., He, H., Yu, Y. *J. Phys. Chem.* C, 2009, *113*, 4426.
[46] Topsøe, N.Y. *Science,* 1994, *265*, 1217.
[47] Topsøe, N.Y., Dumesic, J.A., Topsøe, H., *J. Catal.,* 1995, *151*, 241.
[48] Lietti, L., Nova, I., Forzatti, P. *Top. Catal.*, 2000, *11/12*, 111.
[49] Lietti, L., Forzatti, P., Bregani, F. *Ind. Eng. Chem. Res.*, 1996, *35*, 3884.
[50] Lietti, L., Forzatti, P., Berti, F. *Catal. Lett.*, 1996, *41*, 35.
[51] Wachs, I.E., Deo, G., Weckhuysen, B.M., Andreini, A., Vuurman, M.A., Boer, M., Amiridisz, M.D. *J. Catal.*, 1996, *161*, 211.
[52] Chen, J.P., Yang, R.T. *Appl. Catal. A,* 1992, *80*, 135.
[53] Alemany, L., Lietti, L., Ferlazzo, N., Forzatti, P., Busca, G., Giamello, E., Bregani, F. *J. Catal.*, 1995, *155*, 117.

In: Cerium
Editors: Aleksey Izyumov and Gulnaz Plaksin
ISBN: 978-1-62257-670-8

Chapter 6

EFFECT OF INTERFACIAL INTERACTIONS ON MODIFICATION OF CERIUM OXIDE AND ITS REDUCTION IN COMPOSITE MATERIALS UNDER HYDROGEN CONTAINING ATMOSPHERE

Elena Yu. Konysheva*
School of Chemistry, University of St Andrews, St Andrews, Fife, UK
Ural Federal University, Ekaterinburg, Russia

ABSTRACT

Ce-based materials have attracted much attention because of their applications as electronic ceramics, membranes, electrodes in fuel cells, and catalysts. The redox activity of cerium cations, which is the basis of its application as the catalysts, was usually observed in undoped cerium oxide at elevated temperatures and low oxygen partial pressure. One of the questions remained to be answered is whether cerium oxide in a composite material would reduce during long-term exposure to hydrogen containing atmosphere. If it is, how wide extend would be. To explore this aspect, the redox activity of the composite systems $La_{0.6}Sr_{0.4}CoO_3$ - CeO_2 (LSCCe), $La_{0.95}Ni_{0.6}Fe_{0.4}O_3$ - CeO_2 (LNFCe), and $La_{0.8}Sr_{0.2}MnO_3$ - CeO_2 (LSMCe) fabricated through the reactive sintering in air at 1350 °C was studied under hydrogen containing atmosphere. During the reactive sintering the cross-dissolution of cations at the phase interfaces occurs and results in the modification of the initial phases, forming cerium oxides with different degree of oxygen deficiency. The degree of the cross-dissolution at the phase interfaces increases in the following sequence: LSMCe → LSCCe → LNFCe. The results obtained in the present study demonstrate that reduction of modified cerium oxides in composite materials with highly reducible transition metal cations occurs under hydrogen atmosphere at the initial stage of reduction. During long-term exposure to H_2-Ar atmosphere the involvement of cerium oxide in the reduction process seems to depend on its lattice parameter that determines the length of the pathway for oxygen ion transport through the lattice.

* E-mail: elena.konysheva@googlemail.com

Kinetics of the reduction of composite materials at 800 °C has been discussed in more details.

Keywords: cerium oxide; reduction; composites; Mn, Co, Ni and Fe containing perovskites; Thermogravimetric analysis; X-ray powder diffraction; X-ray photoelectron spectroscopy

1. INTRODUCTION

Material designs and composite architectures have unique benefits over conventional single-phase materials. They can provide effective multi-functional properties, thereby advancing the development of solid state electrochemical devices and their effectiveness for solving energy and environmental issues. Composite materials composed of a perovskite phase with transition metal cations and a rare–earth metal oxide with fluorite structure have been considered as electronic ceramics [1, 2], membranes [3, 4], and electrodes for fuel cells [5-10]. $CeO_{2-\delta}$ and $A_{1-x}A'_xB_{1-x}B'_xO_3$ (A,A′ =La, Sr, Ce; B,B′ = Co, Ni, Fe, Mn) perovskite containing materials have been also explored for a wide variety of catalytic applications: NO reduction by H_2 over perovskite-like mixed oxides [11], hydrogenation of ethylene [12] and benzoic acid [13]; oxidation of carbon monoxide [14,15]; oxidation of *methane*, others hydrocarbons and alcohols [15-17]; peroxide decomposition [18]; and formulation of SO_x reduction catalysts [15].

The following redox process (Eq. 1) in cerium oxide is the basis of its application as catalysts

$$Ce^{4+} + 1e^- \leftrightarrow Ce^{3+} \tag{1}$$

It varies with temperature, oxygen partial pressure, size of crystallites, and applied electrical field [19-24]. Reduction of cerium oxide is accompanied by the formation of oxygen vacancies (Eq. 2), which could play an important role in catalytic activity [25, 26].

$$O^X_O \rightarrow 2e^- + V^{\bullet\bullet}_O + ½\,O_2\uparrow \tag{2}$$

The summary process can be expressed as

$$2\,Ce^X_{Ce} + O^X_O \leftrightarrow 2\,Ce'_{Ce} + V^{\bullet\bullet}_O + ½\,O_2\uparrow \tag{3}$$

The family of cerium oxides with different anion deficiency is wide [27-30], for instance: CeO_2 (space group $Fm\bar{3}m$, no. 225, JCPDS 34-0394), Ce_7O_{12} (space group $R\bar{3}$, no. 148, JCPDS 089-8433), and $Ce_{11}O_{20}$ (space group $P\bar{1}$, no. 2, JCPDS 089-8435); although oxygen partial pressure below ~ 10^{-8} atm and high temperature are required to introduce oxygen deficiency in undoped CeO_2 [20,31]. On the other hand, oxygen deficiency in cerium oxide could be introduced under air through doping by cations with lower oxidation state. Depending on doping cations the modified oxygen deficient cerium oxide can retain the fluorite structure as it was observed for $(Ce_{1-x-z}La_xM_z)O_{2-\delta}$ (M=Ca, Sr) [32], $(Ce_{1-x-z}Gd_xPr_z)O_{2-\delta}$ [33], $Ce_xY_{0.2-x}Sc_{0.6}Zr_{3.2}O_{8-\delta}$ [34], $Ce_{0.905}La_{0.028}Sr_{0.019}Co_{0.048}O_{2-\delta}$ (δ =

0.081), $Ce_{0.859}La_{0.042}Sr_{0.028}Co_{0.071}O_{2-\delta}$ (δ = 0.120), and $Ce_{0.808}La_{0.058}Sr_{0.038}Co_{0.096}O_{2-\delta}$ (δ = 0.163) [35]. The change in crystal structure was also observed for doped oxygen deficient cerium oxide fabricated under air. The phase $(CeLa)_{11}O_{20}$ with triclinic structure (space group $P\bar{1}$, no. 2) and Ce_6O_{11} with monoclinic structure (space group *P*21/c, no. 14) were revealed by X-ray powder diffraction (XRD), respectively, in 75 mol% $La_{0.95}Ni_{0.6}Fe_{0.4}O_3$ – 25 mol % CeO_2 [36] and $La_{0.6}Ce_{0.4}CoO_3$ [37] compositions synthesized under air. The presence of $(CeLaSrCo)_{11}O_{20}$ nanophase at the surface of grains in 57 mol% $La_{0.6}Sr_{0.4}CoO_3$ - 43 mol% CeO_2 composition was confirmed by high-resolution transmission electron microscopy [38].

$La_{1-x}Sr_xB_{1-x}B'_xO_3$ (B,B$'$ = Co, Ni, Fe, Mn) perovskites can be easily reduced with the lowering oxygen partial pressure and they are of considerable interest as redox catalysts. In a composite material composed of a perovskite phase with transition metal cations and cerium oxide, a strong reducibility of a phase with transition metal cations can be combined with the reversible addition and removal of oxygen, thereby introducing an oxygen store for oxidation reactions. Reduction behavior of cerium oxide in the composite materials has not been fully understood yet. One of the questions to be clarified is whether cerium oxide in a composite material with highly reducible transition metal cations would undergo reduction at the initial stage and during long-term exposure to hydrogen containing atmosphere and how wide extend would be. Therefore, in the present chapter the reduction behavior of composite materials spontaneously formed during reactive sintering in the following systems $La_{0.6}Sr_{0.4}CoO_3$ - CeO_2 (LSCCe), $La_{0.95}Ni_{0.6}Fe_{0.4}O_3$ - CeO_2 (LNFCe), and $La_{0.8}Sr_{0.2}MnO_3$ - CeO_2 (LSMCe) has been explored under hydrogen containing atmosphere with particular focus on the reduction behavior of undoped and modified cerium oxide, perovskites, and composite materials containing both cerium oxide and perovskite. The modification of initial phases occurs because of the cross-dissolution of Ce, La, Sr, and transition metal cations at phase interfaces during the high-temperature sintering [36, 38]. The reduction behavior of materials can be studied by several complementary techniques: thermogravimetric analysis (TGA), temperature-programmed reduction (TPR), magnetic susceptibility measurements, Fourier-transform infrared (FTIR), UV–VIS diffuse reflectance, X-ray photoelectron spectroscopy (XPS), and X-ray powder diffraction (XRD). In the present chapter, the results obtained by thermogravimetric analysis and X-ray powder diffraction will be mainly discussed. At some stages data collected by temperature-programmed reduction and X-ray photoelectron spectroscopy will be also presented.

2. Reduction of Undoped and Multi-cation Doped Cerium Oxide

The reduction of CeO_2 by hydrogen starts at 200 °C irrespectively of the initial surface area of the cerium oxide (S_{BET} surface areas = 5 - 115 m^2/g) and it includes several steps [39]. According to temperature-programmed reduction (TPR), an intermediate reduction step was observed between 300-350 °C for the low-surface-area powder (S_{BET} = 5 m^2/g) and corresponded to the reduction of the surface and near surface area of grains. The reduction results in the formation of a formal composition $CeO_{1.83}$ for both types of powders. Temperature higher than 650 °C was required to reduce the cerium oxide further [39].

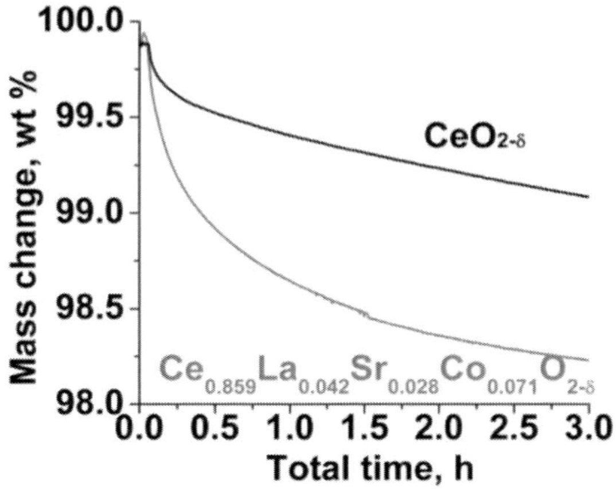

Figure 1. Mass change of CeO_2 and $Ce_{0.859}La_{0.042}Sr_{0.028}Co_{0.071}O_{2-\delta}$ during exposure to H_2-Ar atmosphere at 800 °C.

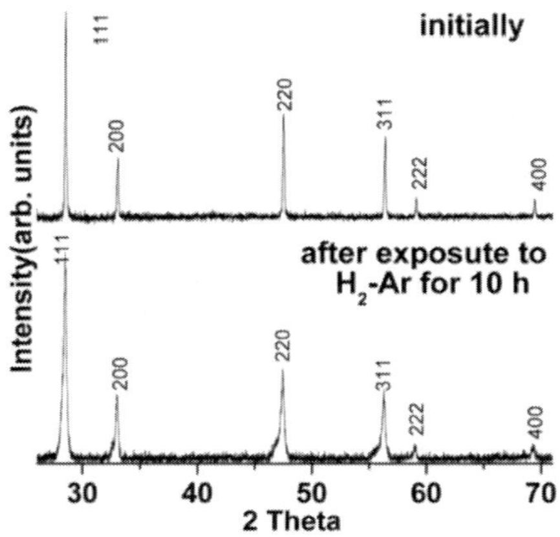

Figure 2. XRD patterns of $CeO_{2-\delta}$ initially and after exposure to H_2-Ar atmosphere at 800 °C for 10 h.

Figure 1 illustrates reduction behavior of undoped cerium oxide under H_2-Ar (5 % H_2-95 % Ar) atmosphere at 800 °C. At the initial stage of reduction (~0.5 h) undoped cerium oxide lost about 0.3 wt %. The weight of cerium oxide gradually decreases by about 1.6 wt % after 10 h that formally corresponds to $CeO_{1.82}$ composition and the appearance of Ce^{3+} cations, oxygen vacancies, and formation of water in the gaseous phase

$$2\,Ce^{X}_{Ce} + O^{X}_{O} + H_2 \leftrightarrow 2\,Ce^{/}_{Ce} + V^{\bullet\bullet}_{O} + H_2O\uparrow \quad (4)$$

The weight loss of $Ce_{0.859}La_{0.042}Sr_{0.028}Co_{0.071}O_{2-\delta}$ was stronger that could be related to the reduction of both Ce and Co cations to lower oxidation state (Eqs.4 - 6):

$$2\,Co^{/}_{Ce} + O^{X}_{O} + H_2 \leftrightarrow 2\,Co^{//}_{Ce} + V^{\bullet\bullet}_{O} + H_2O\uparrow \quad (5)$$

$$Co^{//}_{Ce} + O^{X}_{O} + H_2 \leftrightarrow Co^{////}_{Ce} + V^{\bullet\bullet}_{O} + H_2O\uparrow \quad (6)$$

After 10h the reduced cerium oxide changes color from light-yellow to blue-grey and retains the fluorite structure (Figure 2). However, a broadening of all peaks was observed in the XRD pattern of the cerium oxide, confirming its partial reduction. This indicates also that the bulk of grains is involved in the reduction process in undoped cerium oxide, suggesting the appearance of Ce^{3+} cations and oxygen vacancies not only on the surface of crystallites, but also in the bulk of the grains. The loss of the lattice oxygen within a single-phase chemical compound at a constant temperature could be described through the double exponential decay (Eq. 7) [40] representing the two processes (BULK and nSURF):

$$m(t) = K_0 + K_1^{BULK} \times \exp(-K_2^{BULK} \times t) + K_1^{nSURF} \times \exp(-K_2^{nSURF} \times t) \quad (7)$$

The "BULK" component is related to the mass transport from the internal part or core of grains/crystallites, whereas the "nSURF" component is related to the oxygen ion transport within the near surface area of grains and correlates stronger with the surface processes. Kinetics of the oxygen ion transport from the near surface area and the internal bulk of the same crystallite to the surface could be different because of the redistribution of elements between the bulk and surface of the crystallites as well as differences in an average length for the oxygen ion transport. Corresponding coefficients obtained from the fitting of the data (Fig. 1, 800 °C, H_2-Ar) are presented in Table 1.

The K_1^{BULK} and K_1^{SURF} coefficients obtained for CeO_2 is lower compared to that for $Ce_{0.859}La_{0.042}Sr_{0.028}Co_{0.071}O_{2-\delta}$. XPS analysis does not confirm the presence of Co cations on the surface of grains in $Ce_{0.905}La_{0.028}Sr_{0.019}Co_{0.048}O_{2-\delta}$ and $Ce_{0.808}La_{0.058}Sr_{0.038}Co_{0.096}O_{2-\delta}$ compounds with lower and higher Co content, respectively, than in $Ce_{0.859}La_{0.042}Sr_{0.028}Co_{0.071}O_{2-\delta}$. This implies that the presence of stronger reducible Co cations in the bulk of the grains of the single phase $Ce_{0.859}La_{0.042}Sr_{0.028}Co_{0.071}O_{2-\delta}$ composition could facilitate reduction process.

Table 1. Fitting parameters for the reduction of single phase CeO_2 and $Ce_{0.859}La_{0.042}Sr_{0.028}Co_{0.071}O_{2-□}$ with fluorite structure under H_2-Ar atmosphere at 800 °C

	R^2 parameter	K_0, mg[a] ± 0.009	K_1^{BULK}, mg [a] ± 0.003	K_2^{BULK}, min^{-1} [a] ± 0.001	K_1^{SURF}, mg [a] ± 0.008	K_2^{SURF}, min^{-1} [a] ± 5×10^{-5}
CeO_2	0.999	102.74	0.408	0.165	1.155	0.0038
$Ce_{0.859}La_{0.042}Sr_{0.028}Co_{0.071}O_{2-\delta}$	0.999	105.83	0.999	0.159	1.271	0.0142

[a]The maximum error.

3. Reduction of Perovskites with Transition Metal Cations on B-sites

The reduction behavior of perovskites $A_{1-x}A'_xB_{1-y-z}B'_yB''_zO_{3\pm\square}$ (A, A$'$ = La, Sr and B, B$'$, B$''$ = Ni, Mn, Fe, Co) was studied in H_2-Ar atmosphere by XRD [41] and by TGA during thermal cycling and during isothermal exposure at 800 °C [42]. The following perovskites were explored: $LaCoO_3$ (rhombohedral symmetry: space group $R\bar{3}c$, no. 167) [43], $La_{0.6}Sr_{0.4}CoO_3$ (LSC; rhombohedral symmetry: space group $R\bar{3}c$, no. 167) [44], $La_{0.872}Sr_{0.104}Ni_{0.288}Fe_{0.192}Mn_{0.52}O_{3-\delta}$ (LSMNF48; orthorhombic symmetry: space group Pbnm; standard setting: Pnma, no. 62) [45], $La_{0.8}Sr_{0.2}MnO_{3-\delta}$ (LSM; rhombohedral symmetry: space group $R\bar{3}c$, no. 167) [46], and $La_{0.795}Sr_{0.195}Ce_{0.01}Mn_{0.99}Ce_{0.01}O_{3-\delta}$ (LSMCe02; rhombohedral symmetry: space group $R\bar{3}c$, no. 167) [47]. The nominal composition $La_{0.95}Ni_{0.6}Fe_{0.4}O_{3-\delta}$ (LNF) contains the cation stoichiometric $LaNi_{0.579}Fe_{0.421}O_3$ perovskite with rhombohedral symmetry (space group $R\bar{3}c$, no. 167) and 5.2 mol % NiO (space group $R\bar{3}m$, no. 166) [48].

According to XRD, a rapid reduction of $LaCoO_3$ starts at temperatures as low as 375 °C and accompanied by formation of oxygen deficient brownmillerite $La_3Co_3O_8$ and $La_5Co_2O_5$ phases [41]. In contrast to $LaCoO_3$, $La_{0.6}Sr_{0.4}CoO_3$ is a faster ionic conductor [49-51]. It starts losing weight at the temperature of 300-320 °C that is lower by about 70-80 °C compared to LNF and LSMNF48 [42]. The onset reduction temperature of LNF and LSMNF48 is very close in spite of the difference in the average oxidation state of B cations, the degree of A-site deficiency and Sr-substitution, suggesting the involvement of Ni cations in the reduction process at the initial stage. The weight loss of LSM was observed after about 420 °C and showed slower kinetics.

Figure 3 illustrates the reduction behavior of perovskites during isothermal exposure to H_2-Ar at 800 °C. These experiments were carried out in order to explore in more detail the mechanism and kinetics of reduction. The weight of the samples changes continuously with time. For the perovskites investigated one can distinguish several stages (Figure 3). A slight weight gain (stage I) is followed by a strong weight loss (stage II). Further mechanism and kinetics of reduction (stage III) varies depending upon composition. The strongest weight gain among the perovskites investigated during stage I was observed for the LSC and might indicate direct sorption of hydrogen into the lattice, forming hydride-oxide phases [52 ,53]. Probably, there are a number of active sites at the surface, grain boundaries, or through volume of powders where chemical sorption of hydrogen could take place. Hydrogen atoms have a small size and could penetrate into the perovskite lattice, forming H^- centers near Me^{n+} cations, as was shown for $LaSrCoO_3H_{0.7}$ [52, 53]. On the other hand, the oxide lattice could reduce to form water and, then, the evolved water is reincorporated into the lattice to give a small weight gain. On the other hand, $\{Me^{n+}\text{-}O^{k-}\cdots H^{m+}\}$ intermediate complexes in the area closed to the surface or into the bulk of polycrystallites could form at the initial stage of exposure to H_2-Ar, thereby reducing the oxygen lattice. A sharp weight loss was observed during stage II. It lasts up to 0.5 h. In general, perovskites, depending on their composition, rapidly lose up to 6-12 mol% of the lattice oxygen which is accompanied by phase or structural transformations in the solid or expansion of a crystal lattice [42]. During stage II a strong reduction was observed for LNF and LSC, whereas that of LSMCe02 and LSMNF48 was noticeably lower, implying that LSMNF48 and LSMCe02 containing manganese cations are less reactive with respect to reduction than Co, Ni, and Fe containing perovskites. For LNF and LSMNF48, the weight loss at this stage was slightly higher than expected from the Ni content in the samples if the only process taking place is $Ni^{z+} \rightarrow Ni^{2+}$ (where z ~ 2.8 -3.0). The difference becomes slightly higher with the rise in the gas flow rate and the specific surface area for both LNF and LSMNF48 [42], suggesting reduction of Ni cations to metallic phase (although it was not observed by XRD) and partial reduction of manganese cations.

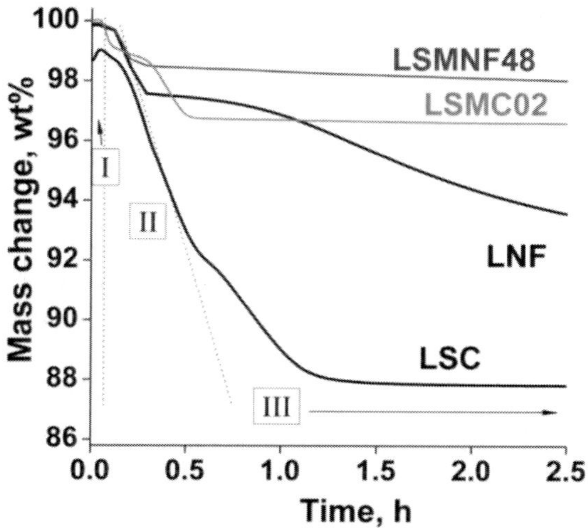

Figure 3. Reduction of perovskites during exposure to H_2-Ar atmosphere at 800 °C. Stages I, II and III are marked on the figure conditionally.

Reduction of manganese cations could also contribute to the weight loss after partial reduction of nickel cations [54], particularly for LSMNF48. A secondary phase ($LaFeO_3$) was detected in the XRD pattern of LNF, whereas the change in symmetry of LSMNF48 from orthorhombic to rhombohedral occurred at this stage. Several consecutive exothermic peaks were observed for LSC, indicating series of fast phase transformations in solid during isothermal exposure to H_2-Ar. According to XPS, both Mn^{4+} and Mn^{3+} cations are present on the B-sites within the perovskite structure of LSMCe02, assuming that the initial strong weight loss could be attributed to the reduction of Mn^{4+} to Mn^{3+} cations on the surface (mainly) and near surface area of crystallites. Water was found to be the main gaseous product formed during reduction.

All compositions showed completely different reduction behavior during long term exposure to H_2-Ar atmosphere (stage III). It seems to be difficult to distinguish stages II and III for LSC because it decomposes completely to La_2O_3, SrO and Co within about 1-2 h. On further reduction of LNF, NiO exsolves from the perovskite structure, forming $LaFeO_3$ phase, which in its turn gradually decomposes to La_2O_3 and metallic phase. Kinetics of the complete reduction of the LNF composition depends on the specific surface area and gas flow rate: the higher the specific surface area and the faster the flow rate, the faster the complete reduction.[42] Further reduction of LSMCe02 occurs through several steps and could be interpreted as complete reduction of Mn^{4+} to Mn^{3+} cations in the bulk of grains followed by further reduction $Mn^{3+} \rightarrow Mn^{2+}$ on the surface and in the bulk of crystallites. Notice that Mn containing perovskites can retain a certain fraction of Mn^{2+} cations on the B-sites before decomposition [55]. After 10 h of exposure to H_2-Ar, a small fraction of perovskite phase was observed in the XRD pattern of LSMCe02. Finally, it decomposes to La_2O_3, SrO and MnO. LSMNF48 shows the slowest reduction. Kinetics of its reduction depends on the gas flow rate. During further 4 h exposure to H_2-Ar at 800 °C, the expansion of the crystal lattice was observed and followed by partial decomposition of the perovskite structure. Higher temperature (~900 °C) is required for the complete reduction of LSMNF48 to La_2O_3, SrO, MnO, and $FeNi_x$.

4. Reduction of Composites Consisted of the Modified Cerium Oxide and Perovskites with Transition Metal Cations on B-sites

4.1. Phase Composition of Composites and Crystal Structure of Constituents

The composites in the *(100-x)* mol % $La_{0.6}Sr_{0.4}CoO_3$ – *x* mol % CeO_2 (LSCCex), *(100-x)* mol % $La_{0.95}Ni_{0.6}Fe_{0.4}O_3$ – *x* mol % CeO_2 (LNFCex), and *(100-x)* mol % $La_{0.8}Sr_{0.2}MnO_3$ – *x* mol % CeO_2 (LSMCex) systems were fabricated through mechanical mixing of a perovskite and CeO_2 in relevant ratios for 1 h followed by firing in air at 1350 °C for 5 h. LSCCe57, LSCCe76, LSMCe57, and LSMCe75 are two-phase compositions [40,56]. They consist of the modified perovskite (space group $R\bar{3}c$, no. 167) and modified cerium oxide (space group $Fm\bar{3}m$, no. 225). The modification of the initial components occurs due to cross-dissolution of Ce, La, Sr, and transition metal cations at the phase interfaces. About 2 mol % CeO_2 can dissolve in the perovskite structure of $La_{0.6}Sr_{0.4}CoO_3$ and $La_{0.8}Sr_{0.2}MnO_3$ [57]. Structural XRD analysis showed that incorporation of La cations into the fluorite structure of CeO_2 in the LSMCe system seems to be negligible since the *a* parameter is comparable for CeO_2, LSMCe57, and LSMCe75 (Figure 4).

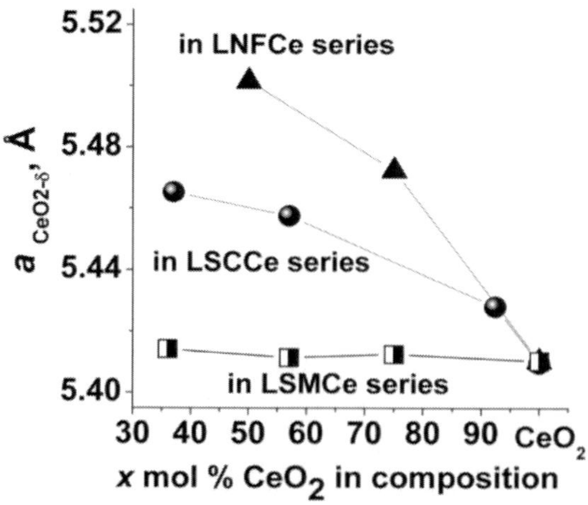

Figure 4. Comparison of the room temperature lattice parameter for the modified cerium oxide with fluorite structure ( ) in the LNFCe, LSCCe, and LSMCe compositions fabricated under air at 1350 oC. Data in the LNFCe series are presented for the main fraction.

In contrast, in the LSCCe system the dissolution of La, Sr, and Co cations in the fluorite structure of cerium oxide results in the expansion of the *a* parameter (Figure 4). The phase compositions in LNFCe50 and LNFCe75 are more complicated. According to structural XRD analysis, at least five phases present in LNFCe50: perovskite phase with rhombohedral symmetry (space group $R\bar{3}c$, no. 167), NiO (space group $R\bar{3}m$, no. 166), and $(CeLa)O_{2-\delta}$ phases (space group $Fm\bar{3}m$, no. 225) with different lattice parameters (major fraction with a = 5.5015(3) Å, which formally corresponds to the dissolution of ~ 30 mol% La and formal oxygen deficiency $\delta = 0.15$; minor fraction with a = 5.4673(7) Å, which formally corresponds to the dissolution of ~ 15 mol% La; and minor fraction with a = 5.4163(18) Å, which lattice parameter is very close to that for CeO_2) [36]. Four phases were revealed in LNFCe75 [36]: perovskite phase with rhombohedral symmetry (space group $R\bar{3}c$, no. 167), NiO (space group $R\bar{3}m$, no. 166), and $(CeLa)O_{2-\delta}$ phases (space group $Fm\bar{3}m$, no. 225) with a = 5.4725(1) Å (major fraction) and with a = 5.4280(2) Å (minor fraction). Dissolution of CeO_2 in LNF is less than 2 mol % [47]. Formation of $(CeLa)O_{2-\delta}$ phases with different lattice parameters in the LNFCe50 and LNFCe75 could occur due to an inhomogeneous distribution of La cations and due to specific dissolution of Ni and Fe cations in the fluorite structure of cerium oxide. Formation of $Ce_{0.9}Ni_{0.1}O_2$ solid solution, $FeCe_2O_4$, and $CeFeO_3$ phases after high temperature treatment was reported in the literature [58-60]. Figure 4 illustrates that $(CeLa)O_{2-\delta}$ phases in LNFCe50 and LNFCe75 have the largest lattice parameter among the systems investigated and suggest the longest pathway for oxygen ion transport from the bulk to the surface of crystallites.

4.2. Reduction of Composites

Perovskites undergo reduction at 800 °C under H_2-Ar atmosphere much stronger compared to undoped cerium oxides or doped by transition metal cations (Figures 1, 3). Therefore, one may expect that the reduction of a perovskite phase would dominate in composite systems with perovskite and cerium oxide phases, whereas the reduction of cerium oxide would be negligible. Figure 5 illustrates the reduction kinetics of composites containing a perovskite phase and cerium oxide (50 mol % or more) at the constant temperature of 800 °C under H_2-Ar atmosphere. The shape of the curves observed for the composite materials in the LSCCe, LSMCe, and LNFCe systems is very similar to that observed for the perovskites (Figures 3 and 5). One can distinguish several stages of the reduction like for perovskites. A small weight gain (stage I) was observed within up to 0.1 h for compositions in the LSCCe system only and it was proportional to the fraction of the modified perovskite phase in the LSCCe composites (Figure 6): the higher the fraction of the perovskite in the composite, the stronger the weight gain (Figures 3, 5). The sharp weight loss (stage II) was revealed for all compositions investigated (Figure 5). At first glance at this stage the weight loss is also proportional to the fraction of the modified perovskite phase in multi-phase compositions (the higher the fraction of the perovskite constituent in the composition, the stronger the weight loss), suggesting a strong contribution of transition metal cation reduction in the perovskite constituents to the total weight loss.

The compositional dependences of the weight loss for the LSCCe, LNFCe, and LSMCe compositions presented on Figures 7 and 8 show the following interesting tendency. We compare (1) the experimental data obtained in the thermogravimetric measurements (presented on Figure 7 as symbols with solid lines) with (2) the expected weight loss due to the reduction of transition metal cations in perovskites only (which are marked as the dash-dot lines) and (3) the expected weight loss due to the simultaneous reduction of transition metal cations in perovskites and Ce cations in CeO_2 with the assumption that ceria in composites contributes to the weight loss as an individual phase and proportionally to its content in the corresponding LSCCe, LSMCe or LNFCe compositions (which are marked as the dot lines).

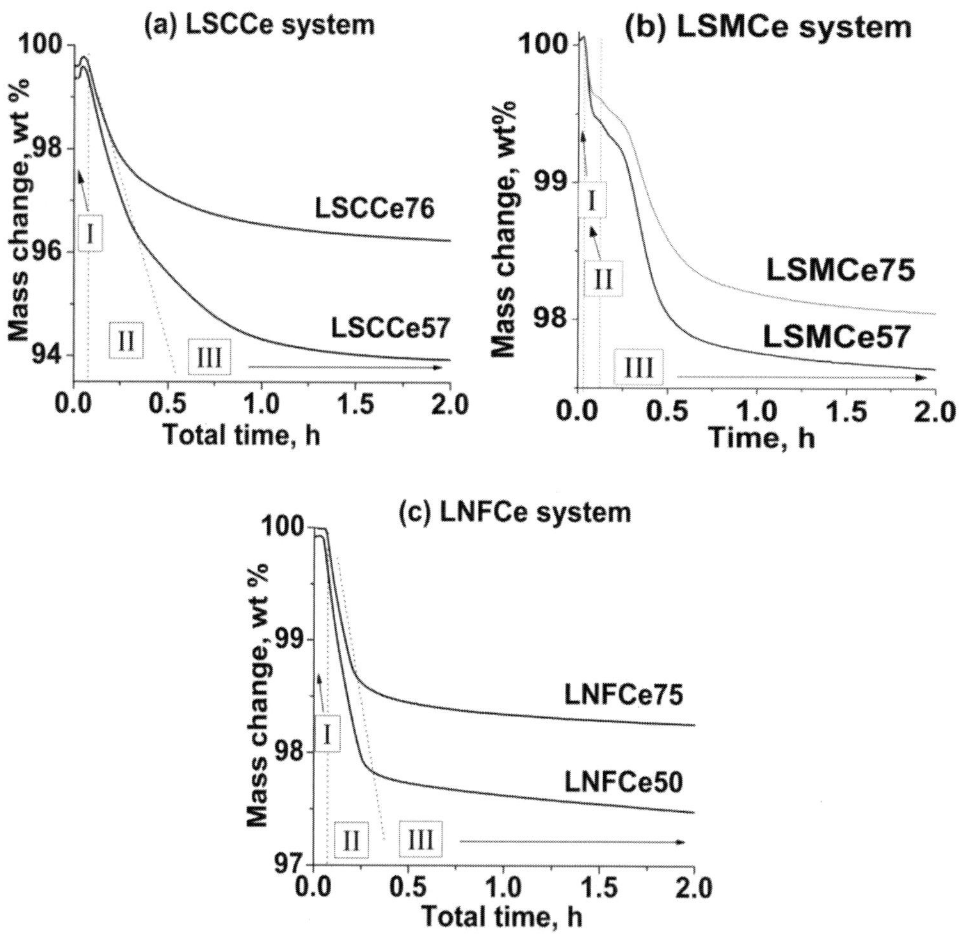

Figure 5. Mass change of compositions in (a) LSCCe, (b), LSMCe, and (c) LNFCe series during exposure to H_2-Ar atmosphere at 800 °C. The borders between stages I, II, and III are mark on figures conditionally. TGA data are presented after the subtraction of the corrections.

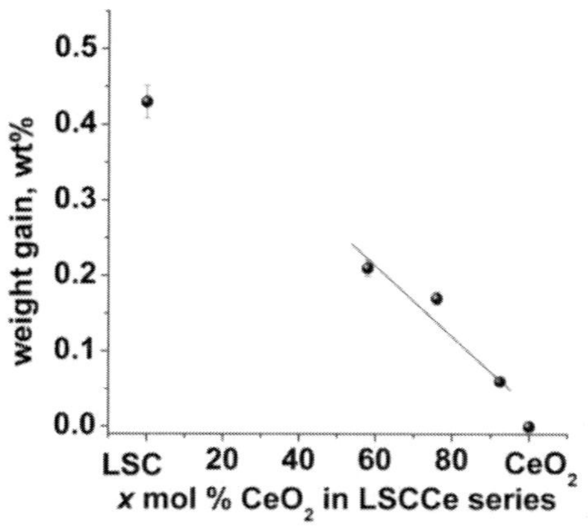

Figure 6. Weight gain in the LSCCe series at 800 °C during initial exposure to H_2-Ar atmosphere.

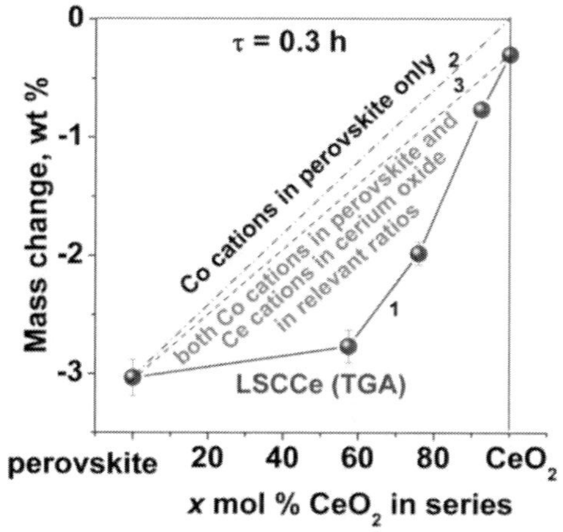

Figure 7. Comparison of (1) the experimental data obtained in the thermogravimetric measurements for LSCCe system after 0.3 h exposure to H_2-Ar atmosphere at 800 °C (symbols with solid line) with (2) the expected weight loss due to the reduction of transition metal cations in perovskites (dash-dot line) and (3) the expected weight loss due to the simultaneous reduction of transition metal cations in perovskites and Ce cations in CeO_2 in proportions corresponding to their fractions in the compositions (dot line).

The same analysis can be carried out with data presented on Figure 8. In fact, the total weight loss of compositions in the LSCCe, LSMCe and LNFCe series does not change linearly with the increase in cerium oxide content. At the initial stage of reduction the experimental weight loss is much stronger than if only the transition metal cations within the perovskite structure or if both a perovskite constituent and cerium oxide in proportions corresponding to their fractions in the compositions would undergo reduction (Figures 7, 8a). It means that the reduction of a perovskite constituent and cerium oxide occurs simultaneously. An additional conclusion is that a strong reduction of a perovskite constituent does not prevent or inhibit the reduction of cerium oxide at the initial stage of reduction. The strongest difference between the total weight loss observed experimentally and the expected weight loss due to the simultaneous reduction of transition metal cations in perovskites and Ce cations in CeO_2 was revealed for the LSCCe compositions (Figures 7, 8a), whereas the smallest difference was detected for the LSMCe compositions. On the one hand, the discrepancy observed could be explained by the dissolution of transition metal cations in the fluorite structure of CeO_2 and their ability to be involved in the reduction cycle. Under H_2-Ar atmosphere, the transition metal cations dissolved in the fluorite structure reduce, resulting in a stronger weight loss, as it was revealed for the single phase $Ce_{0.859}La_{0.042}Sr_{0.028}Co_{0.071}O_{2-\delta}$ (Figure 1). Co cations in oxide and complex oxide phases demonstrate much stronger involvement in the red-ox cycle (even during thermal cycling under air [40]) in contrast to Ni, Mn, and Fe cations. On the other hand, transition metal cations on the surface of grains in perovskite and fluorite phases could reduce to metallic phases, particularly Co and Ni cation, which is difficult to detect by XRD. Metallic phases could be perfect surface centers for hydrogen sorption, thereby facilitating reduction process in both modified perovskite phase and modified cerium oxide.

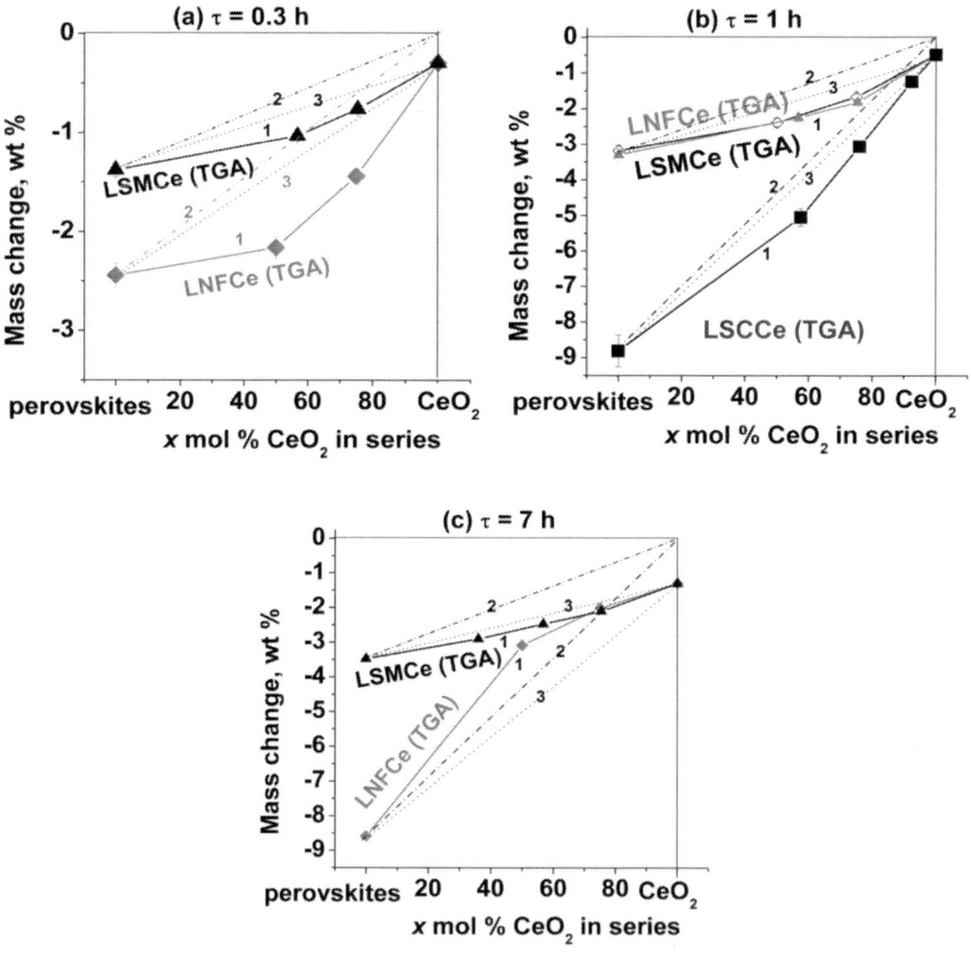

Figure 8. Compositional dependence of the weight loss in the LSCCe, LSMCe, and LNFCe series after exposure to H_2-Ar atmosphere at 800 °C for (a) 0.3 h; (b) 1 h, and (c) 7 h. (1) Experimental data obtained in TGA experiments are marked as solid lines with symbols; (2) the expected weight loss due to the reduction of transition metal cations in perovskites are shown as dash-dot lines; and (3) the expected weight loss due to the simultaneous reduction of transition metal cations in perovskites and Ce cations in CeO_2 in proportions corresponding to their fractions in the compositions marked as dot lines. Error bars are within the symbols.

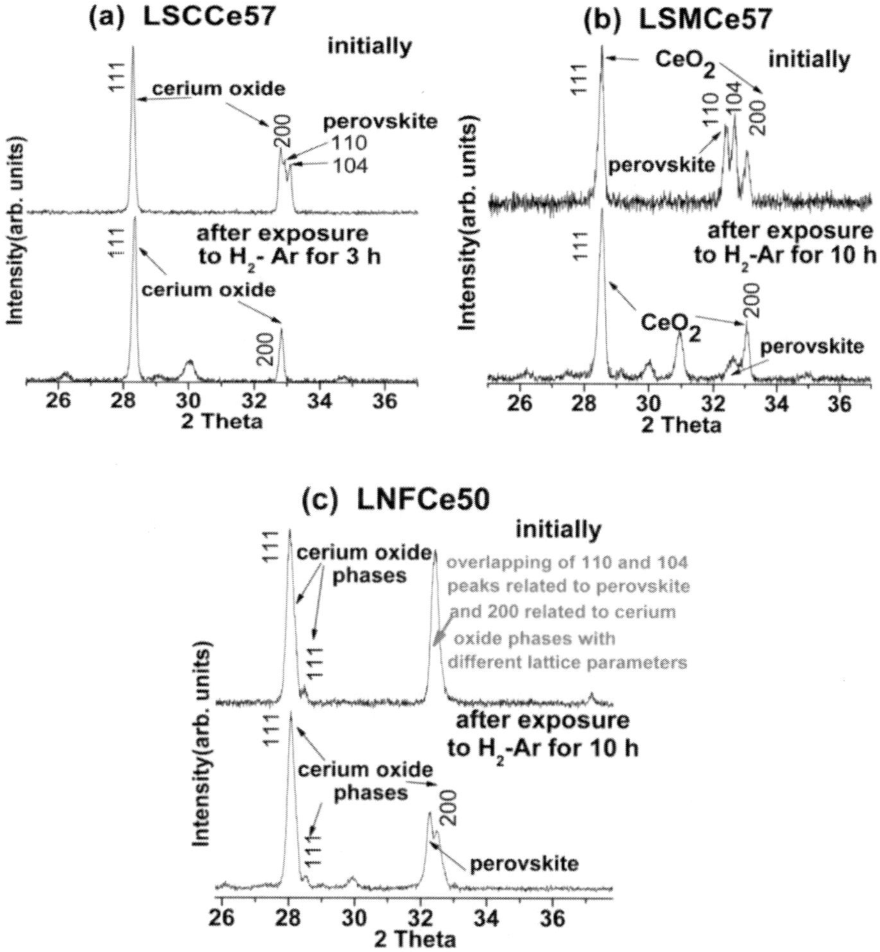

Figure 9. XPD patterns (initial and after reduction under H_2-Ar atmosphere at 800 °C) for (a) LSCCe57 after 3 h , (b) LSMCe57 after 10 h, and (c) LNFCe50 after 10 h.

The difference between the total weight loss observed experimentally and the expected weight loss due to the simultaneous reduction of transition metal cations in perovskites and Ce cations in cerium oxide becomes less for all three systems during further reduction at 800 °C (Figure 8b,c). That is reasonable because with time (stage III) the reduction process is not only limited by sorption of hydrogen on the surface, but also by oxygen ions transport from the bulk crystallites to the surface. During long-term reduction (Figure 8c) the weight loss measured for LNFCe75 was the same if only transition metal cations would be involved in the reduction process, implying that with time the modified cerium oxides become less involved in the reduction process. This was observed only in the LNFCe systems, where $(CeLa)O_{2-\delta}$ phases possess initially the largest lattice parameter and the longest pathway for oxygen ion transport through the crystal lattice with fluorite structure (Figure 4). Formally, the deviation from oxygen stoichiometry (δ) in $(CeLa)O_{2-\delta}$ with a = 5.4725(1) Å (major fraction in LNFCe75) is about 0.075, which is lower compared to $Ce_{0.859}La_{0.042}Sr_{0.028}Co_{0.071}O_{2-\delta}$ (δ = 0.120). Oxygen ion transport through the bulk is proportional to the concentration of oxygen vacancies, their charge and mobility. The latest, in its turn, correlates with the length of the pathway for O^{2-} transport through the crystal lattice. Surprisingly, the weight loss measured for LNFCe50 was even less than expected if only transition metal cations would be involved in the reduction process (Figure 8c). This suggests that the transition metal cations in LNFCe50 do not entirely reduce to the metallic phase and correlates with the XRD data (Figure 9).

According to XRD, the perovskite phase in LSCCe57 composition decomposes completely after 3 h of reduction. A small wide peak at about 32-33 degree 2Θ was observed in the XRD pattern of LSMCe57 after 10 h of reduction, suggesting the existence of a very small fraction of the perovskite phase. The peaks related to the perovskite phase in LNFCe50 underwent some changes, but still one can clearly distinguish peaks related to the perovskite phase (Figure 9c). Non-complete reduction of transition metal cations could be determined by several reasons.

First, as it was shown for the LSCCe system, fast oxygen ion transport through the perovskite lattice is important for a fast reduction of transition metal cations within the crystal lattice. Products of reduction and decomposition of perovskite are located on the top or at the surface of grains, thereby delaying fast oxygen ion transport from the bulk to the surface of crystallites. Another reasons resulting in the non-complete reduction of LNFCe compositions could be that Ni and Fe cations in the non-substituted cation stoichiometric perovskites ABO_3 demonstrate different reducibility. At 1000 °C metallic phase appears in $LaNiO_3$ at about 10^{-11} atm, where that in $LaFeO_3$ at about 10^{-17} atm [54], implying only partial reduction of Fe cations under our conditions, in particular, in the core of grains. The presence of the modified cerium oxide could also result in a detrimental effect similar to that from the products of reduction. In LSCCe system the presence of the modified cerium oxide changes kinetics of oxygen sorption and desorption under air [40]. After reduction up to 10 h, the peaks related to the modified cerium oxide phase(s) in the LSCCe, LSMCe, and LNFCe systems retain their shape and positions (Figure 9) in contrast to the undoped cerium oxide (Figure 2). Probably, the change in oxygen stoichiometry of ceria finely distributed through the volume of multi-phase compositions cannot be precisely monitored with XRD. Another explanation could be that only surface of ceria particles in multi-phase compositions is involved in the reduction process.

CONCLUSION

Reduction behavior of the $La_{0.6}Sr_{0.4}CoO_3$ - CeO_2, $La_{0.95}Ni_{0.6}Fe_{0.4}O_3$ - CeO_2, and $La_{0.8}Sr_{0.2}MnO_3$ - CeO_2 systems has been explored under hydrogen containing atmosphere. Reduction of multi-phase composite materials containing the modified perovskite and modified cerium oxide is the multi-step process including simultaneous reduction of the perovskite phase and cerium oxide. Perovskite phases could lose lattice oxygen, retaining their original perovskite structure; they could change their crystal symmetry or decompose to intermediate phases, oxides or pure metals/intermetallides depending of chemical origin of transition metal cations on the B-site. The presence of structural defects and their mobility in the oxygen sublattice are important factors determining thermochemical stability of perovskites. On long-term exposure H_2-Ar atmosphere, the perovskite containing three transition elements and Sr on the B and A sublattices, respectively, showed better thermochemical stability in hydrogen containing atmosphere. The results presented in the chapter demonstrate that reduction of cerium oxide at the initial stage of exposure to H_2-Ar atmosphere at 800 °C takes place regardless of whether it is present as an individual phase or as a second phase in a composite system containing perovskites with highly reducible transition metal cations into the B sublattice. On long-term exposure to H_2-Ar atmosphere, degree of the involvement of cerium oxide in the reduction process in multi-phase systems seems to depend on the initial lattice parameter of the modified cerium oxide that in its turn correlates with the length of the pathway for oxygen ion transport through the fluorite crystal lattice.

REFERENCES

[1] Yamada, T.; Sandu, C.S.; Gureev, M.; Sherman, V.O.; Noeth, A.; Muralt, P.; Tagantsev, A.K.; Setter N. *Adv. Mater.* 2009, *21*, 1363-1367.

[2] Wang, X.H.; Lu, W.Z.; Fan, G.F.; Zhou, D.X. *Key Eng. Mater.* 2007, *336-338*, 275-278.

[3] Yin, X.; Hong, L.; Liu, Z. *J. Membrane Sci.* 2006, *268*, 2-12.

[4] Kharton, V.V.; Kovalevsky, A.V.; Viskup, A.P.; Figueiredo, F.M.; Yaremchenko, A.A.; Naumovich, E.N.; Marques, F.M.B. *J. Eur. Ceram. Soc.* 2001, *21*, 1763-1767.

[5] Mogensen, M.; Lindegaard, T.; Hansen, U.R.; Mogensen, G. *J. Electrochem. Soc.* 1994, *141*, 2122-2128.

[6] Irvine, J.T.S. in *Perovskite Oxide for Solid Oxide Fuel Cells*; Ishihara, T.; Ed.; Fuel Cells and Hydrogen Energy; Springer Science + Business Media: New York, NY, 2009, pp 167-182.

[7] Zhu, X.; Lü, Z.; Wei, B.; Chen, K.; Liu, M.; Huang X.; Su, W. *J. Power Sources.* 2009, *190*, 326-330.

[8] Harta, N.T.; Brandon, N.P.; Daya, M.J.; Lapena-Rey, N. *J. Power Sources* 2002, *106*, 42-50.

[9] Konysheva, E.; Bonaccorso, A. D.; Irvine, J.T.S. *Russ. J. Electrochem.* 2010, *46*, 805-810.

[10] Antolini, E.; Perez, J. *Int. J. Hydrogen Energ.* 2011, *36*, 15752- 15765.

[11] Ferri, D *Catal. B: Environmental* 1998, *16*, 339-345.

[12] Petunchi, J.O.; Nicastro, J.L.; Lombardo, E.A. *J. Chem. Soc..*; Forni, L.; M.A.P. Dekkers, M.A.P.; Nieuwen, B.E. *Appl. Chem. Commun.* 1980, 467–468.

[13] Cheng, D.G.; Chong, M.; Chen, F.; Zhan, X. *Catal. Lett.* 2008, *120*, 82-85.

[14] Voorhoeve, R.J.H.; Remeika, J.P.; Freeland, P.E.; Matthias, B.T. *Science* 1972, *177*, 353–354.

[15] Trovarelli, A. *Catal. Rev. Sci. Eng.* 1996, *38*, 439-520.

[16] Lombardo, E.A.; Ulla, M.A. *Res. Chem. lntermed.* 1998, *24*, 581–592.

[17] Sin, A.; Kopnin, E.; Dubitsky, Y.; Zaopo, A.; Aricò, A.S., Gullo, L.R.; La Rosa, D.; Antonucci, V. *J. Power Sources* 2005, *145*, 68-73.

[18] Falcon, H.; Carbonio, R.E.; Fierro, J.L.G. *Catalysis* 2001, *203*, 264-272.

[19] Tuller, H.L.; Nowick, A.S. *J. Phys. Chem. Solids* 1977, *38*, 859-867.

[20] Tuller, H.L.; Nowick, A.S. *J. Electrochem. Soc.* 1979, *126*, 209-217.

[21] Blumenthal, R.N.; Sharma, R.K. *J. Solid State Chem.* 1975, *13*, 360-364.
[22] Tschöpe, A.; Sommer, E.; Birringer, R. *Solid State Ionics* 2001, *139*, 255–265.
[23] Tschöpe, A.; Birringer, R. *J. Electroceramics* 2001, *7*, 169–177.
[24] Gao, P.; Kang, Z. Fu, W.; Wang, W.; Bai, X.; Wang, E. *J. Am. Chem. Soc.* 2010, *132*, 4197–4201.
[25] Sayle, T.X.T.; Parker, S.C.; Catlow, C.R.A. *J . Chem. Soc. Chem. Commun.* 1992, 977-978.
[26] Yang, Z.; Yu, X.; Lu, Z.; Li, S.; Hermansson, K. *Phys. Lett. A* 2009, *373*, 2786–2792.
[27] International Centre for Diffraction Data, Database PDF-2 (accessed July 2010).
[28] Yashima, M.; Kobayashi, S.; Yasui, T. *Solid State Ionics* 2006, 177, 211-215.
[29] Eyring, L. in *Synthesis of Lanthanide and Actinide Compound*; Morss, L.R.; Meyer., G.; Eds.; Kluwer Academic Publisher: Netherland, 1991; pp 187-224.
[30] Zinkevich, M.; Djurovic, D.; Aldinger, F. *Solid State Ionics* 2006, *177*, 989-1001.
[31] Hull, S.; Norberg, S.T.; Ahmed, I.; Eriksson, S.G.; Marrocchelli, D.; Madden, P.A. *J. Solid State Chem.* 2009, *182*, 2815-2821.
[32] Yamamura, H.; Katoh, E.; Ichikawa, M.; Kakinuma, K.; Mori, T.; Haneda, H. *Electrochemistry* 2000, *68*, 455-459.
[33] Maricle, D.L.; Swarr, T.E.; Karavolis, S. *Solid State Ionics* 1992, *52*, 173-182.
[34] Preis, W.; Waldhäusl, J.; Egger, A.; Sitte, W.; Carvalho, E.De; Irvine, J.T.S. *Solid State Ionics* 2011, 19, 148–152.
[35] Konysheva, E.; Irvine, J.T.S. *Solid State Ionics* 2011, *184*, 27–30.
[36] Konysheva, E.; Irvine, J.T.S. *Solid State Chem.* 2011, *184*, 1499-1504.
[37] Wen, Y.; Zhang, C.; He, H.; Y. Yu, Y. Teraoka, Catal. Today 126 (2007) 400–405.
[38] Konysheva, E.; Blackley, R.; Irvine, J.T.S. *Chem. Mater.* 2010, *22*, 4700-4711.
[39] Laachir, A.; Perrichon, V.; Badri, A.; Lamotte, J.; Catherine, E.; Lavalley, J.C.; Fallah, J.E.; Hilaire, L.; Normand, F.L.; Quéméré, E.; Sauvion, G.N.; Touret, O. *J. Chem. Soc., Faraday Trans.* 1991, *87*, 1601-1609.
[40] Konysheva, E.; Francis, S.M.; Irvine, J.T.S.; Rolle, A.; Vannier, R.N. *J. Mat. Chem.* 2011, *21*, 15511-15520.
[41] Radovic, M.; Speakman, S.A.; Allard, L.F.; Payzant, E.A.; Lara-Curzio, E.; Kriven, W.M.; Lloyd, J.; Fegely, L.; Orlovskaya, N. *J. Power Sources* 2008, *184*, 77–83.
[42] Konysheva, E.; Irvine, J.T.S. *Chem Mater.* 2009, *21*, 1514-1523.
[43] Mizusaki, J.; Tabuchi, J.; Matsuura, T.; Yamauchi, S.; Fueki, K. *J. Electrochem. Soc.* 1989, *136*, 2082-2088.
[44] Sonntag, R.; Neov, S.; Kozhukharov, V.; Neov, D.; ten Elshof, J.E. *Physica B* 1998, *241-243*, 393-396.
[45] Konysheva, E.; Irvine, J.T.S.; Besmehn, A. *Solid State Ionics* 2008, *179*, 1432–1435.
[46] Mamak, M.; Metraux, G.S.; Petrov, S.; Coombs, N.; Ozin, G.A.; Green, M.A. *J. Am. Chem. Soc.* 2003, 125, 5161-5175.
[47] Irvine, J.T.S. *ECS Transactions* 2008, *13(26)*, 115-122.
[48] Konysheva, E.; Suard, E.; Irvine, J.T.S. *Chem. Mater.* 2009, *21*, 5307–5318.
[49] Teraoka, Y.; Nobunaga, T.; Okamoto, K.; Miura, N.; Yamazoe, N. *Solid State Ionics* 1991, *48,* 207-212.
[50] Ishigaki, T.; Yamauchi, S.; Mizusaki, J.; Fueki, K.; Tamura, H. *J. Solid State Chem.* 1984, *54*, 100-107.

[51] Ishigaki, T.; Yamauchi, S.; Kishio, K.; Mizusaki, J.; Fueki, K., J. Solid State Chem. 1988, *73*, 179-187.

[52] Hayward, M. A.; Cussen, E. J.; Claridge, J. B.; Bieringer, M.; Rosseinsky, M. J.; Kiely, C. J.; Blundell, S. J.; Marshall, I. M.; Pratt, F. L. *Science* 2002, *295*, 1882-1884.

[53] Bridges, C. A.; Darling, G. R.; Hayward, M. A.; Rosseinsky, M. J. *J. Am. Chem. Soc.* 2005, *127*, 5996-6011.

[54] Nakamura,T.; Petzow, G.; Gauckler, L.J. *Mat. Res.Bull.* 1979, *14*, 649-659.

[55] Poulsen, F. W. *Solid State Ionics* 2000, *129*, 145-162.

[56] Konysheva, E.; Irvine, J.T.S.; Besmehn, A. *Solid State Ionics* 2009, *180*, 778-783.

[57] Konysheva, E.; Francis, S.M.; Irvine, J.T.S. *J. Electrochem. Soc.* 2010, *157*, B159-B165.

[58] Pound, B.G. *Solid State Ionics* 1992, 52, 183–188.

[59] Belov, B.F.; Goroh, A.V.; Demchuk, V.P.; Dorschenko, N.A.; Somojlenko, Z.A. *Inorg. Mater.* 1983, 19, 231–234.

[60] Robbins, M.; Wertheim, G.K.; Menth, A.; Sherwood, R.C. *J. Phys. Chem. Solids* 1969, 30, 1823–1825.

In: Cerium
Editors: Aleksey Izyumov and Gulnaz Plaksin

ISBN: 978-1-62257-670-8

Chapter 7

PROSPECTS FOR THE USE OF CERIUM AS FERTILIZER FOR SOILS IN RUSSIA

Yu. N. Vodyanitskii*
Lomonosov Moscow State University, Department of Soil Science, Moscow, Russia

A. T. Savichev
Geological Institute, Russian Academy of Sciences, Moscow, Russia

ABSTRACT

The study of lanthanides in soils is not only of theoretic and soilgenetic importance but also of applied value, because lately lanthanides-containing substances and wastes are used as microfertilizers. The interest in rare earth elements in soils increased in the 1990s, when they found wide use (and cerium in particular) as microfertilizers for some crops in China, which increased the yield and quality of grain. Ce-containing fertilize usage will be baseless without information about Ce-content in soils. Revelation of regions with negative Ce-anomalies is the most important, where Ce-containing fertilize usage is the most effective. Neither positive nor negative geochemical anomalies of natural or technogenic rare earth metals have been revealed in Russia until now.

The recent progress in the study of rare earth elements was related to the use of the expensive method of inductively coupled plasma mass spectroscopy (ICP MS). The expensive method of neutronactivation analysis using a nuclear reactor with the use of a gamma spectrometer allows determining the different dispersed elements in soils even at low concentrations.

We used for determining Ce contents the X-ray radiometric modification of the energydispersive X-ray fluorescence method, in which the sample is excited by a high-energy radioisotope source ^{241}Am rather than by the X-ray tube radiation. This method has made it possible to obtain data on Ce geochemistry in soils of humid landscapes quite expressively and cheaply.

* E-mail: yu.vodyan@mail.ru.

Podzolic soils are specified by substantial negative cerium anomalies. Our studies of podzol and podzolic soil in Pinega (Arkhangelsk oblast), soddy-podzolic soil in town Chusovoi (Perm region) revealed a negative cerium anomalies. This situation is typical for taiga zone, where lanthanides are removed from soil horizons.

In Buryatia, Russia study the effectiveness of introducing lanthanides into the soil is produced. There are many negative Ce-anomalies in Russia. Ce-deficit in these soils may be removed by addition the Ce-containing wastes, including phosphogypsum.

INTRODUCTION

For years, lanthanides (Ln) are regarded as biologically inert elements, though known for their high chemical reactivity (Jollliff et al., 1989; Hidaka et al., 1992; Liu et al., 1993). In particular, the lanthanides are actively interacting with the protein binding of calcium (Furie et al., 1973). Lanthanides activate the biological cycle of nitrogen, crucial in this case have the redox conditions of the environment: in oxidizing conditions are enhanced nitrification, and recovery they are enchanced ammonification (Zhu et al., 2002).

Lanthanides are contributed to the accumulation and nutrient fractionation of some microorganisms (Tsuruta, 2005). In soils they played the active role by iron-oxidizing chemolithotrophic bacteria. Among them is widespread bacterium *Gallionella ferruginea,* which can accumulate lanthanides in quantities thousands of times greater than their contents in the rocks, and millions of times - their concentration in the soil and ground water (Anderson et al., 2003).

In recent years, we have an active effect of lanthanides on plants. Some plants accumulate a significant amount of the lanthanides, and their content in different parts of plants varies greatly. Agricultural chemists have identified the positive effect of low doses of the lanthanide series on the development of plants. At the same time, the effect was not widespread, due to insufficient knowledge of the properties of Ln-deficient soils. Without an understanding of the role of the soil, expand the scope of Ln-fertilizer outside of China, where they are used on large areas, it will be difficult. This applies is particularly to Russia with a huge variety of soils.

FERTILIZER CERIUM AND THE IMPACT ON CROP YIELDS

The physiological need for plants in the lanthanides is not proven. But some of the lanthanides exhibit a high biological activity. The most studied light lanthanides are: La, Ce, Nd with high ckarke. The role of lanthanides in the physiology of plants is largely determined by their relationship to calcium and phosphorus.

Light lanthanides have the physical and chemical affinity with Ca^{2+}. This is due to the proximity of ionic radii (for Ca^{2+} 0.099 nm, La^{3+} - 0.110 nm, Ce^{3+} - 1.102 nm). Large-scale substitution of Ca^{2+} and lanthanides is harm for the development of plants (Ishikawa et al., 1996; Brown et al., 1990).

Small amounts of lanthanides promote biomass production and growth of vascular plants (Wahid et al., 2000). Ln-mixed fertilizers are now widely used in China to improve the nutrition of plants (Xu et al., 2002). This is largely due to the fact that China ranks first in the world for the extraction of lanthanides.

Lanthanides in agriculture of China are used for 30 years, since 1972 (Tyler, 2004, Pang et al., 2002). The usual dose of lanthanides does not exceed 0.23 kg / ha per year (Xu et al., 2002). Lanthanides bring in a soluble form as nitrates, chlorides, or mixtures of amino acids. In 1995 the volume of produced Ln-fertilizer was able to cover the area of 16-20 million hectares (Pang et al., 2002).

Light lanthanides (La, Ce) are made each year by dusting the seed or foliar feeding. Direct soil application gives a weak (or zero) result, it is not recommended, especially in heavy soils, where the lanthanides made quickly secured the solid phase and become unavailable to plants.

While in the laboratory at a moderate concentration of La in solution are often fixed increase in crop yields, but if you transfer experiments in the field the positive effect was not noted. This situation is observed in Ultisole and submerged soils (China), with a high sorption capacity, where the bulk of La sorbed. When the dose of lanthanum increased dramatically (over 10 mg La kg^{-1}) the yield of plants decreased (Xie et al., 2001). This shows the difficulty of determining the optimal doses of fertilizers containing lanthanides.

The effectiveness of lanthanide micronutrients is examined in the laboratories of Buryatia, Russia. Scientists use the lanthanide sulphates and zeolites saturated with lanthanides. Unlike the practice in China, Buryatia scientists introduce lanthanides into the soil when they are depleted in mobile forms in laboratory experiments. Various biochemical parameters were changed. The most noticeable effect is increased soil biological activity, which is reflected by the mineralization of organic matter. Soil nitrogen regime changes through increased nitrification and nitrate accumulation. We emphasize that the role of lanthanides in the activation of biological processes is not unique. The same effect on the decomposition of cellulose and nitrogen regime in the soil have small amounts of other trace elements: zinc, and cobalt. Probably due to improved nitrogen nutrition in the experiments with the introduction of lanthanide point increase in yield: 11-16% - corn, 24-29% - peas.

After foliar feeding the leaves absorb the lanthanides, which are able to migrate to the root (Xu et al., 2002). The degree of concentration varies in different parts of individual plants and for different lanthanides. Admission to the pepper plants were the same for light and heavy lanthanides, but the roots of canola in the concentration of light by 1-2 orders of magnitude higher than the heavy lanthanides (Cao et al., 2000). Solution of 0.088 mg Ce^{3+} l^{-1} increased the length of the roots of corn doubled (Diatloff et al., 1995a).

It is important that cerium (III) positively influences the content of chlorophyll in spinach leaves (Hong et al., 2002). It is known that the introduction of elements of variable valence of Fe, Mn (as Ce) eliminates chlorosis of plants in calcareous soils. Therefore, it would be desirable to study the composition of soils, which increases the cerium content of chlorophyll in plants. If it is confirmed that the cerium treatment eliminates chlorosis of plants in calcareous soils, it will allow more meaningful to apply the Ce-fertilizers. We emphasize that the lanthanides with the constant valence, however, inhibit the synthesis of chlorophyll (Bakou et al., 1992; Ono, 2000).

The negative impact of calcareous soils on the availability of lanthanide indirectly was confirmed in other experiments. It is shown a negative statistical relationship between pH and the content of lanthanides in the roots of bentgrass *Agrostis capillaris* (Tyler et al., 2001). Physical and chemical characteristics of the medium pH and E_H affect the mobility of the lanthanides greatly. Decrease of pH increases the mobility of lanthanides, as well as many other cations. At the same time has a value of Ln-minerals, which determine the difference in the solubility of lanthanides in different soils. In Swedish Cambisoil the concentration of lanthanides in the soil solution is proportional to their content in the soil. It reaches a maximum (3-5 mg l^{-1}) for cerium, neodymium and lanthanum. Only these high-clarke light lanthanides affect plants really. The content of the others heavy lanthanides (Tb, Ho, Tm, Yb, Lu) is minimal: <0.1 mg l^{-1} (Tyler, Olsson, 2001). Gradual decline in concentration in the solution with increasing pH is typical for Ce.

The second important observation is related to the role of soil moisture. The effectiveness of foliar feeding of lanthanides increases in soil moisture deficits (Tyler, 2004), when the soil mobility of lanthanides and their availability to plants decreases.

The toxicity of the lanthanides is determined by their effect on bacteria in the solution, for example, *Vibrio Fischeri* (Weltje et al., 2004). When compared with other heavy metals toxicity of Lu^{3+} is higher than Cd^{3+} or Zn^{2+}, and comparable to the toxicity of Cu^{2+}. The heaviest lanthanide (Lu) is 200 times more toxic than the lightest (La). This is consistent with the known regulation of growth negative biological effect with increasing atomic number. In this respect, the use of fertilizers is justified for the easiest and cheapest of the lanthanides: La and Ce.

Attempts are being made to introduce the Maximum Permissible Concentration (MPC) for soil of lanthanides. In (Weltje et al., 2004; Struijs et al., 1997) MPC considered as the sum of the background concentration of lanthanides with the background and the Maximum Permissible Addition (MPA):

$$MPC = MPA + C_{background.}$$

The value of MPA are based on the transition of the lanthanide in the soil solution and accountig element relative toxicity to living organisms. For the European Union's the proposed value of MPA for cerium is 44 mg kg^{-1}. This high value indicates a low risk of cerium in the soil and, indirectly confirms its beneficial properties when making fertilizer at a dose of 44 mg kg^{-1}. For well-studied soil Dutch background value of $C_{background.} = 9$ mg kg^{-1}. Maximum Permissible Addition is almost 5 times the background content of cerium. For the cerium in the Dutch soils MPC = 9 + 44 = 53 mg kg^{-1} (Weltje et al., 2004; Struijs et al., 1997).

Like other heavy metals, high concentrations of lanthanides in solution lead to toxic reactions of plants. The concentrations of 0.5-25 mg l^{-1} Ce^{3+} in the culture medium inhibited the initial growth of roots, reduced the mass of wheat roots, decreased the content of Ca, Mg, K, Cu, Zn in wheat. The degree of risk increased with increasing concentration of Ce in the medium and with increasing exposure time (Xu et al., 2002). High doses have caused damage to foliage and decreased weight of the roots by spraying solutions of Ce.

Reasons for Lack of Ce in the Soil

Among the Ln-containing minerals is dominated by silicates, carbonates and phosphates. Among the main mineral of cerium oxide - cerianit $(CeTh)O_2$. Minerals have different resistance to weathering, for example, in the acidic environment of lanthanide phosphates and carbonates are less stable than the oxides. The different solubility of Ln-minerals determines the content in the soils of different regions.

In calcareous neutral loamy soils of southern Ontario, Canada was determined coefficient $K_d = C_{soil}: C_{solution}$ (Sheppard et al., 2009). All Ln-containig minerals in the soils were highly stable: for them K_d = 17000-44000. The minerals enriched by other metals are much less stable: for chromium K_d = 1200, for the uranium - 480, for zinc - 400, for molybden - 120 for barium 92. The high stability of Ln-minerals is associated with a neutral reaction of these soils. Characteristically that the stability of the lanthanide cation is positively correlated with pH, whereas the stability of the easily soluble metal is positively correlated with the content of clay particles.

Much of weathering and leaching of lanthanides is controlled by soil microorganisms and organic compounds. There are many microorganisms in the large volume of sorbed lanthanides (Andres et al., 2000). Lanthanides phosphate mineral surfaces are covered with cells, bacteria and fungal hyphae. Plant roots secrete oxalic acid, which is a strong solvent of phosphates. Oxalate dominates in weakly acidic organic compounds in soil and groundwater. Another reason for enhanced release of the lanthanides is due to the formation of complexes with siderophores on mineral surfaces, and these complexes enter the soil solution and made available to the cells of bacteria. Soil bacteria *Arthrobacter* sp. is capable of dissolving hornblende, releasing the heavy lanthanides from Ho to Lu (Brantley et al., 2001).

The behavior of lanthanide is influenced by organic compounds containing a variety of negatively charged groups. They adsorb trivalent cations of these metals in large numbers (Wu et al., 2001a). The mobility of the lanthanides is largely determined by the active forms of organic matter. For example, 46-74% of the variation of the lanthanides in soil solution in the humus horizon of Swedish Cambisol (soils with a pronounced metamorphic illuvial horizon) is determined by the content of organic matter (Tyler, Olsson, 2002). The concentration of dissolved organic carbon determines the content of lanthanides in water (Dia et al., 2000) and in the rivers (Land et al., 1999; Neal et al., 2000; Astrom et al., 2003). On the other hand, the stable form of raw humus serve as a barrier for the movement of the lanthanides.

The fate of lanthanides is affected by the origin and age of the soil, climate, human impact. It has a value of several factors: the degree of enrichment of the soil by different lanthanides, the charge of particles in solution, pH, E_H - especially for cerium, the extent of hydromorphic soils, the composition of primary minerals (Cao et al., 2001; Price et al., 1991). Soil formation is accompanied by a transformation of the parent lanthanide minerals. In different soils, this transformation proceeds differently. The transformation of the parent lanthanide minerals is most active in soils of humid regions. The main trend is the loss of the lanthanides from the upper soil horizons. Podzols in temperate cold regions eluvial horizon E is strongly depleted in lanthanides, especially La and Ce (Minarik et al., 1998; Ohlander et al., 2000; Tyler, 2004). Experiments on the dissolution of the experimental soil on granite in the acidic environment showed that the dissolution begins with allanite, apatite, and then captures, in the end - Ln-containing feldspars (Harlavan et al., 2002; Harlavan et al., 2009). Lanthanides affect the nutrient status of plants, forming Ln-phosphates, such as stable as Fe-phosphates.

In low and high weathering soils pieces of New South Wales (Australia), weathered granite in different ways. The dissolution of low weathering allanite and apatite accompanied by the formation of light lanthanide phosphates in soils. Over time, secondary lanthanide phosphates are dissolved and the accumulated oxide of cerium (IV) tserianit (Taunton et al., 2000a; 2000b).

Lanthanides are removed from the upper horizons is due to the destruction of unstable minerals and they are accompanied by a relative enrichment of elements in the stable minerals under natural conditions humid climate. This trend is typical for relatively young podzols in cold conditions in Sweden (Tyler, 2004).

From the eluvial horizon in podzols Khibiny, Russia cerium leached significantly. We determined the relative (compared with the normalized content of other lanthanides) the content of cerium, which is expressed as an index of its abnormality (Savichev, Vodynitskii, 2012). In soils of the province of Lovozero, Khibiny studied the content of lanthanides in the background area and the zone of geochemical anomalies caused by shallow loparite, mineral-rich lanthanides. Indicators of the cerium anomaly was similar at both sites. But there were differences in the distribution of lanthanides in the peat and in the parent material. The rock relative concentration of cerium - Ce * value of the anomaly (1.14-1.53) was higher than in the peat, where Ce * = 0.69-1.44. Probably, sphagnum moss, which constitutes the basis of peat accumulates has less cerium than the other lanthanides. Similar results were obtained when calculating the degree of cerium anomaly in the humus horizon of Swedish soils light texture (Tyler, Olsson, 2002). The value of Ce * = 1.53 is probably due to specific nutrient accumulation of cerium.

Modern Methods of Diagnostics of Ce in the Soil

The recent progress in the study of rare earth elements was related to the use of the expensive method of inductively coupled plasma mass spectroscopy (ICP MS).

The expensive method of neutron-activation analysis using a nuclear reactor with the use of a gamma spectrometer allows determining the different dispersed elements in soils—Hf, La, Ce, Sm, Eu, Yb, Lu, Th, and U—even at low concentrations. However, the identification of Pr and Nd by this method is complicated because of the short lifetime of these elements.

X-ray fluorescence is the simplest and least expensive method for studying the heavy metals in soils. The contents of rare (Zr and Nb) and rare earth (Y) elements are determined by this method. However, when a Mo, Rh, or Ag anode of the X-ray tube and a common voltage of 35–40 kV are used, other rare earth elements can be identified only from the *L*-lines. Their intensities are lower than those of the *K*-lines by several times, and the weak *L*-lines are overlapped by the strong *K*-lines of macroelements. Niobium with $Z = 41$ is the last element reliably identified by conventional X-ray fluorescence analysis, because the scattered lines of the anode material are located father in the energy spectrum, and the elements with atomic numbers higher than that of the anode material are not excited.

A radically different situation is observed when the X-ray radiometric modification of the energy dispersive X-ray fluorescence method is used, in which the sample is excited by a highenergy radioisotope source ^{241}Am rather than by the X-ray tube radiation. This source is the most suitable for this purpose. An advantage of this method is that the *K*-lines of the heavy metals are actively excited in this case and are not overlapped by the lines of macroelements. Other advantages are the low background radiation compared to X-ray tubes, the high stability, and the small size.

Instead of the traditional sample excitation by emanation from an X-ray tube, excitation by isotope source ^{241}Am with energy of 59 keV and activity of $3.7 \cdot 10^{10}$ s^{-1} was used for a Tefa-6111 analyzer. Preparation of the samples for the X-ray radiometric method was simple: powdered samples were placed in polyethylene dishes with 5-μm-thick Mylar bottoms. The mass of the powdered samples was about 8 g (exact weight of the sample is not required for this analysis). The *K*-lines of Ba, La and Ce are clearly seen (Savichev, Vodynitskii, 2009).

We studied metrological characteristics of the new approach using a set of 30 standard samples of soils and rocks. The limit of element determination was 4 mg/kg at storage time of 800 s. The accuracy of determination corresponds to the third category (according to the requirements of the Scientific Council on analytic methods at the lower limit of determinable concentrations 8 mg/kg), to the second category at concentrations about 100 mg/kg, and to the first category at the concentrations about 500 mg/kg.

THE SPREAD OF NEGATIVE CE ANOMALIES IN SOILS RUSSIA

In the sandy podzol in the content of cerium in Pinega (Arkhangelsk region) soils is significantly lower (in the soil - 48 mg kg^{-1} in the earth's crust – 66 mg kg^{-1}) in the upper horizon of 6-10 mg Ce kg^{-1} (Vodyanitskii et al., 2011). Its low maintenance required light granulometric composition podzol. Sandy podzols represent a strong negative anomaly of cerium.

In the upper horizon of podzolic soil Pinega obsession with cerium above: 18-19 mg kg-1, than in a sandy podzol. This is due to the heavy grading of the soil. Podzolic soils of northern taiga is a negative anomaly of cerium.

Situation is changing in the southern taiga in the Urals region. In the sod-podzolic soil in Chusovaya content of cerium above. In the upper levels it reaches 23-30 mg kg^{-1}, although it does not reach the level of Clarke.

Thus, the deficit of cerium is detected in the soils of humid climate, especially in light soils initially depleted cerium. In these soils can be expected the most effect of Ce-fertilizers. The average content of cerium in the podzolic soils is below the clarke range.

The similar results of rare earth metals are obtaned in the soils of Sweden. In the eluvial horizons of podzolic soils strongly depleted in rare earth metals, while they enriched illuvial (Nagao et al., 1998; Cao et al., 2001; Price et al., 1991). This distinction is particularly enhanced during the transition from the total content of rare earth metals to their mobile (extracted 0.2 M HCl) forms (Price et al., 1991). Depletion of the eluvial horizon matter is associated with many factors: the destruction of clay minerals and (hydr)oxides of Fe and Mn, as the carrier phases of these metals, as well as dissolving the phosphates of rare-earth metals.

CONCLUSION

Biological activity of lanthanides is expressed in different ways. Lanthanides interact with proteins binding calcium, activates the biological nitrogen cycle. Chemolithotrophic iron-oxidizing bacterium *Gallionella ferruginea* accumulate lanthanides in quantities thousands of times higher than their content in the soils.

Deficiency of cerium in the soil is due to the instability of many lithogenic minerals (carbonates, phosphates and cerium) in humid climates. The active forms of organic matter affect its dissolution. In addition, cerium as a metal with variable valence is unstable in a reducing environment.

China is widely used as a fertilizer Ln-containing dusting seed and foliar fertilizer in soils deficient lanthanides - with a low total content and/or low access to them. This is particularly true for calcareous soils experiencing Ce-chlorosis. The role of soil moisture: the effectiveness of foliar feeding of Ln increases in the deficit of soil moisture, soil when the mobility of lanthanides and their availability to plants decreases.

In Buryatia, Russia study the effectiveness of introducing lanthanides into the soil is produced. There are many negative Ce-anomalies in Russia. Ce-deficit in these soils may be removed by addition the Ce-containing wastes, including phosphogypsum.

References

Andres Y., Thouand G., Boualam M., Mergey M. Factors influencing the biosorption of gadolinium by micro-organisms and its mobilization from sand // *Appl. Microbiol. Biot.* 2000. V. 54. P. 262-267.

Anderson C.R., Pederson K. In situ growth of Gallionella biofilms and partitioning of lanthanides and actinides between biological material and ferric oxyhydroxides // *Geobiologyl.* 2003. V. 1. P. 169-178.

Astrom M., Corin N. Distribution of rare elements in anionic, cationic and particulate fractions in boreal humus-rich streams affected by acid sulfate soils // *Water Res.* 2003. V. 37. P. 273-280.

Bakou A., Buser C., Dandulafis G., Brudvig G., Ghaotakis D.F. Calcium binding sites of photosystem II as probed by lanthanides // *Biochem. Biophys. Acta.* 1992. V. 1099. P. 67-77.

Brantley S.L., Liermann L., Bau M., Wu S. Uptake of trace elements and rare earth elements from hornblende by soil bacterium // *Geomicrobiol. J.* 2001. V. 18. P. 37-61.

Brown P.H., Rathjen A.H., Graham R.D., Tride D.E. Rare earth elements in biological system // Handbook on the physics and chemistry of rare earth. *Elsevier Sci. Pab. Amsterdam.* 1990. V. 13.

Cao X.D., Chen Y., Gu Z.M., Wang X.R. Determination of trace rare earth elements in plant and soil by inductively coupled plasma mass spectrometry // *Int. J. Environ. An. Ch.* 2000. V. 76. P. 295-309.

Cao X.D., Chen Y., Wang X.R., Deng X.H. Effect of redox potential and pH value on the release of rare elements from the soil // *Ghemosphere.* 2001. V. 44. P. 655-661.

Dia A., Gruau G., Olivie-Lauquet G., Riou C., Molenat J., Curmi P. The distribution of rare earth elements in groundwater: Assessing the role of source-rock composition, redox changes and colloidal particles // *Geochim. Cosmochim. Acta.* 2000. V. 64. P. 4231-4151.

Diatloff E., Smirh F.W., Asher C.J. Rare-earth elements and plant growth. Three. Responses of corn and mungbean to low concentrations of cerium in dilute, continuously flowing nutrient solutions // *J. Plant Nutr.* 1995. V. 18. P. 1991-2003.

Furie B., Eastlake F., Schechter AN, Anfinsen CB The interaction of the lanthanide ions with staphylococcal nuclease // *J. Biol. Chem.* 1973. V. 248. P. 5821-5825.

Harlavan Y., Erel Y. The release of Pb and REE from granitoids by the dissolution of accessory phases // *Geochim. Cosmochim. Acta.* 2002. V. 66. P. 837-848.

Harlavan Y., Erel Y., Blum JD The coupled release of REE and Pb to the soil labile pool with time by weathering of accessory phases, Wind River Mountains, WY // Geochim. *Cosmohim. Acta.* 2009. V. 73. P. 320-336.

Hidaka H., Holliger P., Shimizu H., Masuda A. Lanthanide tetrad observed in the Oklo and ordinary uraninites and its application for their forming processes // *Geochem. J.* 1992. V. 26. P. 337-346.

Hong F.S., Wang L., Meng X.X., Wei Z., Zhao G.W. The effect of cerium (III) on the chlorophyll formation in spinach // *Biol. Trace Elem. Res.* 2002. V. 89. P. 263-276.

Ishikawa S., Wagatsuma T., Ikarashi T. Comparative toxicity of Al^{3+} , Yb^{3+} , and La^{3+} to root tip cells differing in tolerance to high Al^{3+} in terms of ionic potentials of dehydrated trivalent cations // *Soil Sci. Plant Nutr.* 1996. V. 42. P. 613-625.

Jollliff B.L., Papike J.J., Shearer C.K. Inter- and intro-crystal REE variations in apatite from Bob Ingersoll Pegmatite, Black Hills, South Dakota // Geochim. *Cosmohim. Acta.* 1989. V. 53. P. 429-441.

Land M., Ohlander B., Ingri J., Thunberg J. Solid speciation and fractionation of rare earth elements in a spodosol profile from northern Sweden as revealed by sequential extraction // *Chem. Geol.* 1999. V. 160. P. 121-138.

Liu C-.Q., Masuda A., Okada A., Yabuki S., Zhang J., Fan Z. A geochemical study of loess and desert sand in northern China: Implication for continental crust weathering and composition // *Chem. Geol.* 1993. V. 106. P. 359-374.

Minarik L., Zigova A., Bendl J., Skrivan P., Stchastny M. The behavior of rare elements and Y during the rock weathering and soil formation in the Ricany granite massif, Central Bohemia // *Sci. Total Environ. J.* 1998. V. 215. P. 101-111.

Nagao S., Rao R.R., Killey R.W.D., Young J.L. Migration behavior of Eu(III) in sandy soil in the presence of dissolved organic materials // *Radochim. Acta.* 1998. V. 82. P. 205-211.

Neal C., Jarvie H.P., Whinton B.A., Gemmell J. The water quality of River Wear, north-east England // *Sci. Total Environ.* 2000. V. 251/252. P. 153-172.

Ohlander B., Ingri J., Schonberg H. Ghange of Sm-Nd isotope composition during weathering of till // *Geochim. Cosmochim. Acta.* 2000. V. 64. P. 813-820.

Ono T. Effects of lanthanide substitution at Ca^{2+}-site on properties of the oxygen evolving center of photosystem II // *J. Inorg. Biochem.* 2000. V. 82. P. 85-91.

Pang X., Li D.C., Peng A. Application of rare earth elements in the agriculture of China and its environmental behavior in soil // *Environ. Sci. Pollut. Res.* 2002. V. 9. P. 143-148.

Price R.C., Gray C.M., Wilson R.E., Fray F.A., Tayler S.R. The effects of weathering of rare element, Y and Ba abundances in Tertiary basalts from southeastern Australia // *Ghem. Geol.* 1991. V. 93. P. 245-265.

Savichev A.T., Vodyanitskii Yu.N. Improvenent of the X-ray radiometric method of lanthanides diagnostics in soils // *Eurasian Soil Sc.* 2012. V. 45. № 7. P. 744-753.

Savichev A.T., Vodaynitskii Yu.N. Determination of barium, lanthanium and cerium cocntents in soils by the X-ray radiometric method // *Eurasian Soil Sc.* 2009. V. 42. № 13. P. 1461-1469.

Sheppard SC, Grant CA, Drury CF Trace elements in Ontario soils – mobility, concentration profiles, and evidence of non-point-source pollution // *Can. J. Soil Sci.* 2009. V. 89. P. 489-499.

Struijs J., Van de Meet D., Peijnenburg W., Van den Hoop M., Grommentuijn T. Added risk approach to derive maximum permissible concentration for heavy metals: how to take into account natural background levels? // *Ecotox. Environ. Saf.* 1997. V. 37. P. 112-118.

Taunton A.E., Welch S.A., Banfield J.F. Microbial controls on phosphate and lantanide distributions during granite weathering and soil formation // *Ghem. Geol.* 2000. V. 169. P. 371-382.

Taunton A.E., Welch S.A., Banfield J.F. Geomicrobial controls on light rare element, Y and Ba distributions during granite weathering and soil formation // *J. Alloy Compd.* 2000. V. 303. P. 30-36.

Tsuruta T. Separation of rare elements by microorganisms // *J. Nucl. Radiochem.* 2005. V. 6. P. 81-84.

Tyler G. Rare earth elements in soil and plant systems – A review // *Plant Soil.* 2004. V. 267. P. 191-206.

Tyler G. Vertical distribution of maior, minor, and rare elements in Haplic Podzol // *Ceoderma.* 2004. V. 119. P. 277-290.

Tyler G., Olsson T. Concentrations of 60 elements in the soil solution as related to soil acidity // *Eur. J. Soil. Sci.* 2001. V. 52. P. 151-165.

Tyler G., Olsson T. Conditions related to solubility of rare and minor elements in forest soil // *Soil Sci.* 2002. V. 165. P. 594-601.

Tyler G., Olsson T. Plant uptake of major and minor elements as influenced by soil acidity and liming // *Plant Soil.* 2001. V. 230. P. 307-321.

Vodyanitskii Yu.N., Goryachkin S.V., Savichev A.T. Distribution of rare earth (Y, La,Ce) and other hravy metals in profile of podzolic soil groupe // *Eurasian Soil Sc.* 2011. V. 44. № 5. P. 546-555.

Wahid P.A., Valiathan M.S., Kamalan N.V., Eapen J.T., Vijayalakshmi S., Prabhu R.K., Mahalingam T.R. Effect of rare earth elements on growth and nutrition of coconut palm and root competition for these elements between the palm and Calotrops gigantea // *J. Plant Nutr.* 2000. V. 23. P. 329-338.

Weltje L., Verhoof L.R.C.W., Verweij W., Hamers T. Lutetium speciation and toxicity in microbial bioassay: testing the free-ion model for lanthanides // *Environ. Sci. Technol.* 2004. V. 38. P. 6597-6604.

Wu Z.H., Luo J., Guo H.Y., Wang X.R., Yang C.S. Adsorption isotherms of lanthanum to soil constituents and effects of pH, EDTA and fulvic acid on adsorption of lanthanum onto goethite and humic acid // *Chem. Spec. Bioavailab.* 2001. V. 13. P. 75-81.

Xie Z.B., Zhu J.G., Chu H.Y., Zheng Q., Zhang Y.L., Chao Z.H. Influence of exogenous lanthanum on fertility parameters of Red Soil and Paddy Soil // *J. Rare Earth.* 2001. V. 19. P. 229-232.

Xu X.K., Zhu W.Z., Wang Z.J., Witkamp G.J. Distribution of rare earth and heavy metals in field-grown maize after application of rare earth-containing fertilizer // *Sci. Total Environ.* 2002. V. 293. P. 97-105.

Zhu J.G., Chu H.Y., Xie Z.B., Yagi K. Effects of lanthanum on nitrification and ammonification in three Chinese soils // *Nutr. Cycl. Agroecosyst.* 2002. V. 63. P. 309-314.

In: Cerium
Editors: Aleksey Izyumov and Gulnaz Plaksin
ISBN: 978-1-62257-670-8

Chapter 8

CERIUM(IV) IN AN ACIDIC SOLUTION: A 'NON-INNOCENT' OXIDANT

***Masaki Yoshida*[1,2] *and Shigeyuki Masaoka*[1,3*]**
[1]Institute for Molecular Science, Myodaiji, Okazaki, Japan
[2]Department of Chemistry, Faculty of Science, Kyushu University, Higashi-ku, Fukuoka, Japan
[3]PRESTO, Japan Science and Technology Agency (JST), Kawaguchi, Saitama, Japan

ABSTRACT

Cerium is the most stable lanthanide in the (+4) oxidation state. Consequently, cerium(IV) is commonly used as a one-electron oxidant in aqueous organic synthesis, aqueous polymer synthesis, and water oxidation. Since cerium(IV is a fairly strong Lewis acid, the hydrated ion $[Ce(H_2O)_n]^{4+}$ is easily hydrolyzed and polymerized except at very low pH condition, leading to the loss of oxidizing power. Thus $[Ce(H_2O)_n]^{4+}$ ion exists only in concentrated perchloric acid solution. On the other hand, it has been pointed out in many previous reports that cerium(IV) has specific reactivity in acidic media. In cerium(IV)-driven water oxidation, no oxygen evolves when cerium(IV) is replaced with other strong oxidants, although most of the ruthenium-based O_2-evolving catalysts used cerium(IV) as an oxidant. In acidic aqueous media, cerium(IV) oxidizes aldehydes, ketones and alcohols at the α-carbon atom via the radical-like H-atom abstraction mechanism. In fact, electronic structures of some cerium(IV) complexes are recognized as an 'open-shell singlet state'. These indicate the 'non-innocent' character of cerium(IV) in the oxidation. This chapter focuses on the detailed electronic structure and the specific reactivity of cerium(IV) in the acidic solution.

* E-mail: masaoka@ims.ac.jp, Fax: +81-564-59-5589

INTRODUCTION

Oxidation is one of the most important chemical reactions in both nature and manufacturing. In order to accommodate the current needs for the development of forward-looking technology, numerous oxidation reactions have been developed by researchers all over the world. In order to oxidize substrates effectively, it is necessary to know the detailed properties of not only the catalysts and substrates but also the oxidants. However, the reactivities of oxidants have often been ignored, while reactivities of the substrates and catalysts have been investigated continuously. Cerium(IV) is well-known one-electron oxidant, due to the stability of its (+4) oxidation state. Although most of the lanthanide(IV) ions are too unstable to isolate under mild conditions, cerium(IV) is kinetically stable even in an aqueous solution [1]. Because of their solubility in acidic aqueous media, $(NH_4)_2[Ce^{IV}(NO_3)_6]$ (cerium(IV) ammonium nitrate, CAN) and $(NH_4)_4[Ce^{IV}(SO_4)_4]$ (cerium(IV) ammonium sulfate, CAS) are widely used as oxidants in the aqueous organic synthesis, aqueous polymer synthesis, and water oxidation. The reactivity and oxidizing ability of cerium(IV) are largely dependent on the solution condition (e.g., concomitant anion and solution pH).

This chapter summarizes the reactivity of cerium(IV) in an aqueous media, especially, some of its anomalies in acidic reaction conditions. Moreover, based on the fundamental properties of cerium(IV) as a metal ion, the relevance of the specific reactivity of cerium(IV) and the ambiguous oxidation states of a Ce center and ligands, so-called 'non-innocent' [2] character, will be discussed.

ANOMALIES IN THE OXIDATION OF WATER BY CERIUM(IV)

Oxidation of water to molecular oxygen ($2H_2O \rightarrow O_2 + 4H^+ + 4e^-$) is an essential part of the artificial solar energy conversion and storage, and has attracted considerable attention in recent years. In the study of water oxidation, cerium(IV), oxone, and hypochlorite are used as typical oxidants [3]. In particular, CAN has been most commonly used as a 'simple one-electron oxidant' over the past three decades, and many O_2-evolving catalysts based on metal complexes were discovered by employing cerium(IV) as an oxidant. However, several anomalies have been pointed out in the reactivities of cerium(IV) recently.

First, the rate of O_2 evolution catalyzed by Ru-based catalysts is first order to both the catalyst and cerium(IV) concentrations, giving the following rate law:[4]

$$v_{O2} = d[O_2]/dt = k[\text{Catalyst}][Ce^{4+}]$$

This rate law suggests that the catalyst and cerium(IV) form an encounter complex at the transition state in the rate-determining step, eventhough cerium(IV) is a one-electron oxidant. Second, in the case of a Ru-based catalyst, no oxygen evolves when $(NH_4)_2[Ce^{IV}(NO_3)_6]$ (1.70 V vs. NHE) is replaced with another strong oxidant, i.e., $K_2S_2O_8$ or oxone, whose redox potential is sufficiently high to promote O_2 evolution from water (1.96 or 1.85 V vs. NHE, respectively) [4b]. Third, another important factor in cerium(IV)-driven water oxidation is unveiled by Berlinguette and co-workers. In their experiment, an oxygen atom derived from $[Ce(NO_3)_6]^{2-}$ is incorporated into the final dioxygen product to a small extent [5].

This indicates that the nitrate anion coordinated to cerium(IV) can provide an active oxygen atom and participates in the O-O bond formation step as a minor pathway, which is also supported by the detection of small amount of NO_2 under catalytic conditions [5] These specific reactivities cannot be explained clearly if cerium(IV) is a simple one-electron oxidant.

ANOMALIES IN THE AQUEOUS ORGANIC SYNTHESIS AND AQUEOUS POLYMER SYNTHESIS

In the study of organic and polymer syntheses, cerium(IV) is also commonly used as a one electron oxidant due to its strong oxidizing ability. On the other hand, it has been assumed in many literatures [6, 7] that a hydrated cerium(IV) ion provides an oxygen atom with a highly radical-like character. In the highly acidic media, cerium(IV) oxidizes aldehydes, ketones and alcohols at the α-carbon atom via the radical-like H-atom abstraction mechanism [6]. In this reaction, H atom on the α-carbon atom is selectively abstracted by cerium(IV). This selectivity is further confirmed by the isotopic labeling experiment for the oxidation of alcohol [7]. Thus, it has been assumed in many literatures that a hydrated cerium(IV) ion provides an oxygen atom with a highly radical-like character.

However, it has been generally thought that compounds of Ce^{4+} exhibit simple diamagnetism with a $4f^0$ configuration (i.e., $[Xe](6s)^0(4f)^0$ for Ce(IV)). Why does cerium(IV) show such a radical-like character in an aqueous acidic solution?

BASIC PROPERTIES OF CERIUM(IV) AS A METAL ION

Since cerium(IV) is not only an oxidant but also a metal ion, the reactivity of cerium(IV) is dramatically affected by the ligand. In order to better understand the cerium(IV)-based oxidation reaction, we should understand the complicated properties of cerium(IV) as a metal ion. The formal potential of the Ce(IV)/Ce(III) couple is reported to change upon changing the acid employed as follows: $E_0 = +1.44$ V vs. NHE (1 M H_2SO_4), +1.61 V (1 M HNO_3), and +1.70 V (1 M $HClO_4$) [8]. This order can be well-correlated with the order of the complexation constants of Ce^{4+} with SO_4^{2-}, NO_3^- and ClO_4^-. For instance, the sulfate anion forms a much more stable complex [8] with Ce^{4+} than the nitrate anion [9], leading to the substantial decrease in the formal potential of the Ce(IV)/Ce(III) redox couple. Because the perchlorate anion cannot form a stable complex with Ce^{4+}, high oxidizing ability is achieved in an aqueous perchloric acid, whereas the stability of the cerium(IV) complex with the sulfate or the nitrate is high enough for isolation and crystallization [10]. In addition to such anion dependency, the oxidizing ability of cerium(IV) is also affected by the solution pH. Cerium(IV) is well-known not only as a strong oxidant, but also as a fairly strong Lewis acid [11]. The first and second hydrolysis equilibrium constants (pK_{h1}, pK_{h2}) for cerium(IV) are determined by Yu and O'Keefe as follows [12]:

$Ce^{IV}(OH_2)_n \rightarrow Ce^{IV}(OH)(OH_2)_{n-1} + H^+$ $pK_{h1} = -0.7$

$Ce^{IV}(OH)(OH_2)_{n-1} \rightarrow Ce^{IV}(OH)_2(OH_2)_{n-2} + H^+$ $pK_{h2} = 0.717$

Moreover, such hydroxocerium(IV) ions form a μ-oxo cerium(IV) dimer (e.g., Ce^{IV}(μ-O)Ce^{IV}) [13] or oligomer [14]. Hardwick and Robertson deduce that cerium(IV) exists in an aqueous solution as a mixture of Ce^{4+}, $Ce(OH)^{3+}$, and Ce^{IV}(μ-O)Ce^{IV}, with following equilibria:

$2Ce^{IV}(OH) \rightarrow Ce^{IV}(\mu\text{-}O)Ce^{IV} + H_2O$

In a higher acid concentration, Ce^{4+} and $Ce(OH)^{3+}$ are the dominant ions. Such mononuclear cerium(IV) ions are reported as more reactive than the μ-oxo dimer [15]. On the other hand, $Ce^{IV}(OH)_2$ easily forms a less-reactive μ-oxo dimer or oligomer in comparison with $Ce^{IV}(OH)$, leading to additional complications. Indeed, the formal potential of the Ce(IV)/Ce(III) changes upon changing the acid concentration as follows: E_0 = +1.50 V vs. NHE (0.1 M HNO_3), +1.61 V (1.0 M HNO_3) [16]. At a higher pH region (pH > 2-3), cerium(IV) forms a stable precipitate, i.e., CeO_2 [12]. Therefore, oxidation by cerium(IV) should be carried out in a lower pH region [17].

'Non-Innocent' Character of Cerium(IV) Complexes

Recently, the authors revealed that the hydroxocerium(IV) ion, i.e., $Ce^{IV}(OH)$, has a singlet biradical character [4b], which is the major species in the pH range of $-0.7 < pH < 0.717$.[12] Figure 1 shows the molecular structure, together with the spin density distribution, computed for the open-shell singlet state of one of the possible hydroxocerium(IV) ions, i.e., $[Ce^{IV}(OH)(NO_3)_5]^{2-}$.

The spin density distribution also shows that one spin occupies the 4f orbital of the cerium ion, while the other spin is primarily located over the O(hydroxide) atom. This reveals that this model ion can be described more accurately as $[Ce^{III}(\bullet OH)(NO_3)_5]^{2-}$ and thereby the spins localized at the Ce^{III} and O(hydroxide) atoms are antiferromagnetically coupled. These results indicate the singlet biradical character of the hydroxocerium(IV)ion [4b] Is the singlet biradical character unusual in the case of cerium(IV) complexes? In fact, some cerium(IV) complexes posses such a singlet biradical character [18] Neumann and Fulde investigated the detailed electronic structure of $Ce(C_8H_8)_2$ (cerocene, where $C_8H_8^{2-}$ = cyclooctatetraene dianion), unveiling that the compound is better described as $Ce^{III}(C_8H_8^{2-})(C_8H_8^{\bullet -})$, in which one spin is localized on the cerium ion ($4f^1$) and the other spin on one of the ligands (Figure 2) [18a]. This is now understood as an 'open-shell singlet state', in which two spins within the molecule are antiferromagnetically coupled. Now this electronic structure of cerocene is recognized as a 'non-innocent' character. The singlet biradical character of the hydroxocerium(IV) ion indicates that a highly reactive hydroxyl radical (HO•) is formally given over the cerium ion. This is because the spin density is also slightly localized on the O(nitrate) atom, indicating the slightly radical-like character of nitrate. If this is the case,

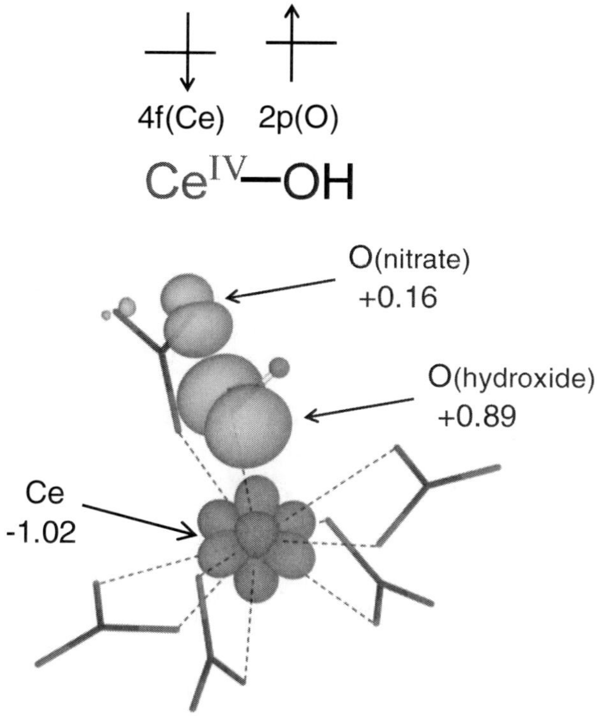

Figure 1. The spin density distributions of $[Ce^{IV}(OH)(NO_3)_5]^{2-}$ in the open-shell singlet state. The values of the spin density of the oxygen atom and cerium atoms are depicted with α and β spins having positive and negative signs, respectively.

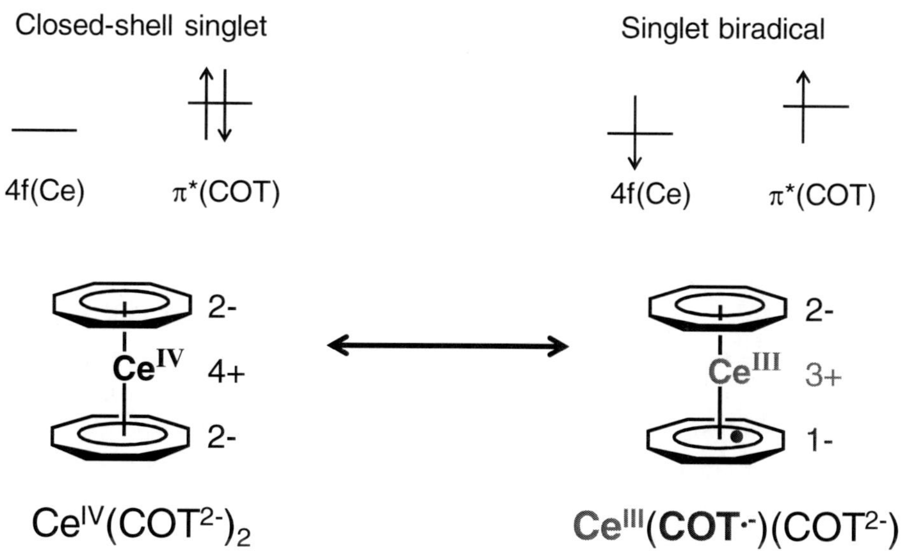

Figure 2. Singlet biradical character of cerocene.

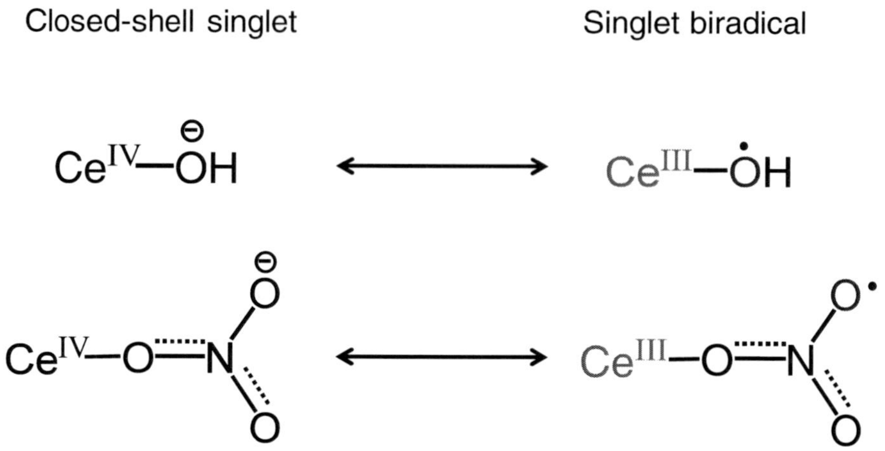

Figure 3. Possible resonance structures of CAN.

cerium(IV) does not only function as an oxidant but also a radical. This is further supported by the fact that cerium(IV) can easily generate free radical by photo-excitation [19]. Cerium(IV) complexes with an O-donor ligand show a broad absorption band in the near-UV to visible region, corresponding to the charge transfer band from the O atom to the metal [20]. Evans and Uri reported that the photo-excitation of this absorption band induces the generation of a free hydroxyl radical, leading to oxygen evolution [19a]. Considering such electronic configurations of the cerium(IV) complexes, it is quite reasonable that the hydrated cerium(IV) ion (i.e., CAN) has radical-like O atom as shown in Figure 3, that is, the hydrated cerium(IV) ion is a 'non-innocent' oxidant.

Radical-like Character of Cerium(IV) Reveals Its Specific Reactivities

This radical-like character of cerium(IV) clearly explains the specific reactivities. Cerium(IV) does not only function as an oxidant but also a radical. In the case of cerium(IV)-driven water oxidation, cerium(IV) provides an efficient path towards the radical-radical coupling between the catalyst and the cerium(IV) (Figure 4).[4,5] As mentioned above, cerium(IV) shows some specific reactivities as follows: (i) the catalyst and cerium(IV) form an encounter complex [4]. (ii) no O_2 evolves when cerium(IV) is replaced with another strong oxidant [4b], (iii) an O atom derived from NO_3^- is incorporated into O_2 to a small extent [5]. This interpretation is quite reasonable for such specific reactivities of cerium(IV). This conclusion is further supported by the fact that the rate of O_2 evolution is accelerated by UV irradiation [5b] caused by the photo-excitation of cerium(IV) (vide supra). In the case of aqueous radical polymerization initiated by cerium(IV), this radical-like character explains the oxidizing process by cerium(IV).

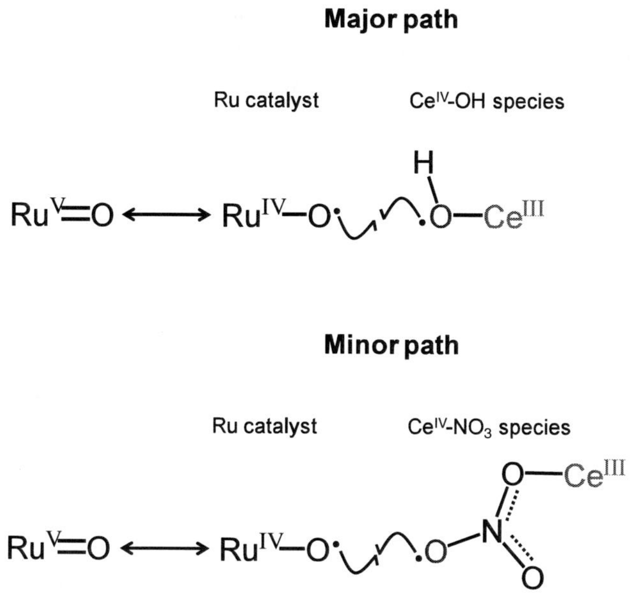

Figure 4. O-O bond formation step in water oxidation promoted by the 'non-innocent' character of the cerium(IV).

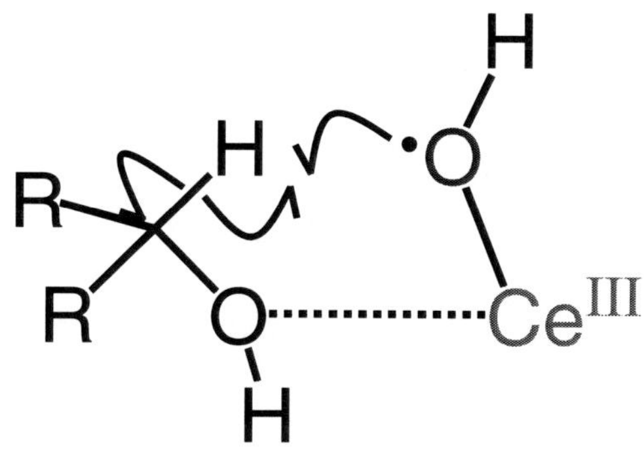

Figure 5. Proposed mechanism of H atom abstraction of alcohol by cerium(IV).

Considering that the cerium(IV)-oxyl radical is highly active toward H-atom abstraction from the organic substrate [21], H-atom abstraction behavior of the hydrated cerium(IV) ion is quite reasonable (Figure 5).

Conclusion

Eventhough cerium(IV) is commonly used as a one-electron oxidant in various studies, its reactivity is largely dependent on the reaction condition. The formal potential of the Ce(IV)/Ce(III) couple is affected by coexisting anions. Due to its Lewis acidity, cerium(IV) does not have high oxidizing ability in neutral or basic solutions. More importantly, cerium(IV) has highly specific reactivities, i.e., radical-like character in an acidic aqueous media. The hydrated cerium(IV) ion bears a radical-like O atom as the ligand, such as the MnO_4^- ion, which oxidizes hydrocarbons via H atom transfer in a non-polar media [22].

Therefore, we should be able to observe its anomalous behavior caused by its 'non-innocent' character in an aqueous media when we chose cerium(IV) as a simple one-electron oxidant. This chapter summarizes such a 'non-innocent' character of cerium(IV) in an aqueous media.

REFERENCES

[1] S. Cotton, *Lanthanide And Actinide Chemistry*, John Wiley and Sons, New York, 2006.

[2] Review: a) M. D. Ward, J. A. McCleverty, *J. Chem. Soc., Dalton Trans.*, 2002, 275. b) A.B.P. Lever, *Coord. Chem. Rev.*, 2010, *254*, 1397. c) W. I. Dzik, X. P. Zhang, B. de Bruin, *Inorg. Chem.*, 2011, *50*, 9896.

[3] Review: a) W. Rüttinger, G. C. Dismukes, *Chem. Rev.*, 1997, *97*, 1. b) M. Yagi, M. Kaneko, *Chem. Rev.*, 2001, *101*, 21. c) I. Romero, M. Rodríguez, C. Sens, J. Mola, M. R. Kollipara, L. Francàs, E. Mas-Marza, L. Escriche, A. Llobet, *Inorg. Chem.*, 2008, *47*, 1824.

[4] a) S. Masaoka, K. Sakai, *Chem. Lett.*, 2009, *38*, 182. b) M. Yoshida, S. Masaoka, J. Abe, K. Sakai, *Chem. Asian J.*, 2010, *5*, 2369. c) A. Kimoto, K. Yamauchi, M. Yoshida, S. Masaoka, K. Sakai, *Chem. Commun.*, 2012, *48*, 239.

[5] a) D. J. Wasylenko, C. Ganesamoorthy, M. A. Henderson, B. D. Koivisto, H. D. Osthoff, C. P. Berlinguette, *J. Am. Chem. Soc.*, 2010, *132*, 16094. b) D. J. Wasylenko, C. Ganesamoorthy, M. A. Henderson, C. P. Berlinguette, *Inorg. Chem.*, 2011, *50*, 3662.

[6] a) J. Shorter, *J. Chem. Soc.* 1962, 1868. b) D. L. Mathur, G. V. Bakore, *Bull. Chem. Soc. Jpn.* 1971, *44*, 2600. c) G. G. Rao, B. M. Rao, *Anal. Chim. Acta*, 1972, *59*, 461.

[7] a) J. S. Litter, *J. Chem. Soc.*, 1959, 4135. b) J. S. Litter, W. A. Waters, *J. Chem. Soc.*, 1960, 2767.

[8] E. Wadsworth, F. R. Duke, C. A. Goetz, *Anal. Chem.*, 1957, *29*, 1824.

[9] a) I. A. Lebedev, B. F. Myasoedov, V. Ya. Frenkel, *Zh. Neorg. Khim.*, 1973, *18*, 3224; b) S. Bayülken, A. S. Saraç, *Turkey J. Chem.*, 1996, *20*, 111.

[10] a) Y. Shan, S. D. Huang, *Acta Crystallogr.*, 1998, *C54*, 1744. b) T. A. Beineke, J. Delgaudio, *Inorg. Chem.*, 1968, *7*, 715.

[11] a) H. G. Offner, D. A. Skoog, *Anal. Chem.*, 1966, *38*, 1520; b) M. Komiyama, N. Takeda, Y. Takahashi, H. Uchida, T. Shiiba, T. Kodama, M. Yashiro, *J. Chem. Soc., Perkin Trans. 2*, 1995, 269. c) P. M. Cullis, E. Snip, *J. Am. Chem. Soc.*, 1999, *121*, 6125.

[12] a) P. Yu, S. A. Hayes, T. J. O'Keefe, M. J. O'Keefe and J. O. Stoffer, *J. Electrochem. Soc.*, 2006, *153*, C74. b) P. Yu, T. J. O'Keefe, *J. Electrochem. Soc.*, 2006, *153*, C80.

[13] a) T. J. Hardwick, E. Robertson, *Can. J. Chem.*, 1951, *29*, 818. b) D. J. Savage, T. W. Kyffin, *Polyhedron*, 1986, *5*, 743. c) A. Ikeda-Ohno, S. Tsushima, C. Hennig, T. Yaita, G. Bernhard, *Dalton Trans.*, 2012, *41*, 7190.

[14] E. L. King, M. L. Pandow, *J. Am. Chem. Soc.*, 1952, *74*, 1966.

[15] a) M. Ardon, G. Stein, *J. Chem. Soc.*, 1956, 104. b) M. Ardon, *J. Chem. Soc.*, 1957, 1811. c) D. L. Mathur, G. V. Bakore, *J. Indian Chem. Soc.*, 1971, *411*, 363. d) M. D. Fernandez, G. Z. Guzman, *Eur. Polym. J.*, 1989, *25*, 1165.

[16] F. Liu, J. J. Concepcion, J. W. Jurss, T. Cardolaccia, J. L. Templeton, T. J. Meyer, *Inorg. Chem.*, 2008, *47*, 1727.

[17] R. A. Binstead, C. W. Chronister, J. Ni, C. M. Hartshorn and T. J. Meyer, *J. Am. Chem. Soc.*, 2000, *122*, 8464.

[18] a) C.-S. Neumann, P. Fulde, *Z. Phys. B* 1989, *74*, 277; b) M. Dolg, P. Fulde, W. Küchle, C.-S. Neumann, H. Stoll, *J. Chem. Phys.* 1991, *94*, 3011; c) C. H. Booth, M. D. Walter, M. Daniel, W. W. Lukens, R. A. Andersen, *Phys. Rev. Lett.* 2005, *95*, 267202.

[19] a) M. G. Evans, N. Uri, *Nature*, 1950, *166*, 602. b) L. Dogliotti, E. Hayon, *J. Phys. Chem.*, 1967, *71*, 3802. c) E. Baciocchi, T. Del Giacco, S. M. Murgia, G. V. Sebastiani, *Tetrahedron*, 1988, *44*, 6651.

[20] A. B. Yusov, V. P. Shilov, *Russ. Chem. Bull., Int. Ed.*, 2000, *49*, 1925.

[21] X.-N. Wu, Y.-X. Zhao, W. Xue, Z.-C. Wang, S.-G. He, X.-L. Ding, *Phys. Chem. Chem. Phys.*, 2010, *12*, 3984.

[22] a) K.A. Gardner, J. M. Mayer, *Science*, 1995, *269*, 1849. b) K. A. Gardner, L. L. Kuehnert, J. M. Mayer, *Inorg. Chem.*, 1997, *36*, 2069.

In: Cerium
Editors: Aleksey Izyumov and Gulnaz Plaksin
ISBN: 978-1-62257-670-8

Chapter 9

STRUCTURE, STABILITY, AND REACTIVITY OF CERIUM(IV) COMPLEXES WITH SOME ALIPHATIC ORGANIC COMPOUNDS

Olga Voskresenskaya*
Joint Institute for Nuclear Research, Dubna,
Moscow Region, Russia

Abstract

An approach to investigation of thermodynamic and kinetic stability of variable-valence metal complexes is developed. Concepts of the generalized inverse problems of quantitative instrumental analysis and the complex formation equilibria are introduced. The procedure for this investigation is regarded as a sequential procedure for solving these two inverse problems. The classical methods for investigating the complex formations in solution are analyzed in terms of these inverse problems. Their generalizations are given to determine together with the formation constants also redox-decomposition rate constants of variable-valence metal complexes, and the rate law of numerous redox processes for which the rate-determining step is the decay of an intermediate complex. These generalized methods are applied to a comparative study of the complexation and intramolecular redox decomposition of cerium(IV) hydroxo complexes with some dicarboxylic, oxycarboxylic acids, and aliphatic polyatomic alcohols in acidity dependence (pH 0.7–3.4) of a sulfate medium at the ionic strength $I = 2$, temperatures of 10.0–43.0°C by UV-Vis spectroscopy, and pH measurements. The region of existence, composition, ligand speciation, stability constants of the complexes $[CeOH(H_{-x}R)_n]^{3-nx}$, rate constants, and activation energies of the intramolecular redox decomposition were determined for these complexes on a sulfate background. The rate law for the redox process that occurred in these studied systems is derived on the basis of calculated constants and the set of all the chemical and algebraic equations describes the equilibria within these systems. The quantitative model for this process is created. The existence of the inverse linear correlation between the logarithms of stability constants and logarithms of the redox-decomposition rate constants for $[CeOHH_{-2}R]^+$ complexes with dibasic ligands $R = H_2L$, $R = H(OH)L$,

*E-mail address: voskr@jinr.ru

and $R = L(OH)_2$ was established. It is shown that the kinetic stability for the corresponding bis-complexes $[CeOH(H_{-2}R)_2]^-$ does not differ from the kinetic stability of the mono-complexes. The constancy of the reactivity for the complexes $[CeOHH_{-x}R]^{3-x}$ with $x = 2, 3, 4$ in the sulfate medium and its independence from their thermodynamic stability are also demonstrated. We discuss the computational aspects of solving the regarded generalized problems.

PACS 82.30.Lp, 82.60.Hc, 82.20.Pm
Keywords: decomposition reactions, equilibrium constants, rate constants

1. Introduction

1.1. Study and applications of tetravalent cerium compounds

The study in the interactions of cerium(IV) with organic substrates has been an active area for investigation [1-8]. It is rather important in relation to the problem of stabilization for unstable oxidation states (OSs) of metals [8, 9, 10], extensively studied electron transfer reactions [11, 12, 13], and numerous applications of cerium(IV) in various domains of chemistry as an one-electron oxidizer [1-4], a catalyst for numerous reactions [5], and a complex-forming agent [6, 7, 8].

Cerium compounds possess various pharmacological properties. Apart from direct antiseptic effects, cerium helps to prevent postburn sepsis. The antineoplastic potential of cerium compounds is currently being re-explored in experimental settings. These compounds are used in nuclear medicine and toxicology [14]. Numerous studies have examined the neurobehavioral, neurodevelopmental, and neurological effects of cerium citrate complexes; the hepatic effects of cerium chloride and cerium dioxide; and the cardiovascular effects of cerium. Available information on the renal effects of cerium chloride, the hematological effects of cerium citrate and cerium ethylenediaminetetraacetate complexes $[Ce(H_{-x}EDTA)_n]^{4-nx}$ were examined; also the genotoxic effects of cerium oxide and cerium nitrate, as well as noncancer effects of cerium sulfide/chloride/oxide/nitrate and cerium citrate [15].

Cerium complexes are widely used in analytical practice [16]. Photoluminescence of cerium complexes with organic compounds applies to their quantitative determination [17]. Cerium(IV) is also used as a test reagent for alcohols [18], which form red 1:1 complexes upon coordination to cerium(IV) [19]. The oldest is the application of cerium(IV) salts as oxidizing agents in redox titrations. The titrimetric methods based on cerium(IV) are known as 'cerimetry' or 'cerate oxidimetry' [20].

Cerium(IV) solutions are used extensively for determination of organic compounds. For instance, it is possible to oxidize malonic, lactic, malic, and tartaric acid by cerium(IV). At present, there are many straightforward and indirect methods, kinetic methods among them [21], for determining inorganic and organic compounds based on this one-electron oxidizer. However, the kinetics and mechanism of organocerium reactions still remains to be studied. They have very diversified features, and any attempt to classify them is justified.

From a technological point of view, that is of great importance, is the separation of rare-earth elements (REE) and processes of radical polymerization of various monomers using cerium copounds; the use of cerium(IV) in the practice of organic synthesis [22, 23]. For example, cerium is used in organic synthesis as an one-electron oxidant in the oxidation of phenols and hydroquinones to quinones [3, 22]; in liquid-phase oxidation of aromatic aldehydes, alcohols, hydrocarbons [23] and phenylpolyethylene glycol [3]; as well as in destruction of EDTA and terephthalic acid [24].

Of special usefulness is the ability of cerium(IV) salts to initiate radical polymerization of vinyl monomers (acrylamide, acrylonitrile, etc.) [25]. For instance, the systems cerium(IV)– alcohols [26, 27] and cerium(IV)– oxalic acid [28] can be used as a source of radicals in the radical polymerization of acrylonitrile. The mixture separation of REE is based on the ability of cerium to form compounds in which it is in OS +4. This led to a pronounced difference between the properties of cerium compounds and those with REE in OS +3 which improves the efficiency method for their separation [29, 30]. However, the problem of cerium stabilization in this (anomalous) OS, especially in complexation with oxycarboxylic acids, has been studied insufficiently.

1.2. The Problem of the Stabilization for Unstable Oxidation States

The stabilization of metals in their anomalous OSs is an interesting theoretical and applied problem (see, e.g., [31]). This problem has recently been reexamined due to the accumulation of information on unusual OSs and the development of a number of topical issues in science materials [8]. Particular aspects for the stabilization of OSs through coordination were considered in [32]. The state of the art in this field and some basic aspects were also analyzed in [8].

Thus, in particular, the following features were observed: an anomalously high redox resistance in compounds with REE ion configurations f^0, f^7, and f^{14}; the highest stability in compounds with the configurations f^0 (the case of tetravalent cerium); the preferred stabilization of the highest metal OSs by high-electronegativity ligands, which increase the apparent π-electron density on the metal atom (fluoride, oxide ions); and the priority of the relationship between the properties of the central ion and ligands.

The last statement is illustrated by earlier studies [29] whose results are summarized in [30]. Those studies concerned the redox interaction rates in the systems containing variable-valence cations (manganese(III), vanadium(V), cerium(IV)), and organic ligands (complexones, oxycarboxylic acids) as functions of the stability constants of complexes, reducing the ability of ligands, their density, and other factors.

These studies led to the conclusions that provided that the metal-ligand bonding is mostly covalent (e.g., in manganese(III) and vanadium(V) complexonates), the reducing ability of the ligand is the key factor in the redox interaction rate in system; provided that the bond is mostly ionic (as, e.g., in cerium(IV) complexonates), the stability of the complexes is the key factor. No simple relationship could be revealed between the stability of the complexes and a 'conditional rate' and apparent activation energy of the redox interaction in cerium(IV) complexation with oxycarboxylic acids.

However, clearly, an analysis with the use of the conditional rate in the redox process and its apparent activation energy as the criterion of metal OS stabilization in the complex can only be regarded as a prerequisite for a more correct analysis based on the intracomplex redox decomposition rate constants. Data on these rate constants are either not found in the literature or refer to complexes of hypothetical structures for which the forms of the metal and ligand are unknown.

It is also evident that such a property of an organic ligand as its reducing capacity may play an important part in the case of a bimolecular (free-radical), but not intermediate, mechanism occurring in oxidation of oxycarboxylic acids by cerium(IV) [33]. A study of the problem for the stabilization in the given context is inseparable from a study of the mechanism and kinetics of the redox process occurring in investigated systems.

1.3. Oscillating Belousov–Zhabotinsky Reaction

Over the past several decades, the oxidation of relatively low molecular weight carbonic acids by cerium(IV) has been the subject of numerous kinetic studies. The ceric sulfate oxidation of aliphatic organic compounds that contain a reactive $C - H$ bond, as the citric, malonic acids, and derivatives, is of special interest [34, 35, 36], since these reactions coincide with the so-called 'organic subset' of the cerium-catalyzed oscillatory Belousov—Zhabotinskii (BZ) reaction [37].

The BZ reaction is one of the most studied chemical oscillators because it is a prime example for nonlinear behavior in chemistry. The fascinating oscillating reaction have attracted biologist and chemists alike for more than 30 years, when the BZ reaction became popular [38-42]. In 1950, Belousov endeavored to model catalysis in the Krebs cycle using cerium ions instead of the protein-bound metal ions common in the enzyme of living cells and first observed oscillations in the rate of $[Ce(IV)]/[Ce(III)]$ during the cerium-catalyzed oxidation of citric acid or malonic acid by acidic bromate.

Further oscillatory phenomena have been discovered in the processes of enzymatic catalysis, accompanying the biological phenomena, including the periodic processes of cellular metabolism; waves of activity in cardiac tissue and brain tissue; as well as the processes occurring at ecological system level.

In its classical form, the BZ reaction can be schematically represented, according to [38], by the Scheme 1.

$$(i)\ M^{q+} + R = M^{q-1} + Products; \quad (ii)\ M^{q-1} + BrO_3^- = M^{q+} + Products$$

Scheme 1. A scheme for the classical BZ reaction. Oxidizers M^{q+} can be cerium(IV) and one-electron oxidizers close to it. Dicarboxylic and oxycarboxylic acids[1], saccharides, alcohols, etc. [34] can play the role of reducing agents R.

While the main features of the inorganic mechanism – especially the 'inorganic subset' (ii) – are rather clear [38], there are still some unknown details in the 'organic subset' (i) of the BZ reaction [40]. The detailed mechanism of the BZ reaction is highly complex and thus not yet fully understood, possibly because of the lack of direct experimental data on the involvement of numerous intermediates in the form of active species. The formation of intermediate organocerium(IV) complexes in the subset (i) had been long postulated [34, 38]. However, the later authors made no efforts to identify and characterize these species.

[1]For instance, malonic acid [35, 37, 38, 40] and derivatives [34], oxalic [35, 36], citric [35, 37, 38], malic [35, 37], and tartaric [35] acids.

1.4. Oxidation of Aliphatic Organic Substrates by Cerium(IV)

It is known from kinetic studies that polar organic compounds (alcohols, ketones, oxycarboxylic acids, etc.), $R = HL$, are oxidized by variable-valence metal ions (e.g., Ce(IV) and Mn(III)) according to the one-electron mechanism, which as a rule includes two stages. First, an organic compound enters into the coordination sphere of the ion-oxidizer in an equilibrium reaction, and a redox process then occurs in this complex, which is the rate-limiting stage [1, 4, 6, 7, 43, 44, 45],

$$M^{q+} + HL \overset{\beta_{11}^{ef}}{\rightleftharpoons} HL \ldots M^{q+} \overset{k_{11}}{\rightarrow} M^{(q-1)+} + H^{+} + \cdot L$$

Scheme 2. A typically scheme for the oxidation of polar organic compounds by variable-valence metal ions. Here, $\beta_{11}^{ef} = k_{+1}/k_{-1}$, $k_{+1} \gg k_{-1} > k_{11}$.

However, kinetic data cannot be used to draw conclusions about the structure of this complex; we can only estimate the effective constants of its formation at a fixed pH value. These constants correspond to indefinite forms of the metal and ligand in the complex. Thermodynamic methods for studying the preequilibrium of Scheme 2 [30] cannot be used to determine the kinetic parameters of these complexes.

Whereas kinetic generalizations of thermodynamic methods [6, 7], on the one hand, provide the possibility of directly determining the composition of intermediate complexes, the form in which organic ligands are present in them, and noneffective constants of their formation. On the other hand, they can be used to study the reactivity of these complexes and the kinetics and mechanism of the initial stages of the redox process that occurs in the system.

In the present work, we develop the generalized methods proposed in [6, 7] and apply them to a comparative study the complexation and intramolecular redox decomposition of cerium(IV) hydroxo complexes with some dicarboxylic (oxalic, malonic, succinic, glutaric) and oxycarboxylic (quinic, lactic, malic, citric tartaric) acids, as well as with aliphatic polyatomic alcohols (glycerol, xylitol, and sorbitol) in dependence on acidity (pH 0.7–3.4) of a sulfate medium at the ionic strength $I = 2$ and temperatures of 10.0–43.0°C by UV-Vis spectroscopy and pH measurements.

2. Procedure

2.1. Generalized Methods in Terms of Inverse and Direct Problems

The study on complexation can be represented as a sequential procedure for solving these two problems: (I) 'the problem of quantitative instrumental analysis' (determination of the compound concentration from measuring its property P); and, (II) 'the inverse problem of complexation equilibria' (calculation of the stability constants β_{mn} of the complexes $[M_m R_n]$ on the basis of their equilibrium concentrations c_{mn}), which can be schematically represented as

$$P \overset{I}{\rightarrow} c_{mn} \overset{II}{\rightarrow} \beta_{mn}. \tag{1}$$

Criterion of correctness for the β_{mn} calculation is a constancy of its value with a variation of concentrations the system components, i.e., in other words, a correct account of all the major equilibria that determine the stoichiometry of the complexes $[M_m(H_{-x}R)_n]^2$.

In order to avoid selection of the stoichiometry (m, n, x) which provides the constancy of the β_{mn}, it establishes in intermediate stages of (I) experimental ($m : n$); and, (II) experimental-calculated (x) solution of these problems:

$$P \stackrel{Ia}{\rightarrow} m{:}n, a_{mn} \stackrel{Ib}{\rightarrow} c_{mn} \stackrel{IIa}{\rightarrow} x \stackrel{IIb}{\rightarrow} \beta_{mn}. \tag{2}$$

At stage Ia, from the maximum value of the property P, determine also the intensity factor: $a_{mn} = P_{max}/c_M$ when $c_M < c_L$; $a_{mn} = P_{max}/c_R$ when $c_R < c_M$).

On one hand, in the context of 'direct' (I_+) and 'inverse' (I_-) problems of spectroscopy and other physical research methods, the problem of quantitative instrumental analysis (I) can be treated as a particular case for the problem I_-. On the other hand, the problems $I_\pm$ may be expressed in terms of quantitative instrumental analysis as

$$I_- : P \rightarrow c_{mn}, \quad I_+ : c_{mn} \rightarrow P. \tag{3}$$

As a result, the initial scheme of research (1) takes a form of sequence two 'inverse problems':

$$P \stackrel{I_-}{\rightarrow} c_{mn} \stackrel{II_-}{\rightarrow} \beta_{mn}. \tag{4}$$

The experimental technique for solving the direct problem I_+ in the case of cations in stable OSs is standard and is not included in the description of a research method (4). The case of cations in unstable OS demands for these purposes using the experimental techniques of kinetic studies. It also points to the need to use for the calculation of stability constants the extrapolated to the 'zero' time $\tau = 0$ property value ($P|_{\tau=0} = P^0$) and the initial equilibrium complexes concentration ($c_{mn}|_{\tau=0} = c^0_{mn}$):

$$P^0 \underset{I_+}{\overset{I_-}{\rightleftharpoons}} c^0_{mn} \stackrel{II_-}{\rightarrow} \beta_{mn}. \tag{5}$$

Specificity studies on complexes with cations in unstable OSs also indicates another need to study, together with thermodynamic (β_{mn}), the kinetic stability of the complexes with respect to intracomplex redox-decomposition, which is naturally characterized by a intramolecular redox-decomposition rate constant (k_{mn}). It also demonstrates the advisability of generalization for the problems $I_\pm$, $II_\pm$ which can be represented as follows:

$$P^0, \dot{P}^0 \underset{\widetilde{I}_+}{\overset{\widetilde{I}_-}{\rightleftharpoons}} c^0_{mn}, \dot{c}^0_{mn} \underset{\widetilde{II}_+}{\overset{\widetilde{II}_-}{\rightleftharpoons}} \beta_{mn}, k_{mn}. \tag{6}$$

Here, $\dot{P}^0$ and $\dot{c}^0_{mn}$ are the initial rates of property and the equilibrium concentration changes, correspondingly.

In terms of modeling the chemical processes, sequential solving the two 'inverse problems' such as $\widetilde{I}_-$ and $\widetilde{II}_-$ in (6) can be treated as solving an extended 'inverse equilibrium modeling' problem. By the sequential solving of the two 'forward problems' $\widetilde{II}_+$ and $\widetilde{I}_+$,

[2]Here and throughout, the indices $-x$ $(-i)$ in the formulas $H_{-x(-i)}R$ denotes the number of protons displaced from reducing molecule R in complexation with cerium(IV) and the number of dissociated protons, correspondingly.

this also represents solving the corresponding problem for 'forward modeling' [46]. First of all, our interest here, is in the extended inverse equilibrium modeling $\widetilde{M}_- = (\widetilde{I}_-, \widetilde{II}_-)$:

$$P^0, \dot{P}^0 \xrightarrow{\widetilde{I}_-} c^0_{mn}, \dot{c}^0_{mn} \xrightarrow{\widetilde{II}_-} \beta_{mn}, k_{mn}. \tag{7}$$

Regarding the kinetics of oxidation on organic compounds by variable-valence metal ions, in which the complexes with cations in unstable OS play the role of intermediates, this approach allows: (i) to determine composition of sufficiently stable intermediate complexes and the speciation of organic ligands in them; (ii) calculate values of their specified intramolecular rate constants and stability constants; (iii) deduce the rate law of the studied redox process on the basis of the calculated model parameters and all the basic equations that describe the equilibria in system; (iv) draw a conclusion on the detailed mechanism of its initial stages; and, (v) create the quantitative model for this investigated process. We shall in more detail consider this procedure.

2.2. Experimental

2.2.1. Reagents

In this study, we used cerium sulfate $Ce(SO_4)_2 \cdot 4H_2O$ of analytical grade; dicarboxylic acids (chemically pure) grade: oxalic acid $(Oxal)$ $HOOCCOOH$, malonic acid $(Maln)$ $HOOC(CH_2)COOH$, succinic acid $(Succ)$ $HOOC(CH_2)_2COOH$, and glutaric acid $(Glut)$ $HOOC(CH_2)_3COOH$; double recrystallized oxycarboxylic acids of chemically pure grade: quinic (1,3,4,5-tetrahydroxycyclohexane-1-carboxylic) acid (Quin), lactic $CH_3CH(OH)COOH$ $(Lact)$, citric $C(OH)(COOH)(CH_2COOH)_2$ $(Citr)$, malic $HOOCHC(OH)CH_2COOH$ $(Malt)$, and tartaric $(CH(OH)COOH)_2$ $(Tart)$ acid; glycerol $(Glyc)$ $C_3H_5(OH)_3$ of analytical grade, xylitol $(Csyl)$ $C_5H_7(OH)_5$, and sorbitol $(Sorb)$ $C_6H_8(OH)_6$ (Russian State Standard GOST 20710-75).

The required ionic strength in solutions $I = 2$ was created with amonium sulfate of analytically pure grade. The stock solutions were prepared from precisely weighed samples. The concentration of organic acid solutions was refined by pH-metric titration with a carbonate-free $NaOH$ solution. The cerium(IV) content in the salt and a freshly prepared solution was determined by back-titration with the Mohr salt in the presence of ferroin [47] before and after performing the experiment.

2.2.2. Spectrophotometric and Kinetic Measurements

Kinetic measurements were performed on a SPECORD UV-Vis recording spectrophotometer (1-cm cell) equipped with temperature cell-holder in the extinction measurement mode at a fixed wavenumber ($\lambda \sim 400$ nm) and on a KF-5 photoelectrocolorimeter attached to a rapid kinetic accessory (MEA-4). All solutions were temperature-controlled by a Wobser thermostat. The UV-Vis absorption spectra of the reactant mixtures and the cerium(IV) solutions were recorded over the 240–355 nm and 330–555 nm wavelength ranges taking into account the kinetics of the observed redox process.

The decomposition of the reactions intermediates was measured at 400 nm, where the difference in absorbances of the intermediates and the cerium(IV) ($\Delta D^0 = D^0 – D_M$) with the increase of pH reaches the highest value. The pH measurements in the reactant mixtures and in the ceric sulfate solutions were carried out with a DATA METER precision pH meter. Reactions were initiated by turning a vessel-mixer, into which the initial components were placed; and the progress of reaction was followed spectrophotometrically by monitoring the decrease in optical density of the reaction mixture.

The 'conditional rate' $-(\lg D^0)'_\tau = -\partial \lg D/\partial\tau|_{\tau=0}$ and the first-order rate pseudoconstant $k_\tau = -2.303(\lg D^0)'_\tau$ (s^{-1}) of the redox decomposition of the intermediates were estimated graphically in the semilogarithmic coordinates $\lg D - \tau$ from the slope $(\lg D^0 - \lg D^i)/(\tau^0 - \tau^i) = const$, because the time variation of the logarithm of the optical density obeys the linear law

$$\lg D = \lg D^0 - k_\tau\,\tau/2.303, \tag{8}$$

and calculated by the least-squares method. The plots were linear approximate up to 75 – 90% completion of reaction for most systems studied (e.g., for all systems with oxyacids). In making measurements, the initial optical densities of the reaction mixtures (D^0) were found from (8) by linear extrapolation of the decay curves in the same coordinates to the moment $\tau = 0$. The initial rate $-(D^0)'_\tau = -\partial D/\partial\tau|_{\tau=0} \equiv -\dot{D}^0$ (s^{-1}) was determined from the formula for the logarithmic derivative:

$$(D^0)'_\tau = 2.303\, D^0\, (\lg D^0)'_\tau. \tag{9}$$

2.2.3. Instrumental Analysis

To solve the generalized problem of the instrumental analysis $\widetilde{I}_-: P^0, \dot{P}^0 \rightarrow c^0_{mn}, \dot{c}^0_{mn}$, we used the UV-Vis spectroscopy ($P^0 = D^0, \dot{P}^0 = \dot{D}^0$). To select the 'working' wavelength in complexation experiments with the use of the $D^0 - pH$ method [48], we studied a behavior of the partial derivative $\partial\Delta D^0(pH,\lambda)/\partial pH = f(pH,\lambda)$ as the function of the λ. Because this function attains a maximum at $\lambda = 400$ nm, the major investigations were performed at this wavelength.

The initial equilibrium concentration of the intermediate complex were determined from

$$c^0_{mn} = \begin{cases} \frac{D^0 - D_M}{D^0_{max} - D_M} c_M\,, & c_M < c_R \\ \frac{D^0 - D_M}{D^0_{max} - D_M} c_R\,, & c_R < c_M \end{cases} \tag{10}$$

where c_M and c_R are the initial concentrations of the metal ion and ligand; D_M is the initial optical density of the metal ion solution; and D^0_{max} is the absorbance of the reaction mixture at the instant of time $\tau = 0$, corresponding to a yield of the complex equal unity.

For $[MR_n]$ complexes, the maximum value of property D^0_{max} was determined in double-reciprocal coordinates by the generalized Benesi–Hildebrand method [6, 7, 49]:

$$\frac{1}{\Delta D^0} \approx \frac{1}{\Delta D^0_{max}} + \frac{1}{\Delta D^0_{max}\beta^{ef}_{1n} c^n_R}, \tag{11}$$

where $\Delta D^0_{max} = \Delta\varepsilon_{1n}\, c_M$, $\Delta\varepsilon_{1n} = \varepsilon_{1n} - \varepsilon_M$, $\varepsilon_M = D_M/c_M$, and ε_{1n} is the extinction coefficient of the complexes $[MR_n]$.

For the complexes $[M_mR]$, the values of D^0_{max} and ε_{m1} are determined from the diagrams $\Delta D^0 - c_M/c_R$. Finding the intensity factor ε_{mn} allows to estimate the $\dot{c}^0 = \dot{D}^0/\varepsilon_{mn}$ value given by the measured value of $\dot{D}^0$.

The composition $m:n$ of intermediate complexes $[M_mR_n]$ formed in system at the moment of mixing a reaction mixture of components was established in direct coordinates by the generalized molar ratio and isomolar series methods [6, 7] from an analysis of the ΔD^0–composition ($\Delta D^0 - c_R/c_M$) and $\Delta\dot{D}^0$–composition ($\Delta\dot{D}^0 - c_R/c_M$) diagrams.

2.3. Calculations

This also solved the generalized inverse problem for chemical equilibria ($\widetilde{II}_-$) besides including (i) calculation of the equilibria stability constants; also, (ii) calculation the rate constants for the processes breaking these equilibria; and, (iii) derivation of the rate law on the basis of the calculated constants and algebraic equations that describe the equilibria in systems.

2.3.1. Calculation of the Preequilibrium Stability Constants

The calculation of effective complexation constants β^{ef}_{mn} for the preequilibrium

$$mM^{q+} + nR \overset{\beta^{ef}_{mn}}{\rightleftharpoons} [M_mR_n]^{q+} \tag{12}$$

were performed using the equation

$$\beta^{ef}_{mn} = \frac{c^0_{mn}}{(c_M - c^0_{mn})^m (c_R - nc^0_{mn})^n}. \tag{13}$$

Its dependence on the pH value

$$\log \beta^{ef}_{mn} = \log K_{mn} + nx\, pH. \tag{14}$$

was studied via a slope analysis for the plot of $\log \beta^{ef}_{mn}$ against pH by averaging the $D^0 - pH$ data series nos. 2-4.

For the preequilibria

$$mM^{q+} + nR \overset{K_{mn}}{\rightleftharpoons} [M_m(H_{-x}R)_n]^{mq-nx} + nx\, H^+, \tag{15}$$

$$mM^{q+} + nH_{-x}R \overset{\beta_{mn}}{\rightleftharpoons} [M_m(H_{-x}R)_n]^{mq-nx},$$

the 'true' equilibrium constants were calculated by these formulas, where charges are omitted:

$$K_{mn} = \frac{[M_m(H_{-x}R)_n][H]^{nx}}{[M]^m[R]^n} = \beta^{ef}_{mn}[H]^{nx},$$

$$\beta_{mn} = \frac{[M_m(H_{-x}R)_n]}{[M]^m[H_{-x}R]^n} = \beta^{ef}_{mn} f^n_x, \tag{16}$$

with

$$f_x = 1 + \sum_{i=1}^{x} B_{xi}[H]^i, \quad B_{xi} = \frac{[H_{-x+i}R]}{[H_{-x}R][H]^i},$$
$$i = \overline{1, x}. \tag{17}$$

To confirm the accuracy of these calculations, numerical optimization methods were also used. The thermodynamic (K_{mn}, β_{mn}) and the optical (ε_{mn}) characteristics of these complexes were fitted with a Hooke and Jeeves' pattern search [50] for minimization of a weighted sum in squared deviations

$$\Delta = \sum_m \sum_n \bar{\varepsilon}_{mn} \beta_{mn} [H_{-x}R]^n - \sum_m \sum_n \hat{\bar{\varepsilon}}_{mn} \beta_{mn} [H_{-x}R]^n \quad (18)$$

or, that equivalent, the calculated mean extinction coefficient

$$\hat{\bar{\varepsilon}}_{mn} = \frac{\varepsilon_M[M] + \sum_m \sum_n \varepsilon_{mn}[M_m(H_{-x}R)_x]}{[M] + \sum_m \sum_n [M_m(H_{-x}R)_n]}$$

$$= \frac{\varepsilon_M[M] + \sum_m \sum_n \varepsilon_{mn} \beta_{mn} [H_{-x}R]^n}{1 + \sum_m \sum_n \beta_{mn} [H_{-x}R]^n}$$

from his experimental value $\bar{\varepsilon}_{mn}$.

2.3.2. Calculation of Rate Constants

The kinetics and mechanism of the redox process apparent in this system was studied by analyzing the dependence of its initial rate $-\dot{D}^0$ on equilibrium concentration of hydrogen ions $[H^+]$ and the initial concentrations of a organic compound c_R and metal ion c_M using the generalizations of the $D^0 - pH$, $D^0 - c_R/c_M$, and $D^0 - c_M/c_R$ methods in the case $\dot{D}^0 \neq 0$ which can be named the $D^0, \dot{D}^0 - pH$, $D^0, \dot{D}^0 - c_R/c_M$, and $D^0, \dot{D}^0 - c_M/c_R$ methods, respectively.

The redox decomposition rate constants of $[M_m(H_{-x}R)_n]^{(q-nx)+}$ complexes were calculated from the data of series $\Delta D^0 - c_R/c_M$, $\Delta D^0 - c_R/c_M$, and $\Delta D^0 - pH$ supplemented by the dependences of $-\Delta\dot{D}^0$ on c_R/c_M, $-\Delta\dot{D}^0$ on c_M/c_R, and $-\Delta\dot{D}^0$ on pH with the equilibrium concentrations of the complex (10), derived in the series $\Delta D^0 - c_M/c_R$, $\Delta D^0 - c_R/c_M$, and $\Delta D^0 - pH$.

The analysis of nonlinear dependence of $-\Delta\dot{D}^0$ on c_R/c_M, $-\Delta\dot{D}^0$ on c_M/c_R and $-\Delta\dot{D}^0$ on pH within the series $\Delta D^0, \Delta\dot{D}^0 - c_R/c_M$ $\Delta D^0, \Delta\dot{D}^0 - c_M/c_R$, and $\Delta D^0, \Delta\dot{D}^0 - pH$ was supplemented with analysis of linear dependences

$$\log(-\Delta\dot{D}^0) = \log k_{mn}^{app} + n_c \log c_{mn}^0, \quad (19)$$

$$-\dot{D}^0 = A + k_{mn}^{app}\, c_{mn}^0, \quad (20)$$

which allows, in particular, to graphically estimate the order n_c of redox process with respect to the initial equilibrium concentrations of these complexes c_{mn}^0 and the value of the apparent intracomplex redox decomposition rate constant k_{mn}^{app}. Here, $\Delta\dot{D}^0 = \dot{D}^0 - \dot{D}_M^0 \approx \dot{D}^0$; $\dot{D}_M^0$ is the initial rate of the density changes in the ceric sulfate solutions.

The calculation of the logarithm the apparent rate constant of intramolecular redox decomposition was carried out for each point of the dependence

$$\log k_{mn}^{app} = \log \Delta\dot{D}^0 - \log c_{mn}^0, \quad (21)$$

with the subsequent averaging of the $\log k_{mn}^{app}$ values obtained.

Since the apparent rate constant of redox decomposition of the complex k_{mn}^{app} ($dm^{-3}mol^{-1}s^{-1}$) depend on their extinction coefficient ε_{mn}

$$k_{mn}^{app} = \Delta\,\varepsilon_{mn}\ l\ k_{mn}, \tag{22}$$

the logarithm of the 'true' rate constant k_{mn} (s^{-1}) of the intramolecular redox decomposition was also calculated:

$$\log k_{mn} = \log k_{mn}^{app} - \log \Delta\varepsilon_{mn} l. \tag{23}$$

Here, $\Delta\varepsilon_{mn} = \varepsilon_{mn} - \varepsilon_M$, l is the thickness of the absorbing solution layer.

The k_{mn} values for the system with two complexes also were fitted using the Cauch's gradient method [51] for minimization of the criterion function

$$U = \sum_{i=1}^{N} w_i \left[\hat{D}^i(k_{mn}) - D^i\right]^2,$$

$$\hat{D}^i(k_{mn}) = \sum_m \sum_n D_{mn}^i \exp\left(-k_{mn}\tau^i\right),$$

$$m, n = \overline{1, 2}; \quad i = \overline{1, N}.$$

Here, w_i is the statistical weight of each measurement, $\hat{D}^i(k_{mn})$ are the fitted progress curves, and D^i are the experimental data points.

The calculated $\log k_{mn}$ values were used to evaluate, in accordance with the formula

$$E_{mn} = 2.303\,\mathcal{R}\,\frac{\log k_{mn}^{T_2} - \log k_{mn}^{T_1}}{\Delta T}\,T_1\,T_2,$$

[$\mathcal{R} = 8.314 \cdot 10^{-3}$ (kJ K^{-1} mol^{-1}) is the gas constant; T the absolute temperature (K); $\Delta T = T_2 - T_1$, $T_2 > T_1$], the activation energy E_{mn} (kJ mol^{-1}) of intramolecular redox decomposition of the complexes (Table 1).

The activation parameters of the redox decomposition of the complexes were also assessed graphically from the $\log k_{mn}$ vs. T^{-1} Arrhenius plots. They were also supplemented by linear least-squares fit using the equation

$$\ln k_{mn} = \ln \mathcal{A}_{mn} - \frac{E_{mn}}{\mathcal{R}} \cdot \frac{1}{T}. \tag{24}$$

Here, $\ln \mathcal{A}_{mn}$ is the natural logarithm of the frequency factor $\mathcal{A}_{mn}$ of redox decomposition of the complexes. Because of the high hydrolyzability of the cerium(IV), the measurements were carried out at the ultimate acid boundary of the yield diagram $\alpha_{mn}^0 - pH$, where $\alpha_{mn}^0 = (D^0 - D_M^0)/(D_\infty^0 - D_M^0)$ is the yield for the complex at 'zero' time $\tau = 0$.

2.3.3. Rate Law and Mechanism

The calculated kinetic and thermodynamic parameters were used to determine the law governing the rate of the redox process observed in these systems and corresponding mechanism for his initial steps.

In view of the equivalence in these two ways determine the rate constant k_{mn}, $\dot{D}^0 \to \dot{c}^0 \to k_{mn}$ and $\dot{D}^0 \to k^{app}_{mn} \to k_{mn}$, where $k_{mn} = \dot{c}^0/c^0_{mn} = k^{app}_{mn}/\varepsilon_{mn} = \dot{D}^0/(\varepsilon_{mn}c^0_{mn})$, the procedure used to determine these parameters and rate law can be also represented as follows:

$$\begin{array}{ccc} & \nearrow \beta^{ef}_{mn} \to x \to \beta_{mn} \searrow & \\ D^0, \dot{D}^0 \to c^0_{mn} & & rate\ law \\ & \searrow k^{app}_{mn} \to k_{mn} \to E_{mn} \nearrow & \end{array}$$

Scheme 3. The procedure to determine the model parameters and rate law of the redox process that occur in these investigated systems.

In the framework of these offered kinetic generalizations of the thermodynamic methods, a set of equations that describe the basic equilibria in a system is transformed into a corresponding set of kinetic equations. This allows to determine the rate law of the redox process that violate these equilibria and its associated mechanism. Therefore, such generalized methods can be considered as a certain tool for studying the mechanism of these redox processes.

Table 1. Stability constant, redox-decomposition rate constant, and corresponding activation energy for $[CeOHH_{-3}Malt]$ complex determined by the $D^0, \dot{D}^0 - pH$ technique* [$c_M = 2.14 \cdot 10^{-3}$ M, $c_L = 2.00 \cdot 10^{-3}$ M, $I = 2$, $t = 17.7\,°C$, $\lambda = 400$ nm, $D^0_{\max} = 0.310$ (run no. 1); $c_M = 3.21 \cdot 10^{-3}$ M, $c_L = 3.00 \cdot 10^{-3}$ M, $I = 2$, $t = 27.0\,°C$, $\lambda = 400$ nm, $D^0_{\max} = 0.246$ (run no. 2)]

D^0	D_M	$-\dot{D}^0$, s^{-1}	pH	c^0_{11}, M	$\log k^{app}_{11}$	$\log k_{11}$	$\log \beta^{ef}_{11}$	$\log \beta_{11}$
				Run no. 1				
0.228	0.227	$0.4 \cdot 10^{-4}$	1.67	$0.024 \cdot 10^{-3}$	−0.77	−2.77	0.76	18.31
0.231	0.228	$0.3 \cdot 10^{-4}$	1.79	$0.073 \cdot 10^{-3}$	−0.39	−2.39	1.26	18.45
0.229	0.220	$0.9 \cdot 10^{-4}$	1.90	$0.200 \cdot 10^{-3}$	−0.35	−2.35	1.76	18.61
0.230	0.215	$1.1 \cdot 10^{-4}$	2.05	$0.316 \cdot 10^{-3}$	−0.46	−2.46	2.01	18.64
0.244	0.210	$2.1 \cdot 10^{-4}$	2.27	$0.680 \cdot 10^{-3}$	−0.51	−2.51	2.55	18.34
0.250	0.212	$2.7 \cdot 10^{-4}$	2.42	$0.775 \cdot 10^{-3}$	−0.46	−2.45	2.67	18.01
0.262	0.208	$3.2 \cdot 10^{-4}$	2.38	$1.060 \cdot 10^{-3}$	−0.52	−2.52	3.02	18.48
				Run no. 2				
0.158	0.155	$1.0 \cdot 10^{-4}$	1.68	$0.095 \cdot 10^{-3}$	0.02	−1.80	1.02	18.54
0.160	0.155	$2.0 \cdot 10^{-4}$	1.85	$0.158 \cdot 10^{-3}$	0.10	−1.72	1.26	18.28
0.162	0.150	$3.7 \cdot 10^{-4}$	1.96	$0.360 \cdot 10^{-3}$	0.01	−1.81	1.68	18.38
0.162	0.148	$5.0 \cdot 10^{-4}$	2.06	$0.412 \cdot 10^{-3}$	0.08	−1.74	1.75	18.15
0.167	0.148	$5.1 \cdot 10^{-4}$	2.06	$0.559 \cdot 10^{-3}$	−0.04	−1.86	1.94	18.34
0.185	0.140	$7.8 \cdot 10^{-4}$	2.20	$1.227 \cdot 10^{-3}$	−0.20	−2.02	2.85	18.83

* $\log \beta_{11} = 18.39 \pm 0.18$, $\log k_{11} = -2.57 \pm 0.10$ $(17.7\,°C)$ (run no. 1); $\log \beta_{11} = 18.42 \pm 0.24$, $\log k_{11} = -1.83 \pm 0.10$ $(27\,°C)$, $E_{11} = 133.0 \pm 1.3$ kJ mol^{-1} (run no. 2).

This experiment shows the pseudo-first order plots, in most cases were linear up to $75 - 90\%$ completion in the redox reaction, with the same degree of reaction completion remains valid and the established rate law.

Note, that to the same rate law there can correspond not a unique mechanism of process. In this case, the choice of the mechanism needs the identification of intermediates via which this mechanism is realized. However, these considered methods allow to identify composition of such intermediates.

The concrete examples, showing the possibilities for this considered method regarding the investigation in the reactions mechanism, are given at the end of this chapter.

3. Results and Discussion

3.1. Spectroscopic Evidence for Existence of Intermediate Complexes

Marked initial optical density changes were observed when solutions of cerium(IV) mixed with organic compounds solutions. The intensity of the peak in the spectrum of the $Ce(SO_4)_2$ solution at 315 nm increases; and, simultaneously, the peak is shifted to the shortwave region (λ_{max}~300–310 nm) by the addition of oxalic, malonic, succinic, glutaric, malic, citric, and tartaric acids at pH values from 1.3 to 2.3. These changes in the Ce(IV) spectrum can be attributed to the formation of an intermediate complex within the mixing time of the reagents.

The appearance in the spectrum of a second absorbtion band (λ_{max}~285–290 nm) and the decrease of the intensity of the peak over pH range 2.3-3.5 probably indicates the formation of second complex in the systems $Ce(IV)$–SO_4^{2-}–R, $R = Oxal$, $Maln$, $Succ$, $Citr$, and $Tart$. Availability of an isosbestic point in the obtained absorption spectra at pH values from 2.3 to 3.5 confirms this assumption and can be regarded as evidence for the existence of at least two complexes in the system. A typical spectrum in the example of the $Ce(IV)$–SO_4^{2-}–$Citr$ system is shown in Figure 1.

On these grounds, we can assume that the two intermediate complexes formed in the system at the instant of mixing the reaction mixture components. Further evidence for this explanation is given in the following sections.

3.2. Stoichiometry

3.2.1. Composition of Intermediates

The composition of intermediate complexes $[M_m(H_{-x}R)_n]^{q-nx}$ formed in the systems $Ce^{4+} - SO_4^{2-} - R$ at the instant of mixing the reaction mixture components was studied by the methods of isomolar series and molar ratios. The results obtained indicate that $1:1$ complexes $[M(H_{-x}R)]^{q-x}$ are predominant in all the systems containing dicarboxylic and oxycarboxylic acids as well as polyatomic alcohols over the pH range $pH \lesssim 2.3$.

The formation of these complexes was additionally confirmed by using kinetic analogs of these methods which are particularly appropriate in the cases of low yield or low light absorption by the complexes. The data on the composition of these complexes $[M(H_{-x}R)]^{q-x}$, $R = Oxal$, $Maln$, $Succ$, $Glut$, $Quin$, $Lact$, $Citr$, $Malt$, $Tart$, $Glyc$, $Csyl$, $Sorb$, over this pH range are in agreement with the conclusions of [52] that the $1:1$

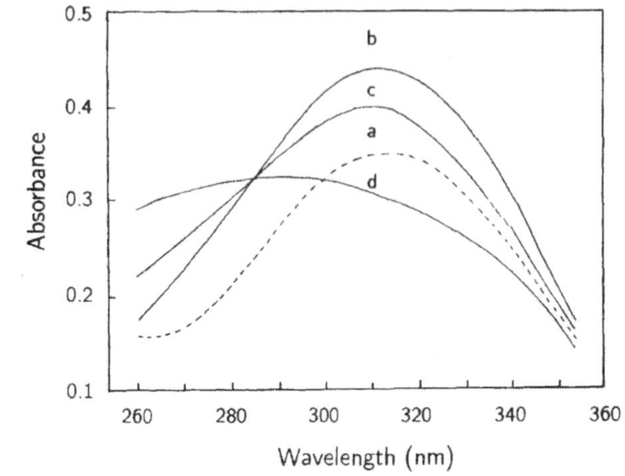

Figure 1. UV-Vis spectra for the reactant mixtures in the absence (a) and in the presence (b-d) citric acid at various pH levels in the $\tau = 0$ moment for the system: $Ce^{4+} - SO_4^{2-} - Citr$: $pH = 2.0$, $c_M = 2.2 \cdot 10^{-4}$ M (a); pH levels of 2.0 (b), 2.3 (c), 3.5 (d), $c_M = 2.2 \cdot 10^{-4}$ M, $c_R = 2.2 \cdot 10^{-4}$ M (b-d); $I = 2$, $t = 25.0°C$.

composition of intermediate organocerium complexes is independent of the organic ligand denticity.

However, the investigation of the Ce^{4+}–SO_4^{2-}–R systems at pH~2.3–3.5 showed a second pH range in the existence of organocerium complexes $[M_m(H_{-x}R)_n]^{q-nx}$. A second type of complex species, $[M(H_{-x}R)_2]^{(q-2x)+}$ ($R = Oxal, Maln, Succ, Citr$) and $[M_2(H_{-x}Tart)]^{2(q-x)+}$, formed at these pH values (Figure 2).

3.2.2. Ligand Speciation

The form in which a coordinated ligand is present in the $[(CeOH)_m(H_{-x}R)_n]^{3-nx}$ ($m, n = 1, 2$) complexes and their stability constants were determined using the $D - pH$ method [48]. As the dominant form of cerium(IV) in the pH range studied was taken the monohydroxo form $CeOH^{3+}$ [53]. The formation of $Ce(OH)_2^+$ occurs to a much smaller extent in comparison with the others and was therefore neglected.

The slope analysis shows, that the multicharged $CeOH^{3+}$ ion displaces 2 protons ($x = 2$) from the ligand molecule in complexation with all the dicarboxylic acids and the polyatomic alcohols, as well as with lactic, quinic, and citric oxycarboxylic acids. Upon complexation of malic and tartaric oxycarboxylic acids, the $CeOH^{3+}$ ion displaces 3 and 4 protons ($x = 3, 4$), respectively (Figure 3). In these cases, similarly to the interaction of oxycarboxylic acids with the cations Sc^{3+}, Ga^{3+}, In^{3+}, and Th^{4+} [54], protons of the oxy groups of oxyacids are displaced because of the high coordination numbers of REE and the effective negative charge on electron-donor oxygen atoms.

For the complex species $[CeOH(H_{-x}R)_2]^{3-2x}$ ($R = Oxal, Citr$) and complex $[(CeOH)_2H_{-x}Tart]^{6-x}$, the number x of protons displaced from the ligand molecule by metal ion equals two and four, respectively.

Since it is known that in the oxidation of oxycarboxylic acids by variable-valence metal ions the intermediates are complexes with charge transfer from the carboxylic groups [55];

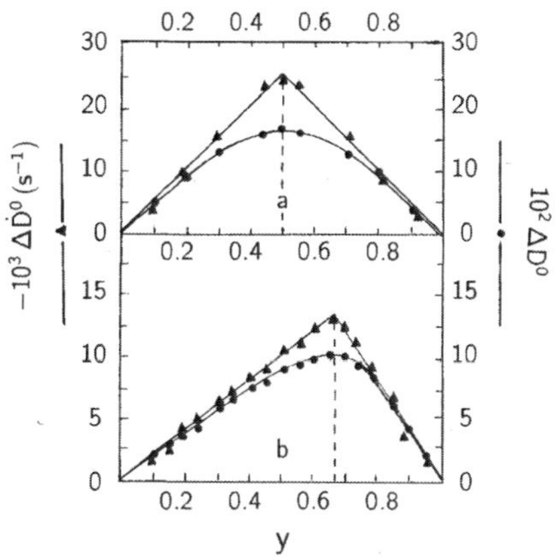

Figure 2. Determination of the composition for the intermediate cerium citrate complexes by the generalized isomolar series method: $pH = 1.4, c_{tot} = 1.0{\cdot}10^{-3}\ M\ (a)$; $pH = 2.4, c_{tot} = 4.0{\cdot}10^{-3}\ M\ (b)$; $t = 25.0°C, I = 2, \lambda = 400$ nm.

the intramolecular transfer in the cerium(IV) intermediate complexes (e.g., with malic, lactic, and tartaric acids) can be represented as

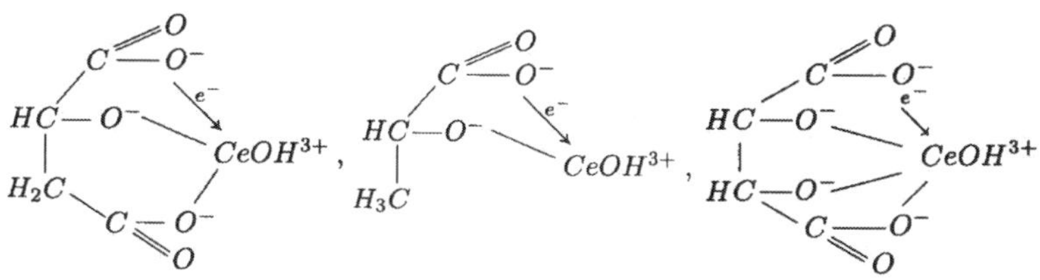

3.2.3. Preequlibrium Scheme

Thus, for all of these investigated systems, the preequlibrium scheme reads:

$$m\,CeOH^{3+} + nH_{-x}R^{x-} \overset{\beta_{mn}}{\rightleftharpoons} [(CeOH)_m(H_{-x}R)_n]^{3m-xn}, \quad (25)$$

$[m, n = 1, 2$ and $x = 2, 3, 4]$ or, which is equivalent in terms of the ionic equilibria theory, can be written as follows

$$m\,CeOH^{3+} + nH_{-x}R^{x-} \overset{\beta_{mn}}{\rightleftharpoons} [(CeO)_m(H_{-x+m}R)_n]^{3m-xn}.$$

In view of the possible dimerization [56], the formulas $[(CeOH)_m(H_{-x}R)_n]^{3m-2n}$ and $[(CeO)_m(H_{-x+1}R)_n]^{3m-xn}$ should be regarded as notations which refer to all of the isomers for the complexes

$$[(CeOH)_{mp}(H_{-x}R)_{pn}]^{(3m-xn)p}$$

$$[(CeO)_{mp}(H_{-x+m}R)_{np}]^{(3m-xn)p}$$

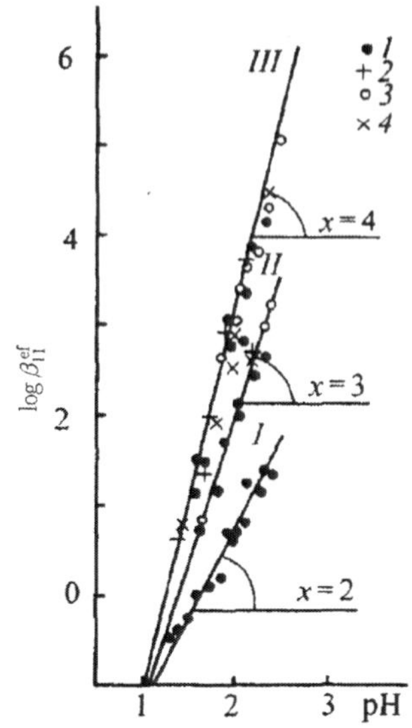

Figure 3. $\log \beta_{11}^{ef}$ vs. pH diagrams for the systems Ce^{4+}–SO_4^{2-}–R. Ligand: (I) $R = Lact$ [$c_M = 6 \cdot 10^{-3}$ M, $c_R = 3 \cdot 10^{-2}$ M]; (II) $R = Malt$ [(1) $c_M = c_R = 2.14 \cdot 10^{-3}$ M; (2) $c_M = c_R = 2.04 \cdot 10^{-3}$ M; (3) $c_M = c_R = 3.21 \cdot 10^{-3}$ M]; (III) $R = Tart$ [(1) $c_M = c_R = 1.22 \cdot 10^{-3}$ M; (2) $c_M = c_R = 2.14 \cdot 10^{-3}$ M; (3) $c_M = 3.02 \cdot 10^{-3}$ M, $c_R = 1.65 \cdot 10^{-3}$ M].

[$p = 1, 2$] which can be either by a mix of species or coincide with one of the major species in the mix. It is noteworthy that these formulas for a sulfate medium should only be regarded as an abbreviated designation of all the possible mixed organocerium-sulfato hydroxo complexes with a variable number of ions for the ionic background

$$\sum_p \sum_r \sum_z [(NH_4)_r(CeOH)_m(SO_4)_z(H_{-x}R)_n]_p^{(3m-xn+r-z)p}.$$

Here, $m, n, p = 1, 2$; r and z can be estimated by the inequalities $r \geq 0; 0 \leq z \leq 3$ [53, 57, 58].

3.3. Stability Constants of the Complexes

For the preequilibrium (25), the equilibrium constant

$$\beta_{mn} = \frac{[(CeOH)_m(H_{-x}R)_n]}{[CeOH]^m[H_{-x}R]^n} = \beta_{mn}^{ef} f_x^n, \qquad f_x = 1 + \sum_{i=1}^{x} B_i[H]^i, \tag{26}$$

was calculated from the data in the series $D^0 - pH$ and the ascending sections of the saturation curves for each point of the dependence $\log \beta_{mn} = \log \beta_{mn}^{ef} + n \log f_x$ with the subsequent averaging of the $\log \beta_{mn}$ values obtained. The calculations were performed using the logarithms of the overall protonization constants $B_i = [R]\,([H_{-i}R][H]^i)^{-1}$ of the organic anion $H_{-x}R^{x-}$ from [59].

For this data, average $\log \beta_{11}$ values found on the sulfate background for the cerium complexes $[CeOH(H_{-2}R)]^+$ with dicarboxylic acids, using the $D^0 - pH$ and molar ratio methods, are $\log\beta_{11} = 7.44 \pm 0.13$ ($R = Oxal$), $\log\beta_{11} = 9.36 \pm 0.12$ ($R = Glut$), $\log\beta_{11} = 9.63 \pm 0.10$ ($R = Succ$), and $\log\beta_{11} = 10.15 \pm 0.09$ ($R = Maln$).

The $\log\beta_{11}$ value for the complex $[CeOHH_{-x}R]^{3-x}$ ($x = 2, 3, 4$) is found in a sulfate medium by averaging data of two to four series for each of the systems, that grows in order $Quin \sim Lact < Citr < Malt < Tart$ with an increasing ligand denticity from $\log\beta_{11} = 14.93{+}0.14$ ($R = Quin$), $\log\beta_{11} = 14.97{\pm}0.18$ ($R = Lact$), and $\log\beta_{11} = 15.87{\pm}0.13$[3] ($R = Citr$) to $\log\beta_{11} = 18.40 \pm 0.21$ ($R = Malt$), and $\log\beta_{11} = 27.83 \pm 0.23$ ($R = Tart$).

For the complexes $[CeOHH_{-2}R]^+$ with polyatomic alcohols these values are $\log\beta_{11} = 22.41 \pm 0.05$ (Glyc), $\log\beta_{11} = 22.46 \pm 0.03$ (Csyl), and $\log\beta_{11} = 22.50 \pm 0.11$ (Sorb). The agreement between these obtained results by different methods (molar ratios (MR), Benesi–Hildebrand, Ardon, D–pH) in determining the stability constants for, e.g., the complex $[CeOH(H_{-2}Sorb)]^+$ indicates that the basic equilibria in solution are taken into account correctly (Table 2).

Table 2. Average stability constants for complex $[CeOH(H_{-2}R)]^+$, $R = Sorb$

pH	$\log\beta_{11}^{ef}$	$\log\beta_{11}$	Method
2.02	0.79	22.65	MR**
2.02	0.79	22.65	Benesi**
2.25	1.30	22.70	MR**
2.25	1.30	22.70	Benesi**
2.25	1.04	22.44	Ardon**
2.45	1.42	22.42	D– pH*
2.55	1.68	22.48	D– pH*
2.60	1.78	22.57	MR*
2.70	1.93	22.43	MR*
2.70	1.94	22.44	Ardon**
2.70	1.90	22.40	D– pH*
2.85	2.18	22.38	D– pH*
2.98	2.53	22.47	D– pH*
3.05	2.58	22.38	D– pH*
3.40	3.37	22.47	D– pH*
$\log\beta_{11} = 22.50 \pm 0.11$			

Note: $^*\lambda = 400$ nm, Specord UV VIS;
$^{**}\lambda_{ef} = 400$ nm, KF-5.

[3] In literature, most of the studies assume that only the carboxylic groups of citric acid participate in complexation of cerium(IV) [6, 60] and tetravalent actinides $An(IV)$ [61] with citrate. However, recent structural studies on thorium(IV) and uranium(IV) complexes with citric acid [62] suggest that citrate is bound to the cation both, by its carboxylate site to the central carbon atom and by its hydroxyl group in alpha position. Accounting for this option in the present paper, the following values of the logarithms for the overall protonization constants were applied to the calculations for citric acid: $\log B_1 = 14.60$, $\log B_2 = 17.45$.

The $\log\beta_{mn}$ values for the cerium bis-complexes with oxalic and citric acids $[CeOH(H_{-2}R)_2]^-$ ($R = Oxal$, $Citr$) as well as for binuclear cerium hydroxotartrato complex $[(CeOH)_2H_{-4}Tart]^{2+}$ are $\log\beta_{12} = 12.57 \pm 0.52$, $\log\beta_{12} = 30.38 \pm 0.27$, and $\log\beta_{21} = 28.63 \pm 0.29$, respectively.

Obtained values for stability constants β_{mn} in systems with two complexes ($R = Tart$, $Citr$, $Oxal$) are consistent with the corresponding values of $\hat{\beta}_{mn}$, and fit when using the Hook–Jeeves algorithm. For instance, $\log\beta_{11} = 15.87 \pm 0.13$, $\log\beta_{12} = 30.38 \pm 0.27$ and $\log\hat{\beta}_{11} = 15.94$, $\log\hat{\beta}_{12} = 30.61$ for $[CeOH(H_{-2}Citr)_n]^{3-2n}$ ($n = 1, 2$). This additionally confirms the results of this present work.

3.4. Intramolecular Rate Constants and Activation Energies

The logarithms of apparent redox decomposition rate constants $\log k_{11}^{app}$ (21) for the complexes $[CeOHH_{-x}R]^{3-x}$ ($x = 2, 3, 4$) grows in the order of lactic, quinic, malic, citric, and tartaric acids from -0.61 ± 0.06 to 0.11 ± 0.04 ($20°C$). However, the logarithms of true rate constants, $\log k_{11}$, calculated using expression (23), are constant and equal to -2.44 ± 0.05 for all the complexes ($20°C$). This magnitude of $\log k_{11}$ is close to the size of $\log k_{21} = -2.47 \pm 0.06$ ($20°C$) found for the binuclear cerium hydroxotartrato complex $[(CeOH)_2H_{-4}Tart]^{2+}$, as well as to the value of $\log k_{12} = -2.40 \pm 0.06$ ($20°C$) for the complex $[CeOH(H_{-2}Citr)_2]^-$. Coincidence in the $\log k_{mn}$ values for these complexes $[CeOH(H_{-2}Citr)_n]^{3-2n}$ ($n = 1, 2$) are demonstrated in Table 3.

The activation energy of intramolecular decomposition also takes a constant, within experimental error, value $E_{11} = 131.0 \pm 2.5$ kJ mol^{-1} for $Quin$, $Lact$, $Citr$, $Malt$, and $Tart$. This value close in magnitude to the activation energy of electron transfer, e.g., in the system $Fe(bipy)_3^{2+} - Co(CN)_5N_3^{3-}$ ($E = 127.0 \pm 2.0$ kJ mol^{-1}) [63]. The observed rise in the rate pseudoconstants which went from $Lact$ to $Malt$ [29, 43] and from $Malt$ to $Tart$ [29, 33] is only apparent and reflects the increase in the extinction coefficients of the complexes for the series in question.

The logarithm of the redox decomposition rate constant $\log k_{11} = -1.71 \pm 0.08$ for the complex $[CeOHH_{-2}Oxal]^+$, within the experimental error, coincides with $\log k_{12} = -1.74 \pm 0.10$ of the bis-complex $[CeOH(H_{-2}Oxal)_2]^-$.

The corresponding $\log\hat{k}_{mn}$ values obtained when using the Cauch's method are $\log\hat{k}_{11} = -1.77$ and $\log\hat{k}_{12} = -1.80$ ($20°C$). The activation energy takes the size 81.17 ± 1.8 kJ mol^{-1}. The magnitude of $\log k_{11}$ for the complexes $[CeOHH_{-2}R]^+$ with other dicarboxylic acids are $\log k_{11} = -2.00 \pm 0.04$ ($R = Maln$), $\log k_{11} = -2.01 \pm 0.05$ ($R = Succ$), and $\log k_{11} = -2.02 \pm 0.03$ ($R = Glut$).

Table 3. Redox-decomposition rate constants for the $[CeOH(H_{-2}Citr)_n]^{3-2n}$ ($n = 1, 2$) complexes are determined by the D^0, $\dot{D}^0 - c_R/c_M$ technique [$I = 2$, $t = 25.0°C$, $\lambda = 400$ nm; $n = 1$: $c_M = 2.8 \cdot 10^{-3}\ M$, $pH = 2.23$, $D^0_{\max} = 0.720$, $D_M = 0.195$; $n = 2$: $c_M = 2.5 \cdot 10^{-3}\ M$, $pH = 2.65$, $D^0_{\max} = 0.430$, $D_M = 0.112$]

$\log(-\Delta\dot{D}^0)$	$10^3\ c_R$, M	$\log k_{11}^{app}$	$\log k_{11}$	$\log(-\Delta\dot{D}^0)$	$10^2\ c_R$, M	$\log k_{12}^{app}$	$\log k_{12}$
−2.85	0.82	0.41	−1.99	−2.72	0.15	0.25	−2.01
−2.73	1.23	0.46	−1.94	−2.62	0.25	0.31	−1.95
−2.56	1.65	0.45	−1.95	−2.60	0.30	0.35	−1.91
−2.49	2.07	0.44	−1.96	−2.57	0.35	0.25	−2.01
−2.35	2.49	0.46	−1.94	−2.54	0.40	0.26	−2.00
−2.41	2.91	0.41	−1.99	−2.52	0.50	0.23	−2.03
−2.38	3.30	0.40	−2.00	−2.44	0.60	0.32	−1.94
−2.36	3.72	0.40	−2.00	−2.40	0.70	0.33	−1.93
−2.35	4.14	0.40	−2.00	−2.39	0.75	0.33	−1.93
−2.35	4.56	0.37	−2.03	−2.40	1.00	0.27	−1.99
−2.37	4.98	0.35	−2.05	−2.41	1.50	0.32	−1.94
−2.35	5.40	0.39	−2.01	−2.40	2.00	0.26	−2.00
−2.36	5.80	0.37	−2.03	−2.41	2.50	0.22	−2.04
−2.36	6.22	0.38	−2.02	−2.40	2.65	0.23	−2.03
−2.36	6.64	0.35	−2.05	−2.41	2.75	0.22	−2.04
−2.34	7.06	0.35	−2.05	−2.37	3.00	0.23	−2.03
−2.35	7.48	0.36	−2.04	−2.41	3.15	0.26	−2.00

$\log k_{11}^{app} = 0.40 \pm 0.04$, $\log k_{11} = -2.00 \pm 0.04$; $\log k_{12}^{app} = 0.26 \pm 0.06$, $\log k_{12} = -2.00 \pm 0.06$.

The values of $\log k_{11}$ for the complexes with dicarboxylic acids are higher than the corresponding $\log k_{11}$ values for cerium(IV) complexes with α-oxycarboxylic acids formed with the displacement of the alcohol hydroxyl proton and are larger by two orders of magnitude than the $\log k_{11} = -3.75 \pm 0.06\ (20°C)$ value for the complexes $[(CeOH)H_{-2}R]^+$ of the same type with aliphatic polyatomic alcohols ($R = L(OH)_k, k = 3, 5, 6$). A comparison of the kinetic data on this series of organic substances show that the relative destabilizing effect of the carboxy group in the cerium complexes $[(CeOH)H_{-2}R]^+$ with dicarboxylic, α-oxycarboxylic acids, and polyatomic alcohols (Table 4).

Table 4. The averaged stability constants and rate constants of the complexes $[(CeOH)_m(H_{-x}R)_n]^{3m-xn}$ identified in this study $[I = 2, t = 20°C]$

$[(CeOH)_m(H_{-x}R)_n]^r$	$\log\beta_{mn}$	$-\log k_{mn}$
$[CeOH(H_{-2}Oxal)]^+$	7.44 ± 0.13	1.71 ± 0.08
$[CeOH(H_{-2}Glut)]^+$	9.36 ± 0.12	2.02 ± 0.03
$[CeOH(H_{-2}Succ)]^+$	9.63 ± 0.10	2.01 ± 0.05
$[CeOH(H_{-2}Maln)]^+$	10.15 ± 0.09	2.00 ± 0.04
$[CeOH(H_{-2}Quin)]^+$	14.93 ± 0.10	2.50 ± 0.06
$[CeOH(H_{-2}Lact)]^+$	14.97 ± 0.18	2.45 ± 0.05
$[CeOH(H_{-2}Citr)]^+$	15.87 ± 0.13	2.40 ± 0.06
$[CeOH(H_{-3}Malt)]$	18.40 ± 0.21	2.42 ± 0.05
$[CeOH(H_{-4}Tart)]^-$	27.83 ± 0.23	2.43 ± 0.06
$[CeOH(H_{-2}Glyc)]^+$	22.41 ± 0.05	3.72 ± 0.10
$[CeOH(H_{-2}Csyl)]^+$	22.46 ± 0.03	3.81 ± 0.08
$[CeOH(H_{-2}Sorb)]^+$	22.50 ± 0.11	3.71 ± 0.09
$[CeOH(H_{-2}Oxal)_2]^-$	12.57 ± 0.52	1.74 ± 0.10
$[CeOH(H_{-2}Citr)_2]^-$	30.38 ± 0.27	2.40 ± 0.04
$[(CeOH)_2H_{-4}Tart]^{2+}$	28.63 ± 0.29	2.47 ± 0.06

3.5. Correlation Between Thermodynamic and Kinetic Stability of the Complexes

The obtained values for $\log\beta_{11}$ and $\log k_{11}$ (Table 4) allow us to establish a direct ('positive') linear correlation between thermodynamic ($\log\beta_{11}$) and kinetic ($-\log k_{11}$) stability of the complexes $[CeOH(SO_4)_zH_{-2}R]^{1-2z}$ ($z \leq 3$) with dibasic aliphatic ligands of the type $R = H_2L$ ($Oxal$, $Maln$, $Succ$, $Glut$), $R = H(OH)L$ ($Lact$, $Quin$, $Citr$), and $R = L(OH)_2$ ($Glyc$, $Csyl$, $Sorb$) (Figure 4,I). This correlation demonstrates the stabilizing action of the oxy groups for ligands in complexes with predominantly an ionic character of bonds[4]. In other words, increased stabilization of cerium OS +4 in the complex with the considered ligands can be associated with an increase in the electronegativity of the oxygen atom for the functional group of the ligand, from which occurs a transfer of electron density on the central ion, for this series of ligands.

[4]For complexes with the predominantly covalent type of metal-ligand bonds ($M^{q+} = Mn^{3+}$) can be observed a negative correlation dependence [30].

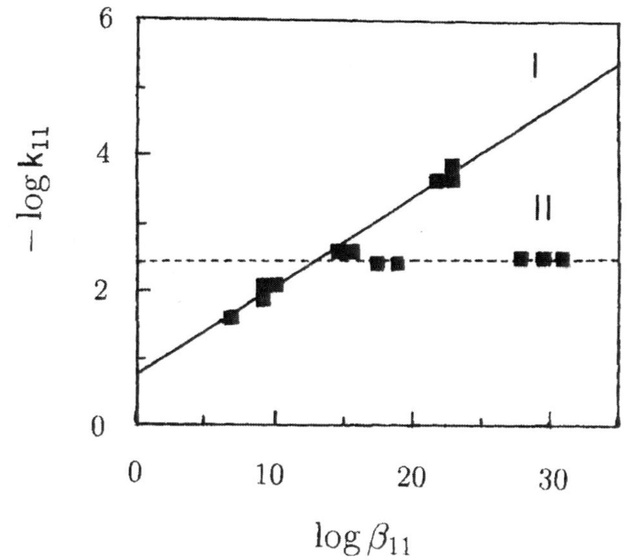

Figure 4. $-\log k_{11}$ versus $\log \beta_{11}$ correlation dependences: (I) for complexes of the type $[CeOH(H_{-2}R)]^{+}$ and (II) for complexes of the type $[CeOH(H_{-x}R)]^{3-x}$ ($x \geqslant 2$).

The correlation equation for $x = 2$ has the form $-\log k_{11} = 0.80 + 0.13 \log \beta_{11}$. It should, however, be used with care because of the statistical spread of values. The results obtained in this work lead us to suggest the existence of the corresponding correlations between the electrochemical and optical characteristics of $[(CeOH)H_{-x}R]^{3-x}$ complexes [8]. Studies of such correlations are currently at the initial stage but are well worth the effort.

An increase in the basicity of oxycarboxylic acids above two ($x > 2$) and, as a consequence, an increase in the thermodynamic stability of complexes does not influence their reactivity (Figure 4, II). The constancy of the reactivity of the complexes $[CeOH(SO_4)_z H_{-x}R]^{3-2z-x}$ with $x = 2, 3, 4$ ($R = Lact, Malt, Tart$), as well as the absence of observable effect of relative stabilizing cerium OS +4 in the bis-complexes $[CeOH(SO_4)_z(H_{-2}R)_2]^{-2z-1}$ ($R = Oxal, Citr$) when the second organic anion penetrates into inner coordination sphere, can be explained in two alternative ways.

On one hand, it is possible to assume, according to [53, 58], a direct interaction between trisulfato-hydroxy cerate ion with the organic substrates without the substitution of sulfate anion from the inner coordination sphere. In this case, it is probable that no more than two functional groups of organic ligands can penetrate into the inner coordination sphere of the cerium(IV) complexes in the sulfate medium since the cerium(IV) ion is eight-nine coordinate [64]. The functional groups of organic ligands, remaining in outer sphere, do not render influence on the rate of the intramolecular electron transfer.

An another explanation for the lack of this stabilizing effect may be an approximate compensation in this effect by a corresponding destabilizing effect which can accompany the displacement of sulfate anion from the inner coordination sphere. Arguments in favor of the substitution of sulfate anion by organic ligands in the cerium(IV) complexes will be given later.

3.6. Reaction Orders and Rate Law

3.6.1. Reaction Orders

Plots of $\log D$ vs. time, for various initial concentrations of cerium(IV) and organic compounds (oxycarboxylic and dicarboxylic acids) were found to be linear indicating good pseudo-first-order kinetics for the decomposition of the intermediate complexes. The typical decay curves showing the decomposition of the intermediates are presented in Fig. 5.

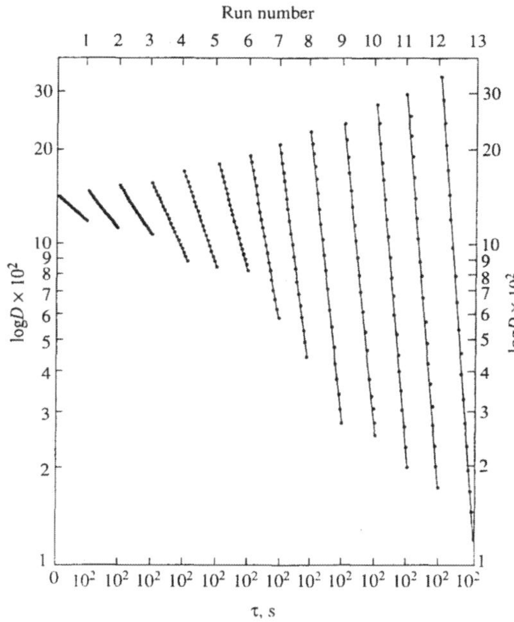

Figure 5. First-order rate curves $\log D^i$ vs. τ^i ($\tau^{i+1} - \tau^i = 5s$, $i = 1, \ldots, 20$) for the cerium(IV)-quinic acid system: (1-13) $c_M = 1.26 \cdot 10^{-3}\ M$, $c_R = 1.87 \cdot 10^{-4}$ to $1.33 \cdot 10^{-3}\ M$, $pH = 2.64$, $I = 2$, $t = 24.5°C$, and $\lambda = 400$ nm.

The typical data of Tables 1 and 2 show also that $k_{11} = -\Delta\dot{D}^0/(c_{11}^0 \cdot \Delta\varepsilon_{11}\, l)$ was unaffected, when the concentrations of the intermediate complexes with organic compounds were varied, indicating first-order dependence on the concentrations of these intermediates. Eq. (19) gives the value of reaction order n_c from 0.95 to 1 for all studied intermediate complexes. The first order with respect to concentrations of the cerium(IV) complexes with the oxycarboxylic and dicarboxylic acids ($n_c = 1$) coincides with the first order with respect to time ($n_\tau = 1$) and represent the true reactions order of the investigated redox process.

The equality to zero for the intercepts of the plots $-\dot{D}^0$ vs. c_{11}^0 in Figure 6 is the evidence that the bimolecular reaction has no contribution to the process in the range of the pH values studied. Therefore, the production of the chelate intermediates $[CeOH(H_{-x}R_n]^{3-xn}$ is a necessary stage of the redox process observed in the systems with the aliphatic organic acids and polyatomic alcohols.

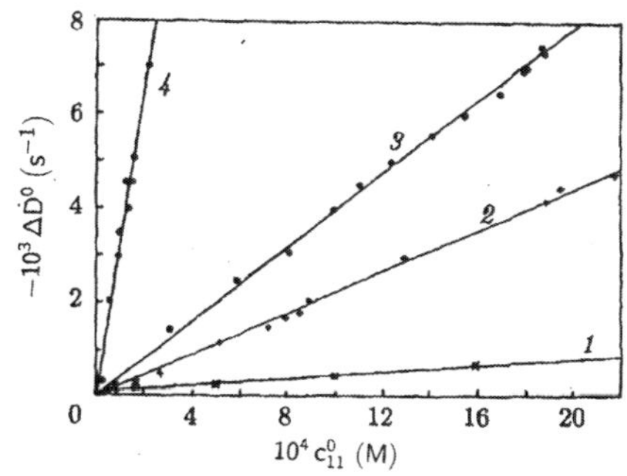

Figure 6. $-\dot{D}^0$ vs. c^0_{11} diagrams for the systems Ce^{4+}–SO_4^{2-}–R. Ligand: (1) $R = Sorb$ [$c_M = c_R = 2.14 \cdot 10^{-3}\ M,\ t = 20.0\,°C$]; (2) $R = Lact$ [$c_M = 6.0 \cdot 10^{-3}\ M,\ c_R = 3.0 \cdot 10^{-2}\ M,\ t = 18.0\,°C$]; (3) $R = Malt$ [$c_M = 1.88 \cdot 10^{-3}\ M,\ c_R = 3.00 \cdot 10^{-3}\ M,\ t = 18.0\,°C$]; (4) $R = Tart$ [$c_M = c_R = 2.14 \cdot 10^{-3}\ M, t = 28.0\,°C$]; $I = 2$, $\lambda = 400$ nm.

3.6.2. Rate Law Over the Range pH $\lesssim$ 2.3

The equation for the initial rate of the redox process occurring in the range $0.6 < pH < 2.3$ can be represented as:

$$-\Delta\dot{D}^0 = \Delta\varepsilon_{1n} l k_{1n} \cdot c^0_{1n}\,. \tag{27}$$

When given the account of expressions (16), Eq. (27) can be rewritten as follows:

$$-\frac{dc_M}{dt} = k_{1n}K_{1n}\frac{[CeOH^{3+}][R]}{[H^+]^x} = k_{1n}\beta_{1n}[CeOH^{3+}][H_{-x}R^{x-}], \tag{28}$$

where $R = Oxal, Maln, Succ, Glut; Quin, Lact, Citr, Malt, Tart; Glyc, Csyl, Sorb$.

This equation establishes the type of the functional dependence of the rate on the equilibrium concentration of hydrogen ions in solution. It is in agreement with the general form of the rate equation $-d[Ce(IV)]/dt = k^{ef}_{(n_c=2)}[Ce(IV)][R]$ [43] and the inverse dependence of the rate pseudoconstant $k^{ef}_{(n_c=2)}$ on the $[H^+]$ observed in several studies (see e.g., [43]). According to (28), the latter dependence can be described by the equation $k^{app}_{(n_c=2)} = k_{1n}K_{1n}/[H^+]^x$.

3.6.3. Rate Law Over the Range pH $\gtrsim$ 2.3

Assume that the cerium(IV) is completely bound in the organic complexes and assume for definiteness that $c_M \le c_R$. Then, e.g., for the system $Ce^{4+} - SO_4^{2-} - Citr$, the rate law can be approximated by the empirical equation

$$\begin{aligned}-\Delta\dot{D}^0 = \sum_n k^{app}_{1n}c^0_{1n} \cong klc_M\sum_n \Delta\varepsilon_{1n}\alpha^0_{1n} &= k\,l\,c_M\left[\Delta\varepsilon_{11}(1-\alpha^0_{12}) + \Delta\varepsilon_{12}\alpha^0_{12}\right]\\ &= k\,l\,c_M\Delta\bar{\varepsilon}_{1n},\end{aligned} \tag{29}$$

where $n = 1, 2$ and $k_{11} \simeq k_{12} \equiv k$.

Equation (29) allows the decrease in the initial rate and apparent rate constant of the redox decomposition of cerium citrato complexes for $\alpha^0_{11} > 0.7$ (Fig. 7b) to be interpreted as arising from a decrease in the mean extinction coefficient of these complexes.

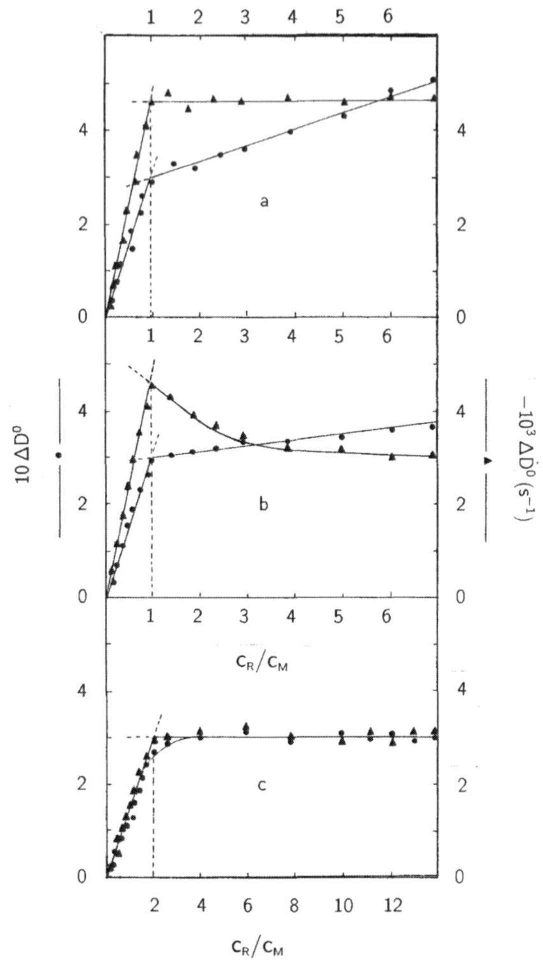

Figure 7. Determination of the intermediate complexes compositions by the generalized molar ratios technique at different pH levels: $\Delta D^0, \Delta \dot{D}^0$ vs. composition diagrams at $pH = 2.23$, $c_M = 2.8 \cdot 10^{-3}\ M$ (a); $pH = 2.34$, $c_M = 2.8 \cdot 10^{-3}\ M$ (b); and $pH = 2.80$, $c_M = 2.8 \cdot 10^{-3}\ M$ (c); $t = 25.0°C$, $I = 2$, $\lambda = 400$ nm.

The decrease is presumably caused by an increment in the yield of the weakly absorbing bis-complex ($\varepsilon_{11} \sim 269 > \varepsilon_{12} \sim 192$) in accordance with

$$\varepsilon_{11} \geq \varepsilon_{11}(1 - \alpha_{12}^0) + \varepsilon_{12}\alpha_{12}^0 \geq \varepsilon_{12}\,. \tag{30}$$

Thus, the slow down of the rate of the redox process apparent for $\alpha_{11}^0 > 0.7$ only seems to be but arises from the decrease in the mean extinction coefficient of the mixture of cerium hydroxocitrato complexes.

A completely analogous situation takes place for the system Ce^{4+}–SO_4^{2-}–$Tart$ with $c_R \leq c_M$ and $\varepsilon_{11} > \varepsilon_{21}$ that contain mononuclear and binuclear cerium hydroxotartrato complexes.

The concentration rate equation for the systems with two complexes, which is independent of the optical density, can be represented as follows:

$$-\frac{dc_M}{dt} = k_{11}[MH_{-4}Tart] + k_{21}[M_2H_{-4}Tart] \tag{31}$$

$$= \Big(\beta_{11}k_{11} + \beta_{21}k_{21}\,[M]\Big)[M][H_{-4}Tart] = \Big(K_{11}k_{11} + K_{21}k_{21}[M]\Big)\frac{[M][Tart]}{[H]^4}\,;$$

$$-\frac{dc_M}{dt} = k_{11}[MH_{-2}R] + k_{12}[M(H_{-2}R)_2] \tag{32}$$

$$= \Big(k_{11}\,\beta_{11} + k_{12}\beta_{12}[H_{-2}R]\Big)[M][H_{-2}R] = \Big(k_{11}K_{11} + k_{12}K_{12}\frac{[R]}{[H]^2}\Big)\frac{[M][R]}{[H]^2}\,,$$

where $M = CeOH^{3+}$, $R = Citr$ or $Oxal$; charges are omitted.

The redox process in these systems occurs through the formation of the two intermediate complexes which have almost identical reactivity: $\log k_{11} = -2.60 \pm 0.04 \sim \log k_{21} = -2.64 \pm 0.06$ (18 °C) for $R = Tart$; $\log k_{11} = -2.01 \pm 0.03 \sim \log k_{12} = -2.02 \pm 0.06$ (25 °C) for $R = Citr$, and $\log k_{11} = -1.76 \pm 0.09 \sim \log k_{12} = -1.80 \pm 0.08$ (22 °C) for $R = Oxal$.

3.7. Detailed Mechanism

Equations (31) and (32) can be rewritten in terms of reactive species. In this way, e.g., for $R = Citr$ we have:

$$\begin{aligned} -\frac{dc_M}{dt} &= \left(k_{11}K_{113} + k_{12}K_{122}\frac{[H_{-2}Citr]}{[SO_4]}\right)[CeOH(SO_4)_3]\,[H_{-2}Citr] \\ &\approx \Big(k_{11}\beta_{11} + k_{12}\beta_{12}[H_{-2}Citr]\Big)[CeOH(SO_4)_3]\,[H_{-2}Citr]\,. \end{aligned} \tag{33}$$

The preequilibrium (25) and the equilibrium constant (26) in this case reads:

$$[CeOH(SO_4)_3]^{3-} + nH_{-2}Citr^{2-} \overset{K_{1nz}}{\rightleftharpoons}$$

$$[CeOH(SO_4)_z(H_{-2}Citr)_n]^{2i-2n-3} + (3-z)SO_4^{2-},$$

$$n = 1, 2;\; z = 3, 2; \tag{34}$$

$$K_{1nz} = \frac{[CeOH(SO_3)_z(H_{-2}Citr)_n][SO_4]^{3-z}}{[CeOH(SO_4)_3][H_{-2}Citr]^n}. \tag{35}$$

As observed in [43] and in some other works (e.g., [33]) inverse dependence of apparent rate constants from a concentration of the sulphate anion confirms this rate law.

Equations (32) and (33) can also be written as:

$$-\frac{dc_M}{dt} = \Big(k_{11} + k_{12}\kappa_{12}[H_{-2}Citr]\Big)[CeOH(SO_4)_3H_{-2}Citr]$$

$$= \Big(k_{n=1} + k_{n=2}[H_{-2}Citr]\Big)[CeOH(SO_4)_3H_{-2}Citr], \tag{36}$$

where

$$\kappa_{12} = \frac{\beta_{12}}{\beta_{11}} = \frac{[CeOH(SO_4)_3(H_{-2}Citr)_2]}{[CeOH(SO_4)_3H_{-2}Citr][H_{-2}Citr]}, \tag{37}$$

$k_{n=1}$ (s^{-1}) and $k_{n=2} = \kappa_{12}k_{12}$ ($l\,mol^{-1}\,s^{-1}$) are the first-order and second-order rate constants, respectively; and $[H_{-2}Citr]$ is the concentration of an incoming ligand. In this form, these equations are formally identical with the kinetic equation for the substitution of a ligand in the complex [65]. This ligand here is the sulfate anion.

Assuming that at low $[H^+]$ (below 1N) the trisulfato-hydroxy cerate ion is the main reactive species ($[Ce]_{tot} \approx [CeOH(SO_4)_3]$), we get

$$K_{1nz} \approx \frac{c^0_{1n}\, f_2^n\, [SO_4]^{3-z}}{(c_M - c^0_{1n})(c_R - nc^0_{1n})^n} = \beta^{ef}_{1n}\, f_2^n\, [SO_4]^{3-z} =$$

$$= \beta_{1n}\left(\frac{c_{SO_4}}{1 + B_{1SO_4}[H^+]}\right)^{3-z}, \tag{38}$$

where c_{SO_4} is the initial concentration of the sulphate anion in the solutions, $B_{1SO_4} = [HSO_4^-]/([SO_4^{2-}][H^+])$.

Equations (34), (35), and (33) reduces to the (25), (26), and (32), correspondingly, when $z = 3$. If $z = 2$, $c_{SO_4} = 0.67$ mol dm^{-3}, and $\log B_{1SO_4} = 0.92$, then $\log K_{113} = \lg \beta_{11}$ and $\log K_{122} \approx \log \beta_{12} - 0.2$. Therefore, in the case of exact equality $K_{122} = \beta_{12}[SO_4]$ in (38) the equality (33) also will exact.

Consequently, the detailed mechanism could be described by Scheme 4.

$H_2O \overset{K}{\longleftrightarrow} H^+ + OH^-$	pK	15.74
$SO_4^{2-} + H^+ \overset{B_{1so_4}}{\longleftrightarrow} HSO_4^-$	$\log B_{1so_4}$	1.08
$H_{-2}Citr^{2-} + H^+ \overset{B_1}{\longleftrightarrow} H_{-1}Citr^-$	$\log B_1$	14.60
$H_{-2}Citr^{2-} + 2H^+ \overset{B_2}{\longleftrightarrow} Citr$	$\log B_2$	17.45
$Ce^{4+} + SO_4^{2-} \overset{\kappa_{1so_4}}{\longleftrightarrow} [Ce\,SO_4]^{2+}$	$\log \kappa_{1so_4}$	4.73
$[Ce\,SO_4]^{2+} + SO_4^{2-} \overset{\kappa_{2so_4}}{\longleftrightarrow} [Ce(SO_4)_2]$	$\log \kappa_{2so_4}$	2.93
$[Ce(SO_4)_2] + SO_4^{2-} \overset{\kappa_{3so_4}}{\longleftrightarrow} [Ce(SO_4)_3]^{2-}$	$\log \kappa_{3so_4}$	1.58
$[Ce(SO_4)_3]^{2-} + OH^- \overset{\beta_{hso_4}}{\longleftrightarrow} [CeOH(SO_4)_3]^{3-}$	$\log \beta_{hso_4}$	13.90
$[CeOH(SO_4)_3]^{3-} + H_{-2}Citr^{2-} \overset{K_{113}}{\longleftrightarrow} [CeOH(SO_4)_3H_{-2}Citr]^{5-}$	$\log K_{113}$	15.87
$[CeOH(SO_4)_3H_{-2}Citr]^{5-} \overset{k_{11}}{\longrightarrow} [Ce(SO_4)_2]^- + PRODUCTS$	$\log k_{11}$	−2.40
$[CeOH(SO_4)_3]^{3-} + 2H_{-2}Citr^{2-} \overset{K_{122}}{\longleftrightarrow} [CeOH(SO_4)_2(H_{-2}Citr)_2]^{5-} + SO_4^{2-}$	$\log K_{122}$	30.38
$[CeOH(SO_4)_2(H_{-2}Citr)_2]^{5-} \overset{k_{12}}{\longrightarrow} [Ce(SO_4)_2]^- + PRODUCTS$	$\log k_{12}$	−2.40

Scheme 4. Detailed mechanism for cerium(IV) oxidation of citric acid over the range $pH \geqslant 2.3$ in aqueous sulfate media.

In terms of $[Ce^{4+}]$ and $[Citr]$ the rate law becomes

$$-\frac{dc_M}{dt} = \left\{ \left(k_{11}K_{113} + k_{12}K_{122}\left(1 + B_{1SO_4}[H]\right) \frac{[Citr]}{c_{SO_4}\left(1 + B_{21}[H] + B_{22}[H]^2\right)} \right) \right.$$

$$\left. \times \frac{\kappa_{1SO_4}\kappa_{2SO_4}\kappa_{3SO_4}\beta_{hSO_4}K\,[H_20]}{\left(1 + B_{21}[H] + B_{22}[H]^2\right)[H]} \left(\frac{c_{SO_4}}{1 + B_{1SO_4}[H]} \right)^3 \right\} [Ce][Citr]. \quad (39)$$

The obtained model parameters of the investigated systems can be regarded as a quantitative basis for the 'forward modeling' to test these models. One example of such test is given in Table 5.

Table 5. The residuals Δ^i between the experimental D^i and fitted $\hat{D}^i$ values of optical density in the system $Ce^{4+} - SO_4^{2-} - Oxal\ (pH = 1.9)$

τ^i, s^{-1}	D^i	$\hat{D}^i$	Δ^i
005	0.580	0.584	0.004
010	0.525	0.521	−0.004
015	0.480	0.468	−0.012
020	0.410	0.423	0.013
025	0.380	0.383	0.003
030	0.350	0.349	−0.001
035	0.320	0.319	−0.001
040	0.290	0.292	0.002
045	0.270	0.268	−0.002
050	0.250	0.247	−0.003
055	0.225	0.227	0.002
060	0.210	0.209	−0.001
065	0.195	0.193	−0.002
070	0.180	0.178	−0.002
075	0.165	0.165	0.000
080	0.150	0.152	0.002
085	0.140	0.141	0.001
090	0.130	0.130	0.000
095	0.120	0.121	0.001
100	0.115	0.112	−0.003
105	0.105	0.103	−0.002
110	0.095	0.096	0.001
115	0.085	0.089	0.004
120	0.075	0.074	−0.001

Note: calculated with the values of $\log \hat{k}_{11} = -1.77$ and $\log \hat{k}_{12} = -1.80$

The criterion for an acceptable fit in the model was minimal systematic deviations $\Delta^i = \hat{D}^i - D^i$ in the fitted progress curves $\hat{D}^i$ $(i = \overline{1, N})$ from the experimental data points D^i

($i = \overline{1, N}$), i.e., the minimum value of the criterion function $\min U = \sum_{i=1}^{N} w_i(\hat{D}^i - D^i)^2$, where w_i is the statistical weight of each measurement.

Until recently there were no general computer programs for solving the generalized inverse problem of these complex formation equilibria. However, very recently a real possibility appeared in the numerical solution of similar problems [66]. One of the key features of this new family of ReactLab software products are proprietary reaction modeling algorithms which enable the fitting of both rate constants and rapid equilibrium constants in the same model. Its predecessor was the well-known SPECFIT/32 package [67] for modeling and nonlinear least-squares fitting for chemical kinetics (not available now).

An another possibility for the simultaneous fitting in stability constants (β_{mn}) and decomposition rate constants ($k_{mn} = k_{mn}^{app}/\varepsilon_{mn}$) is to adapt programs like PSEQUAD [68] to solve this previously mentioned problem. In this case, the initial rate of change by the first property, obtained from the experiment, should be considered as the second measured property $P_2 = \dot{P}^0$. And, the apparent decomposition rate constant must be treated as the second intensity factor $a_2 = k_{mn}^{app}$ which will be fitted, as the first intensity factor $a_1 = \varepsilon_{mn}$ by linear methods. Calculation of stability constants based on the initial rates $\dot{P}^0$ are considered in [69].

Such an approach requires the introduction of weights $\tilde{w}_k$ ($k = \overline{1, 2}$) for each measured property $\tilde{U} = \sum_{k=1}^{2} \tilde{w}_k U_k$ that allows you to automatically agree evaluation of the stability constants obtained from measuring both properties, $P_1 = P^0$ and $P_2 = \dot{P}^0$. In a certain sense it is redundant. However, this calculation procedure can be minimized if we use the second property $\dot{P}^0$ to determine only the corresponding intensity factor k_{mn}^{app} by a linear optimization method, without iterative fitting for the second set of stability constants on the basis of this property. Such a computational procedure no longer requires agreement in the values of stability constants obtained on the basis of two types of measurements and is less time consuming.

One can hope that such modifications of programs to simultaneous determine thermodynamic and kinetic stability of variable-valence metal complexes will appear in the near future.

Conclusion

The classical methods for investigating the complex formation are analyzed in terms of the direct and inverse problems of quantitative instrumental analysis and the complex formation equilibria. A generalization of these methods is presented to determine together with the thermodynamic stability of variable-valence metals complexes also their kinetic stability with respect to redox decomposition.

These generalized methods are applied to the study of cerium(IV) complexes with three groups of organic ligands: dicarboxylic, hydroxycarboxylic acids, and polyatomic alcohols. Identified their composition, the specification of the ligand in the complex; calculated stability constants, rate constants and activation energy of their redox decomposition. On the basis of the calculated parameters and the set of all equations that describe the establishment and violation of equilibria in the systems found the rate law occurring in the systems redox processes and its corresponding mechanism.

Established the positive correlation between thermodynamic and kinetic stability for the complexes $[CeOH(SO_4)_zH_{-2}R]^{1-2z}$ with dibasic ligands of the type $R = H_2L$, $R = H(OH)L$, and $R = L(OH)_2$, which demonstrate the stabilizing effect of hydroxy groups in the ligands complexes. Also considered in this work, organic ligands by their stabilizing action on cerium(IV) through coordination can be arranged in series: $H_{-2}Sorb$, $H_{-2}Csyl$, $H_{-2}Glyc > H_{-4}Tart$, $H_{-3}Malt$, $H_{-2}Lact$, $H_{-2}Quin > H_{-2}Glut$, $H_{-2}Succ$, $H_{-2}Maln > H_{-2}Oxal$.

Increased stabilization of $Ce(IV)$ in the complexes can be associated with increased electronegativity for the oxygen atom in the functional groups of the ligands, from which there is a transfer of electron density on the central ion. The absence of observable effect of relative stabilizing $Ce(IV)$ in the complexes $[CeOH(SO_4)_{z-1}H_{-x}R]^{5-2z-x}$ with $x > 2$, as well as in the bis-complexes $[CeOH(SO_4)_{z-1}(H_{-2}R)_2]^{1-2z}$, can be interpreted as compensation of this effect by a destabilizing effect in the displacement of sulfate anion from the inner coordination sphere in the complex.

A discussion concerning the possibilities of using some software products is addressed to provide the simultaneous determination of the stability constants and redox-decompositions rate constants for variable-valence metal complexes in solutions.

References

[1] Mehdi, H.; Bodor, A.; Lantos, D., et al., Imidazolium ionic liquids as solvents for cerium(IV)-mediated oxidation reactions. *J. Org. Chem.* 2007, *72* (2), 517-524; Balaji, S.; Chung, S.J.; Matheswaran, M., et al., Cerium(IV) mediated electrochemical oxidation process for the mineralization of various organic pollutants in batch and continuous organic feeding modes. *Kor. J. Chem. Eng.* 2007, *24* (6), 1009-1016; Tadon P.K.; Singh, A.K.; Sahgal, S., et al., Oxidation of cyclic alcohols by cerium(IV) in acidic medium in presence of iridium(III) chloride. *J. Molec. Catal.* 2008, *282A* (1-2), 136-143.

[2] Yadav, M.B.; Devra, V.; Rany, A., Kinetics and mechanism of silver(I) catalyzed oxidation of valine by cerium(IV) in acid perchlorate medium. *Ind. J. Chem.* 2010, *49A* 442-447; Wisniewska, J.; Wrzeszcz, G.; Kurzawa, M., et al., The oxidative degradation of dibenzoazepine derivatives by cerium(IV) complexes in acidic sulfate media. *Dalton Trans.* 2012, *41*, 1259-1267; Tandon, P.K.; Khanam, S.Z.; Singh, S.B., Oxidation of vicinal diols by cerium(IV) in aqueous acidic media catalyzed by rhodium(III). *Open Catal. J.*. 2012, *5*, 1-7.

[3] Matheswaran, M.; Balaji, S.; Chung, S.J., et al., Mineralization of phenol by Ce(IV)-mediated electrochemical oxidation in methanesulphonic acid medium: A preliminary study. *Chemosphere* 2007, *69* (2), 325-331; Matheswaran, M.; Balaji, S.; Chung, S.J., et al., Mediated electrochemical oxidation of phenol in continuous feeding mode using Ag(II) and Ce(IV) mediator ions in nitric acid: A comparative study. *Chem. Eng. J.*. 2008, *144* (1), 28-34.

[4] Nayak, B.B; Sahu, S.; Patel, S., et al., Ce(IV) oxidation of alcohols using cetyltrimethylammonium ceric nitrate. *Indian J. Chem.* 2008, *47A*, 1486-1490; Patil,

R.K.; Chimatadar, S.A.; Nandibewoor, S.T., Mechanistic study of cerium(IV) oxidation of antimony(III) in aqueous sulphric acid in the presence of micro amounts of manganese(II) by stopped flow technique. *Transition. Met. Chem.* 2008, *33*, 625-633; Hosahalli R.V.; Savanur A.P.; Nandibewoor, S.T., et al., Ruthenium(III)-mediated oxidation of D-mannitol by cerium(IV) in aqueous sulfuric acid medium: A kinetic and mechanistic approach. *Int. J. Chem. Kinet.* 2010, *42*, 440-452.

[5] Matheswaran, M.; Balaji, S.; Chung, S.J., et al., Studies on cerium oxidation in catalytic ozonation process: A novel approach for organic mineralization. *Catal. Commun.* 2007, *8*, 1497-1501; Najafpour, M.M.; Boghaei, D.M., Heterogeneous water oxidation by bidentate Schiff base manganese complex in the prsence of cerium(IV). *Transition. Met. Chem.* 2009, *34*, 367-372; Playford, H.Y.; Modeshia, D.R.; Barney, E.R., et al., Structural characterization and redox catalytic properties of cerium(IV) pyrochlore oxides. *J. Mater. Chem.* 2011, *23* (24), 5464-5473.

[6] Voskresenskaya, O.; Skorik, N., Determination of the stability constants and rate constants for the intracomplex redox decomposition of cerium(IV) lactate and cerium(IV) malate hydroxo complexes in a sulfate medium. *Russ. J. Inorg. Chem.* 2000, *45* (3), 390-397; Voskresenskaya, O.; Skorik, N.A., The kinetics of cerium(IV) sulfate reaction with citrate and the thermodynamic characteristics of formation of intermediate complexes. *Russ. J. Phys. Chem.* 2009, *83*, 945-950.

[7] Voskresenskaya, O.O.; Skorik, N.A., Composition, stability, and reactivity of cerium(IV) hydroxotartrato complexes in sulfate solutions. *Russ. J. Inorg. Chem.* 2000, *45* (12), 1936-1944; Voskresenskaya, O.O.; Skorik, N.A., Stability constants and rate constants of intramolecular redox decomposition of cerium(IV) complexes with some polyatomic alcohols. *Russ. J. Inorg. Chem.* 2005, *50* (7), 1089-1097.

[8] Kiselev, Yu.M., On the stabilization of oxidation states through coordination. *Russ. J. Inorg. Chem.* 2002, *47* (4), 473-486; Kiselev, Yu.M.; Dobrynina, N.A., *Coordination Chemistry.* Academie: Moscow, 2007; Kiselev, Yu.M., On the methods of stabilizing the oxidation states of transition metals. *Russ. J. Inorg. Chem.* 2007, *52* (11), 1826-1835.

[9] Lavoie, N., *Disubstituted 1,8-Diaminonaphthalene as Supporting Ligands for High Oxidation State Early Transition Metal and Lanthanide Complexes.* Ph.D. Thesis, University of Ottawa, Ottawa, June, 2010; Dubois, M.; Dieudonner, B.; Mesbah, A., et al., Stabilization of Th^{3+} ions into mixed-valence thorium fluoride *J. Solid State Chem.* 2011, *184*, 220-226.

[10] Wong, E.W.Y., *An investigation of redox-active isoindoline-based ligands and their coordination complexes.* Ph.D. Thesis, Simon Fraser University, Vancouver, May 2012; Ingleson, M.J.; Layfield, R.A., N-Heterocyclic carbene chemistry of iron: Fundamentals and applications. *Chem. Commun.* 2012, *48*, 3579-3589; Kleinberg, J., *Unfamiliar Oxidation States and Their Stabilization*; Literary Licensing, LLC: Whitefish, 2012.

[11] Marcus, R.A., Electron transfer reactions in chemistry: Theory and experiment (Nobel Lecture). *Angew. Chem. Int. Ed. Engl.* 1993, *32* (8), 1111-1121; Royea, W.J.; Fajardo, A.M.; Lewis, N.S., Fermi golden rule approach to evaluating outer-sphere electron-transfer rate constants at semiconductor/liquid interfaces. *J. Phys. Chem.* 1997, *101B* (51), 11152-11159; Kuznetsov. A.M.; Ulstrup, J., *Electron Transfer in Chemistry and Biology*; Wiley: Chichester, 1999.

[12] Sethuram, B., *Some Aspects of Electron Transfer Reactions Involving Organic Molecules*; Allied Publishers (P) Ltd.: New Deli, 2003; Swaddle, T.W., Homogeneous versus heterogeneous self-exchange electron transfer reactions of metal complexes: Insights from pressure effects. *Chem. Rev.* 2005, *105*, 25732608; Khoshtariya, D.E.; Dolidze, T.D.; Shushanyana, M., et al., Fundamental signatures of short- and long-range electron transfer for the blue copper protein azurin at Au/SAM junctions. *PNAS*. 2010, *107* (7), 2757-2762.

[13] Mohammed, B.; Subramani, N.K.; Ahmed, A.N., Induced electron transfer reaction of Cobalt(III) complexes by imidazoliniumdichromate in the presence of surfactant medium. *J. Chem. Biol. Phys. Sci.* 2012, *2* (2), 668-676; Amadei, A.; Daidone, I.; Aschi, M., A general theoretical model for electron transfer reactions in complex systems. *Phys. Chem. Chem. Phys.* 2012, *14*, 1360-1370; Jagannadham, V., Electron transfer reactions: A treatise. *Amer. J. Chem.* 2012, *2* (2), 57-82.

[14] Jakupec, M.A.; Unfried, P.; Keppler, B.K., Pharmacological properties of cerium compounds. *Rev. Physiol. Biochem. Pharmacol.* 2005, *153*, 101-111; Moraes, V.; Marczewski, B.; Dias, C.R., et al., Study of gels of molybdenum with cerium in the preparation of generators of ^{99}Mo - ^{99m}Tc. *Braz. Arch. Biol. Technol.* 2005, *48*, 51-56; Park, B.; Martin, P.; Harris, C., et al., Initial in vitro screening approach to investigate the potential health and environmental hazards of Envirox^{TM} - a nanoparticulate cerium oxide diesel fuel additive. *Particle and Fiber Toxicology*. 2007, *4*, 12-21.

[15] *Toxicological Reviev of Cerium Oxide and Cerium Compounds* (CAS No. 1306-38-3). U.S. Environmental Protection Agency: Washington, 2009; http://www.epa.gov/iris (accessed September 2009).

[16] Inczédy, J., *Analytical Applications of Complex Equilibria*; Ellis Horwood: London, 1976.

[17] Wu, X.; Yang, J.; Sun, C., et al., A fluorimetric method for the determination of nucleic acids using Ce(IV) and sodium triphosphate. *Luminescence*. 2005, *20* 41-45; Cui, H.; Zhang, Q.; Myint, A., et al., Chemiluminescence of cerium(IV)-rhodamine 6G-phenolic compound system. *J. Photochem. Photobio.* 2006. *181A*, 238-245; Walash, M.I.; Rizk, M.; Sheribah, Z.A., et al., Development and validation of spectrofluorimetric method for determination of biotin in bulk and pharmaceutical preparations via its oxidation with cerium (IV). *Int. J. Biomed. Sci.* 2010, *6* (3), 268-275.

[18] Duke, F.R.; Smith, G.F., Rapid qualitative test for alcoholic hydroxyl group. *Ind. Eng. Chem. Anal.* 1940, *12*, 201-203; Reid, V.W.; Truelove, R.K., The colorimetric determination of alcohols. *Analyst*. 1952, *77*, 325-328; Lau, O.; Luk, S., Spectrophoto-

metric method for the determination of ethanol in beverages and beer samples using cerium(IV) as reagent. *Int. J. Food Sci. Technol.* 1994, *29* (4), 469-472; Pinyou, P.; Youngvises, N.; Jakmunee, J., Flow injection colorimetric method using acidic ceric nitrate as reagent for determination of ethanol. *Talanta.* 2011, *84* (3), 745-51.

[19] Offner, H.G., *Spectrophotometric studies of some alcohol, hydroxy, and ether complexes of quadrivalent cerium.* Ph.D. Thesis, Stanford University, Stanford, 1957; Hintz, H.L., *The mechanism of cerium(IV) oxidations of cyclic alcohols.* Ph.D. Thesis, Lawence University, Appleton, January 1966; Young, I.B.; Trahanovsky, W.S., Oxidation of organic compounds with cerium(IV). VII. Formation constants for 1:1 cerium(IV)-alcohol complexes. *J. Am. Chem. Soc.* 1969, *91*, 5060-5068.

[20] Smith, G.F., *Cerate Oxidimetry*; G.F. Smith Pub. Co.: Columbus (Ohio), 1942; *Oxidimetry*; Nikolky, B.P., Palchevsky, V.V., Eds.; Chemistry: Leningrad, 1985; Bassett, J.; Denney, R.C.; Jeffry, G.H., et al., *Vogel's Textbook of Quantitative Inorganic Analysis*; Longman: London, 1981; Harris, D.C., *Quantitative Chemical Analysis*; Freeman W.H. and Company: New York and London, 1995.

[21] Mark, H.B.; Rechnitz, G.A., *Kinetics in Analytical Chemistry*; New York: Wiley, 1968; Yatsimirskii, K.B., *Kinetic Methods of Analysis*; Pergamon: Oxford, 1966; Motolla, H.A., *Kinetic Aspects of Analytical Chemistry*; Wiley: New York, 1988; Perez-Bendito, D.; Silva, M., Kinetic Methods in Analytical Chemistry; Ellis Horwood Limited: Chichester, 1988.

[22] Richardson, W.H., Ceric ion oxidation of organic compounds. In *Oxidation in Organic Chemistry, Vol. 5A*; Wiberg, K.B., Eds.; Academic Press: New York and London, 1965; pp 244-277; Cotton, F.A.; Wilkinson, G., *Advanced Inorganic Chemistry*; Interscience: New York, 1972; Ho, T.-L., Ceric ion oxidation in Organic Chemistry. *Synthesis.* 1973, *6*, 347-355; Nair, V.; Balagopal, L.; Menon, R.S., et al., Oxidative intramolecular cyclization reactions of cinnamyl ethers mediated by cerium(IV) ammonium nitrate (CAN): A stereoselective synthesis of 3,4-trans-disubstituted tetrahydrofuran derivatives. *ARKIVOC.* 2003, *2003* (viii), 199-210. http:www.arkat-usa.org/ARKIVOC/arkivoc_articles.asp.

[23] Ho, T.-L., *Organic Synthesis by Oxidation with Metal Compounds*; Plenum Publishing Corporation: New York, 1986; Dakshinamoorthy, A., Cerium(IV) ammonium nitrate: A versatile oxidant in synthetic organic chemistry. *Synlett.* 2005, *19*, 3014-3015; Yao, W.; Chen, Y.; Min, L., et al., Liquid Oxidation of cyclohexane to cyclohexanol over cerium-doped MCM-41. *J. Molec. Catal.* 2006, *246A*, 162-166.

[24] Thiruvenkatachari, R.; Kwon, T.; Jun, J., et al., Application of several advanced oxidation processes for the destruction of terephthalic acid (TPA). *J. Hazard. Mater.* 2007, *142* (1-2), 308-314; Lee, J.-W.; Chung, S.J.; Balaji, S., et al., Destruction of EDTA using Ce(IV) mediated electrochemical oxidation: A simple modeling study and experimental verification. *Chemosphere.* 2007, *68*, 1067-1073; Balaji, S.; Chung, S.J.; Matheswaran, M., Destruction of organic pollutants by cerium(IV) MEO process: A study on the influence of process conditions for EDTA mineralization. *J. Hazar. Mater.* 2008, *150* (3), 596-603.

[25] Dolgoplosk, B.A.; Tinyakova, E.I., *Redox Systems as a Source of Free Radicals*; Science: Moscow, 1972; Bajpai, A.; Simon, J., Redox initiation of copolymerization of styrene with hydroxyl terminated polybutadiene using Ce(IV) as an oxidant. *J. Appl. Polym. Sci.* 2002, *85*, 244-256; Öz, N.; Akar, A., Redox initiation system of ceric salt and dihydroxy poly(dimethylsiloxane)s for vinyl polymerization. *J. Appl. Polym. Sci.* 2006, *102*, 2112-2116; Ozturk, T.; Cakmak, I., Synthesis of block copolymers via redox polymerization process: A critical review. *Iran. Polym. J.* 2007, *16* (8), 561-581.

[26] Mino, G.; Keizerman, S., A new method for the preparation of graft copolymers: Polymerization initiated by ceric ion redox systems. *J. Polym. Sci.* 1958, *31*, 242-243; Mino, G.; Keizerman, S.; Rasmussen, E., The polymerization of acrylamide initiated by ceric nitrate-3-chloro-1-propanol redox systems. *J. Polym. Sci.* 1959, *38*, 393-401; Katai, A.A.; Kulshrestha, V.K.; Marchessault, R.H., Polymerization of acrylonitrile by ceric ion-ethylene glycol redox system. *J. Polym. Sci.* 1963, *2*, 403-407.

[27] Tunca, U.; Serhatli, I.E.; Yagci, Y., Polymerization of acrylamide initiated by the redox system Ce(IV)-4,4-azo-bis(4-cyano pentanol). *Polym. Bull.* 1989, *22*, 483-488; Yagci, C.; Yildiz, U., Redox polymerization of methyl methacrylate with allyl alcohol 1,2-butoxylateblock-ethoxylate initiated by $Ce(IV)/HNO_3$ redox system. *Eur. Polym. J.* 2005, *41*, 177-184.

[28] Sarac, A.S.; Basak, H.; Soydan, A.B., et al., Polymerization of acrylamide by electrolytically generated Ce(IV)-organic acid redox systems. *Angew. Makromol. Chem.* 1992, *198*, 191-198; Sarak, A.S .; Yavuz Ö.; Sezer, E., Electrochemically induced redox polymerization of acrylamide. *J. Appl. Polym. Sci.* 1999, *72* (7), 861-869; Erbil, C., Block copolymerization and chain extension by using PAAm prepolymer initiated with Ce(IV)-malonic acid. *Eur. Polym. J.* 1999, *35*, 1747-1754.

[29] Vakhramova, G.P.; Pechurova, N.I.; Spitsyn, V.I., A study of complexation of cerium(IV) with oxalic acid in a solution of ammonium nitrate. *Vestn. Mosk. Gos. Univ.* 1974, *6*, 682-686; Pechurova, N.I.; Vakhramova, G.P.; Spitsyn, V.I., Investigation of cerium(IV) interaction with citric acid in the sulphate and nitrate media. *Russ. J. Inorg. Chem.* 1974, *19* (6), 2074-2077; Vakhramova, G.P.; Pechurova, N.I.; Spitsyn, V.I., A study on the interaction of cerium(IV) with tartaric acid in the presence of SO_4^{2-}- and NO_3^--ions. *Russ. J. Inorg. Chem.* 1974, *19*, 2722-2725.

[30] *Coordination Chemistry of Rare-Earth Elements*; Spitsyn, V.I., Martynenko, L.I., Eds.; MGU Publisher: Moscow, 1974; Kiselev, Yu.M., New advances in chemistry of highest unstable oxidation states of transition elements. *Russ. J. Inorg. Chem.* 1999, *44* (5), 711-717.

[31] *Coordination Chemistry*; Bailar, J.C., Busch, D.I., Eds.; Reinhold: New York, 1956.

[32] Yatsimirskii, K.B., Unusual oxidation states of d-transition metals in coordination compounds. *Bull. Polish Acad. Sci. Chem.* 1994, *42* (3), 299-305; Yatsimirskii, K.B., Stabilization of unstable oxidation states of d-transition metals in macrocyclic complexes. *Russ. J. Inorg. Chem.* 1991, *36* (8), 2010-2020.

[33] Rao, B.M.; Sastrys, T.P.; Parekh, T., A comparative kinetic study of oxidation of structurally related organic sulphates viz. succinic, malic, and tartaric acids and also of glycerol with cerium(IV) sulfate in sulfuric acid media. *Z. phys. Chem. (Leipzig).* 1983, *264*, 906-920.

[34] Kvernberg, P.O.; Hansen, E.W.; Pedersen B., et al., Oxidation of methylmalonic acid by cerium(IV). Evidence for parallel reaction pathways. *J. Phys. Chem.* 1997, *101A*, 2327-2331; Yu, Y.-O.; Jwo, J.-J., Kinetic study of the Ce(III)-, Mn(II)-, or ferroin-catalyzed Belousov-Zhabotinsky reaction with Allylmalonic Acid. *J. Chin. Chem. Soc.* 2000, *47*, 433-439.

[35] Butler, K.; Steinbock, O.; Steinbock, B., et al., Carbon dioxide production in the oxidation of organic acids by cerium(IV) under aerobic and anaerobic condtions. *Int. J. Chem. Kinetics.* 1998, *30*, 899-902.

[36] Pelle, K.; Wittmann, M.; Lovric K., et al., Mechanistic investigation of the BZ reaction with oxalic acid substrate. *J. Phys. Chem.* 2004, *108A*, 5377-5385; Pereira, J.A.M.; Faria, R.B., Oscillatory bromate-oxalic acid-Ce-acetone-sulfuric acid reaction, in CSTR. *J. Braz. Chem. Soc.* 2004, *15* (6), 976-978; Čevčik, P.; Mišicák, D.; Adamčiková, L., Carbon dioxide production in the oscillating BZ reaction with oxalic acid. *Chem. Pap.* 2006, *60* (1), 1-4.

[37] Belousov, B.P., Periodically acting reaction and its mechanism. In *Collection of Short Papers on Radiation Medicine for 1958, Vol. 147*. Medgiz: Moscow, 1959; pp 145-147; Zhabotinsky, A.M., Periodical oxidation of malonic acid in solution (a study of the Belousov reaction kinetics). *Biofizika*. 1964, *9*, 306-311.

[38] Field, R.J.; Körös, E.; Noyes, R.M., Oscillations in chemical systems. II. Thorough analysis of temporal Oscillations in the bromate-cerium-malonic acid system. *J. Am. Chem. Soc.* 1972, *94*, 8649-8664; Kasperek, G.T.; Bruice, T.C., Observation on an oscillating reaction of potassium bromate, ceric sulfate, and a dicarboxylic acid. *Inorg. Chem.* 1971, *10*, 382-386.

[39] *Oscillations and Traveling Waves in Chemical Systems*; Field, R.J., Burger, M., Eds.; Wiley-Inerscience: New York, 1985.

[40] Y. Gao, Y.; H.-D. Försterling, H.-D.; Noszticzius, Z., HPLC studies on the organic subset of the oscillatory BZ reaction. 1. Products of the Ce^{4+}-malonic acid reaction. *J. Phys. Chem.* 1994, *98*, 8377-8380; Sirimungkala, A.; Försterling, H.-D.; Noszticzius, Z., HPLC Studies on the organic subset of the oscillatory BZ reaction. 2. Two different types of malonyl radicals in the Ce^{4+}-malonic acid reaction. *J. Phys. Chem.* 1996, *100*, 3051-3055.

[41] Epstein, I.R.; Pojman, J.A., *An Introduction to Nonlinear Chemical Dynamics: Oscillations, Waves, Patterns, and Chaos.* Oxford University Press: New York, 1998.

[42] Taylor, A.F., Mechanism and phenomenology of an oscillating chemical reaction. *Progr. React. Kinet. Mech.* 2002, *27*, 247-325; Mikhailov, A.S.; Showalter, K., Control of waves, patterns and turbulence in chemical systems. *Phys. Rep.* 2006, *425*,

79-194; Sobel, S.G.; Hastings, H.M.; Field, R.G., Oxidation state of BZ reaction mixtures. *J. Phys. Chem.* 2006, *110A* (1), 5-7; Zhou, H.; Liang, E.; Zheng, Z., et al., Smart polymers based on Belousov-Zhabotinsky reaction. *Progr. Chem.* 2011, *23* (11), 2368-2376; Lu, X.; Zhao, Y.; Ren, L., et al., Spatiotemporal dynamics of photosensitive BZ reaction. *Progr. Chem.* 2012, *24* (05), 709-721.

[43] Lahid, A.; Alexander, Mc A., Oxidation of malic acid by cerium(IV) in perchloric acid media. *J. Chem. Soc. Dalton Trans.* 1974, *23*, 2521-2530; Mehrotra, R.N.; Ghosh, S., Kinetics of oxidation of citric acid by ceric sulfate in aqueous sulfuric acid. *Z. phys. Chem. (Leipzig).* 1963, *224*, 57-64; Sengupta, K.K.; Aditya, S., Kinetics and mechanism of the oxidation of malonic acid by ceric salts. *Z. phys. Chem.* (N.F.) 1963, *38*, 25-35; Mehrotra, R.N.; Ghosh, S., Kinetics of oxidation of tartaric acid by ceric sulfate in aqueous sulfuric acid. *Z. phys. Chem. (Leipzig).* 1965, *230* (3/4), 231-238.

[44] Tadon, P.K.; Sahgal, S.; Gayatri, P.M., et al., Oxidation of ketones by cerium(IV) in presence of iridium(III) chloride. *J. Molec. Catal.* 2006, *250A*, 203-209; Tadon, P.K.; Dwivedi, P.B.; Singh, S., Liquid-phase oxidation of some aromatic aldehydes, hydrocarons, and alcohols by cerium(IV) catalyzed by iridium(III) in acidic medium. *Synth. Commun.* 2009, *39*, 1920-1928; Hiremath, M.I.; S.T. Nandibewoor, S.T., Kinetics and mechanism of oxidation of gabapentin (neurontin) Aanion by alkaline diperiodatonickelate(IV). *Z. phys. Chem.* 2011, *225*, 95-106.

[45] Denisov, E.T., *Rate Constants of Homolytic Liquid-phase Reactions*; Science: Moscow, 1971; Candlin, J.P.; Taylor, K.A.; Thompson, D.T., *Reactions of Transition Metal Complexes*; Elsevier: Amsterdam-London-New York, 1968; Lappin, G., *Redox Mechanisms in Inorganic Chemistry;* Ellis Horwood: New York, 1994.

[46] Sandu, A.; Liao, W.; Carmichael, G.R., et al., Inverse modeling of aerosol dynamics using adjoints: Theoretical and numerical considerations. *Aerosol Sci. Technol.* 2005, *39*, 677-694; Daescu, D.; Sandu, A.; Carmichael, G.R., Direct and adjoint sensitivity analysis of chemical kinetic Ssystems with KPP. *Atmosph. Environ.* 2003, *37*, 5097-5114; Soetaert, K.; Petzoldt, T., Inverse modeling, sensitivity and Monte Carlo analysis. *J. Statist. Software.* 2010, *33*(3), 1-28.

[47] Kolthof, I.M., et al., *Volumetric Analysis*; Interscience: New York, 1957.

[48] Kostromina, N.A.; Kumok, V.N.; Skorik, N.A., *Chemistry of Coordination Compounds)*, Higher School: Moscow, 1990.

[49] Benesi, H.; Hildebrand, J., A spectrophotometric investigation of the interaction of iodine with aromatic hydrocarbons. *J. Am. Chem. Soc.*. 1949, *71* (8), 2703-2307; Seal, B.; Sil, H.; Mukherjee, D., Independent determination of equilibrium constant and molar extinction coefficient of molecular complexes from spectrophotometric data by a graphical method. *Spectrochim. Acta.* 1982, *38A* (2), 289-292.

[50] Hooke, R.; Jeeves, T.A., Direct search solution of numerical and statistical problems. *Assoc. Computer. Mach.* 1961, *8*, 212-229; Bhatti, M.A., *Practical Optimization Methods: With Mathematical Applications*; Springer: Heidelberg, 2000; Cramer, C.J.,

Essentials of Computational Chemistry: Theories and Models; Wiley-Interscience: New-York, 2004.

[51] Cauchy, A., Methode Generale pour la Resolution des Systemes d'Equations Simultanees. *Comp. Rend. Acad. Sci.* 1847, 536-538; Nazareth, J.L., *The Newton-Cauchy Framework. A Unified Approach to Unconstrained Nonlinear Minimization*, Springer: Berlin and Heidelberg, 1994.

[52] Sarana, N.V., Effect of ligand structure on the stability of cerium(IV) complexes with hydroxyl-containing compounds. *Coll. of Sci. Works of Leningrad Technol. Inst. of Pulp and Paper Industry*. 1969, *24*, 56-59.

[53] Trubacheva, L.V.; Pechurova, N.I., A study of the cerium(IV) hydrolysis in sulfate medium. *Russ. J. Inorg. Chem.* 1981, *26*, 3254-3258.

[54] Skorik, N.A.; Artish, A.S., The stability of scandium, gallium, indium and thorium complexes with anions of some organic acids. *Russ. J. Inorg. Chem.* 1985, *30* (30), 1994-1997.

[55] Duka, G.G.; Batyr, D.G.; Romanchuk, L.S., et al., Photochemical transformation of oxycarboxylic acids in the presence of iron(III). *Coord. Chem.* 1990, *16* (1), 93-106.

[56] Hardwick T.G.; Robertson, E., Ionic species in ceric perchlorate solution. *Canad. J. Chem.* 1955, *29*, 818-827; Yatsimirskii, K.B.; Kostromina, N.A.; Sheka, Z.A., et al., *Coordination Chemistry of Rare-Earth Compounds*. Naukova Dumka: Kiev, 1966; Voskresenskaya, O.; Skorik, N.A., Composition, stability and reactivity of intermediate complexes for some cerium(IV)-organic reactions. *VINITI (Moscow)*. 1998, *100-B98*, 1-7; Ikeda-Ohno, A.; Tsushima, S.; Hennig, C., et al., Dinuclear complexes of tetravalent cerium in an aqueous perchloric acid solution. *Dalton Trans.* 2012, *41*, 7190-7192.

[57] Hardwick, T.G.; Robertson, E., Association of ceric ions with sulphate (a spectral study). *Canad. J. Chem.* 1955, *29*, 828-837; Bachmann K.; Lieser, K.H., Homogene redoxreactionen in sulfationenhaltigen Lösungen. II. System Ce(III)/Ce(IV). *Ber. Bunsgesell. Phys. Chem.* 1963, *67* (8), 810-816.

[58] Singh, R.S.; Jha, P.N., Prasad, R.K., Kinetics of oxidation of phenyl hydrazine with Ce(IV) in sulphuric acid medium. *Proc. Nat. Acad. Sci., India*. 1987, *LVII* (III), 272-278; Binnemans, K., Applications of tetravalent cerium(IV) compounds. In *Handbook on the Physics and Chemistry of Rare Earts. Vol. 36*; Schneider, K.A., Bűnzli, J.-K., Pecharsky, V.K., Eds.; North Holland Publisher, 2006; pp 282-392.

[59] Sillen, L.G.; Martell, A.E., *Stability Constants of Metal-Ion Complexes.* Special publ. *18* (17); Chem. Soc.: London, 1964.

[60] Nebel, D.; Urban G., Potentiometrische Untersuchungen zur Komplexbildung von $Ce^{III}, Ce^{IV}, Th^{IV}, U^{VI}$ und Citrat in wäßriger Lösungen. *Z. phys. Chem. (Leipzig)*. 1966, *233*, 73-84; Pechurova, N.I.; Vakhramova, G.P; Spitsyn, V.I., Study of

cerium(IV) interaction with citric acid in the sulphate and nitrate media. *Russ. J. Inorg. Chem.* 1974, *19* (8), 2074-2077.

[61] Schmeide, K.; Bernhard, G., Spectroscopic study of the uranium(IV) complexation by organic model ligands in aqueous solution. In *Uranium, Mining and Hydrogeology*; Merkel, B.J., Hasche-Berger, A., Eds.; Springer: Berlin and Heidelberg, 2008; pp 591-598; Bonin, I.; Den Auwer, Ch.; Ansoborlo, E., et al., Study of Np speciation in citrate media. *Radiochim. Acta* 2007, *95*, 371-379; Bonin, I.; Cote, G.; Moisy, Ph., Speciation of An(IV) (Pu, Np, U, and Th) in citrate media. *Radiochim. Acta* 2008, *96*, 145-152.

[62] Felmy, A.R.; Cho, H.; Dixon, D.A., et al., The aqueous complexation of thorium with citrate under neutral to basic conditions. *Radiochim. Acta.* 2006, *94*, 205-212; Thuéry, P. Uranyl ion complexation by citric and citromalic acids in the presence of diamines. *Inorg. Chem.* 2007, *46*, 2307-2315; Thuéry, P., Novel two-dimensional uranyl-organic assemblages in the citrate and D(-)-citromalate families. *Cryst. Eng. Comm.* 2008, *10*, 79-85.

[63] Ogunlusi, G.O.; Ige, J.; Oyetunji, O.A., et al., *Transition. Met. Chem.* 2009, *34*, 483-491.

[64] Casari, B.M.; Lander, V., A pseudo-merohedrally twinned rare-earth sulfate: $K_6[Ce(HSO_4)_2\ (SO_4)_4]{\cdot}H_2O$, a novel structure type. *Acta Cryst.*. 2007, *63C* (4), i25-i27.

[65] Langford, C.H.; Gray, H.B., *Ligand Substitution Processes*; W.A. Benjamin, Inc.: Amsterdam-New York, 1965.

[66] Maeder, M.; King, P., Analysis of chemical processes, determination of the reaction mechanism and fitting of equilibrium and rate constants. In *Chemometrics in Practical Applications*; Varmuza, K., Ed.; InTech Publisher: March 23, 2012; pp 41-62. http://www.intechopen.com/books/chemometrics–in–practical–applications; http:// www.jplusconsulting. com; http://www.hi-techsci.com/products/reactlab.

[67] Binstead, R.A.; Zuberbuhler, A.D.; Jung, B., *Specfit, Version 3.0.38 (32-bit Windows).* Spectrum Software Associates: Chapel Hill, NC, 2006. http://www.hi-techsci.com/products/specfit32.

[68] Zékány, L.; Nagypál, I., PSEQUAD: A comprehensive program for the evaluation of potentiometric and/or spectrophotometric equilibrium data using analytical derivatives. In *Computational Methods for the Determination of Formation Constants*; Leggett, D.J., Ed.; Plenum Press: New York, 1985; pp 291-352; Zékány, L.; Nagypál, I.; Peintler, G., *Manual for PSEQUAD. Update 5.10 (last modification: 01/10/2008)*; http://www.staff.u-szeged.hu/ peintler/enprogs.htm#psequad.

[69] Voskresenskaya, O.O.; Skorik, N.A., *Thermodynamic and Kinetic Aspects of the Formation and Redox Decomposition of Cerium(IV) Complexes with a Series of Hydroxyl-containing Organic Ligands*; NTL Publisher: Tomsk, 2011.

INDEX

A

B

C

D

E

F

G

H

I

N

O

P

Q

R

S

T

U

V

W

X

Z